Oxidation of Organic Compounds

Oxidation of Organic Compounds

Volume I. Liquid-Phase, Base-Catalyzed and Heteroatom Oxidations, Radical Initiation and Interactions, Inhibition

Proceedings of the International Oxidation Symposium, arranged by Stanford Research Institute, in San Francisco, Calif. Aug. 28-Sept. 1, 1967

Frank R. Mayo
General Chairman

ADVANCES IN CHEMISTRY SERIES **75**

AMERICAN CHEMICAL SOCIETY

WASHINGTON, D.C.   1968

Copyright © 1968

American Chemical Society

All Rights Reserved

Library of Congress Catalog Card 967-7520

PRINTED IN THE UNITED STATES OF AMERICA

Advances in Chemistry Series
Robert F. Gould, *Editor*

Advisory Board

Sidney M. Cantor

Frank G. Ciapetta

William von Fischer

Edward L. Haenisch

Edwin J. Hart

Stanley Kirschner

John L. Lundberg

Harry S. Mosher

Edward E. Smissman

AMERICAN CHEMICAL SOCIETY PUBLICATIONS

FOREWORD

ADVANCES IN CHEMISTRY SERIES was founded in 1949 by the American Chemical Society as an outlet for symposia and collections of data in special areas of topical interest that could not be accommodated in the Society's journals. It provides a medium for symposia that would otherwise be fragmented, their papers distributed among several journals or not published at all. Papers are refereed critically according to ACS editorial standards and receive the careful attention and processing characteristic of ACS publications. Papers published in ADVANCES IN CHEMISTRY SERIES are original contributions not published elsewhere in whole or major part and include reports of research as well as reviews since symposia may embrace both types of presentation.

CONTENTS

Preface xi

Introduction xiii

LIQUID-PHASE OXIDATIONS

1. Present State and Main Trends of Research on Liquid-Phase Oxidation of Organic Compounds 1
   N. M. Emanuel, Academy of Sciences, Moscow, USSR

2. Absolute Rate Constants for Hydrocarbon Autoxidation. VII. Reactivities of Peroxy Radicals Toward Hydrocarbons and Hydroperoxides 6
   J. A. Howard, W. J. Schwalm, and K. U. Ingold, National Research Council of Canada, Ottawa, Canada

3. Co-Oxidation of Butadiene with Various Hydrocarbons 24
   Dale G. Hendry, Stanford Research Institute, Menlo Park, Calif.

4. Co-Oxidations of Hydrocarbons 38
   Frank R. Mayo, Martin G. Syz, Theodore Mill, and Jane K. Castleman, Stanford Research Institute, Menlo Park, Calif.

5. Kinetic Data on the Radical Oxidation of Petrochemical Compounds 59
   L. Sajus, Institut Francais du Pétrole, 92-Rueil-Malmaison, France

6. Liquid-Phase Oxidation of High Molecular Weight 1-Alkenes 78
   Charles J. Norton and Dennis E. Drayer, Marathon Oil Co., Littleton, Colo.

7. Butene Hydroperoxide 93
   William F. Brill, Princeton Chemical Research, Inc., Princeton, N. J.

8. Gamma-Radiation–Induced Oxidation of 2-Propanol 102
   Gordon Hughes and H. A. Makada, Donnan Laboratories, The University, Liverpool, England

9. Ionic Catalysis in Chain Oxidation of Alcohols 112
   E. T. Denisov, V. M. Solyanikov, and A. L. Alexandrov, Academy of Sciences, Moscow, USSR

10. Liquid-Phase Oxidation of Acrolein, Action of Catalysts 120
    Akira Misono, Tetsuo Osa, and Yasukasu Ohkatsu, University of Tokyo, Tokyo, Japan

11. Autoxidation of Chloroprene 138
   H. C. Bailey, BP Chemicals (U.K.), Ltd., Epsom, Surrey, England

12. Effect of Solvents on Rates and Routes of Oxidation Reactions 150
   G. E. Zaikov and Z. K. Maizus, Academy of Sciences, Moscow, USSR

BASE-CATALYZED AND HETEROATOM COMPOUND OXIDATIONS

13. The Role of Heteroatoms in Oxidation 166
   Cheves Walling, Columbia University, New York, N.Y.

14. Oxidation of Carbanions, Oxidation of Diarylmethanes and Diarylcarbinols in Basic Solution 174
   Glen A. Russell, Alan G. Bemis, Edwin J. Geels, Edward G. Janzen and Anthony J. Moye, Iowa State University, Ames, Iowa

15. Base-Catalyzed Autoxidation of 9,10-Dihydroanthracene and Related Compounds 203
   J. O. Hawthorne, K. A. Schowalter, A. W. Simon, and M. H. Wilt, United States Steel Corp., Monroeville, Pa., and M. S. Morgan, Carnegie-Mellon University, Pittsburgh, Pa.

16. Liquid-Phase Oxidation of Thiols to Disulfides 216
   J. D. Hopton, C. J. Swan, and D. L. Trimm, Imperial College, London, England

17. Oxidation and Chemiluminescence of Tetrakis(dimethylamino)-ethylene. Decay Rates of the Chemiluminescent Intermediate 225
   Carl A. Heller, Michelson Laboratory, China Lake, Calif.

18. Oxidation of Ozonization Products 245
   Dennis G. M. Diaper, Royal Military College of Canada, Kingston, Ontario, Canada

RADICAL INITIATION AND INTERACTIONS

19. Determination of Rate Constants for the Self-Reactions of Peroxy Radicals by Electron Spin Resonance Spectroscopy 258
   J. R. Thomas and K. U. Ingold, Chevron Research Co., Richmond, Calif.

20. Cage Reactions of Acetoxy Radicals 269
   J. C. Martin and Steven A. Dombchik, University of Illinois, Urbana, Ill.

21. Bond Dissociation Energies in the Phenyl Benzoate Molecule and in Related Free Radicals 282
   Peter Gray, Leeds University, Leeds, England

22. Thermochemistry of Oxidation Reactions 288
    S. W. Benson and R. Shaw, Stanford Research Institute, Menlo Park, Calif.

INHIBITION

23. Inhibition of Autoxidation 296
    K. U. Ingold, National Research Council, Ottawa, Canada

24. Action of Aliphatic Amines on Slow Oxidation of Acetaldehyde and Ethyl Ether, and on Decomposition of Organic Peroxides in the Gas Phase 306
    P. W. Jones and D. J. Waddington, University of York, York, England

25. Mechanism of Oxidation Inhibition by Zinc Dialkyl Dithiophosphates 323
    A. J. Burn, British Petroleum Co., Ltd., Middlesex, England

26. A New Method for Determining the Absolute Rate Constants of Autoxidation of Some Hydrocarbons 346
    H. Berger, A. M. W. Blaauw, M. M. Al, and P. Smael, Koninklijke/Shell-Laboratorium, Amsterdam, The Netherlands

Index 359

Oxidation of Organic Compounds

Volume II

Gas-Phase Oxidations, Homogeneous
and Heterogeneous Catalysis,
Applied Oxidations and Synthetic
Processes

Volume III

Ozone Chemistry, Photo and
Singled Oxygen and Biochemical
Oxidations

PREFACE

The International Oxidation Symposium was organized to summarize progress and remaining problems in reactions of organic compounds with oxygen. Original manuscripts were distributed in advance to the participants. The proceedings, presented here in three volumes, include refereed and revised papers, summaries by some section chairmen, and discussion as submitted by participants:

The program was arranged by Dr. Theodore Mill of Stanford Research Institute and by the section chairmen, who had considerable latitude in arranging their sessions. The business and hotel arrangements were made and supervised by L. Thomas Evans and Jean Burnet, respectively, of Stanford Research Institute.

Both the National Science Foundation and the U. S. Army Research Office (Durham) made substantial contributions towards the large expenses of preprinting manuscripts and assisting overseas speakers.

The following companies also contributed substantially to the costs of organizing and conducting the symposium:

Allied Chemical Corp.
ARCO Chemical Co.
Celanese Corp.
Cities Service Co., Inc.
Commercial Solvents Corp.
Ethyl Corp.
W. R. Grace & Co.
Hercules, Inc.
Marathon Oil Co.
Standard Oil (Indiana) Foundation, Inc.
Sun Oil Co.
Universal Oil Products Co.

Smaller contributions were made by:

Air Products and Chemicals, Inc.
Chevron Research Co.
Philip Morris, Inc.
Shell Development Co.
Sinclair Research, Inc.

The scientific success of the symposium depended almost entirely on the written and oral contributions of the participants. We particularly appreciated the able assistance of the section chairmen in arranging their programs and advising us on the manuscripts.

FRANK R. MAYO

Menlo Park, Calif.
March 1968

Introduction

FRANK A. MAYO

Stanford Research Institute, Menlo Park, Calif.

Although this symposium made considerable progress on many problems, it made little or no progress on those which I presented in opening the symposium as being of special interest and importance.

(1) Many years ago, Semenov (13) interpreted autoxidations of alkanes by rearrangement of alkyl radicals from carbon to oxygen of the alkylperoxy radicals. Many others have accepted his idea (4). However, I know of no precedent in organic chemistry for such reactions. Cullis and co-workers (3) have proposed some similar rearrangements as side reactions in autoxidations. In our own work (1) we have found no need for such rearrangements. We can account quite well for the rates and products of our autoxidations by the nonterminating interactions of alkylperoxy radicals to give oxygen and alkoxy radicals, followed by cleavage of these radicals. There is ample precedent for these reactions in low temperature organic chemistry. The problem is: at what temperatures and to what extent do rearrangements of carbon skeletons of alkylperoxy radicals become important in oxidation?

(2) For some time, we have been concerned with the interactions of secondary alkylperoxy radicals. From our own work (6), substantially all interactions of secondary alkylperoxy radicals are terminating in liquid-phase reactions at or below 100°C. However, at least some of these interactions are nonterminating in the gas phase above 100°C., the fraction of nonterminating reactions probably increasing with temperature. We have few good numbers for these relations. We need such numbers to extend the conclusions above, to account for alcohol-ketone ratios, and to correlate the amounts of cleavage accompanying oxidations of straight carbon chains. Data can be obtained either from autoxidation studies or from induced decompositions of secondary hydroperoxides (6).

(3) While the direct cleavage of a carbon-carbon double bond to two carbonyl groups by unzipping of polyperoxides (10, 11) or by ozonolysis is well known, the cleavage of α-methylstyrene to acetophenone and formaldehyde (11), the cleavage of polyisoprene in solution (2, 12), and the

high temperature oxidation of ethylene to formaldehyde (9) proceed by mechanisms which are still obscure. I think that the Knox mechanism (8) for ethylene, where an $HO_2 \cdot$ radical adds to the double bond and the resulting radical then reacts with oxygen, is untenable because the first adduct should decompose readily into an epoxide and an $HO \cdot$ radical (14). In the two liquid-phase cleavages cited, the cleavage step is kinetically identical with one of the chain propagation steps and therefore corresponds to either Reaction 1 or Reaction 2 below, written with α-methylstyrene as an example:

$$R\text{—}O_2\text{—}CH_2\text{—}\underset{Me}{\overset{Ph}{C}}\cdot + O_2 \rightarrow RO_2\cdot + CH_2O + PhAc \qquad (1)$$

$$R\text{—}O_2\cdot + H_2C\text{=}\underset{Me}{\overset{Ph}{C}} \rightarrow R\cdot + CH_2O + PhAc \qquad (2)$$

Filling in the mechanistic details is the important and unsolved problem.

(4) Howard and Ingold (7) proposed participation of a first-order termination reaction in the autoxidation of styrene. If such contributions from first-order terminations are real and widespread in autoxidation, more knowledge about them becomes essential.

(5) In gas-phase oxidations of isobutane at around 350°C., several workers have reported that the products are about 80% isobutylene and 20% of a mixture of several oxygen compounds. Hay, Knox, and Turner (5) have reported that the nature, but not the total amount, of the oxygen compounds depends on the walls of the reaction vessel. They proposed that oxygen compounds arise from wall reactions. I have found that other workers show little enthusiasm for this conclusion, but the right answer is important.

The problems above mostly involve homogeneous oxidations. Another objective of this symposium was to find out how similar are the mechanisms and reactions in homogeneous oxidations to those in heterogeneous catalysis and biological systems. So far it seems that they are not very similar because neither ordinarily involves free radicals. However, the methods used to study biological oxidations have much in common with those used by physical-organic chemists in homogeneous oxidations.

Literature Cited

(1) Allara, D. L., Mill, T., Hendry, D. G., Mayo, F. R., ADVAN. CHEM. SER. **76,** 40 (1968).
(2) Bell, C. L. M., *Trans. Inst. Rubber Ind.* **41,** T202 (1965).
(3) Cullis, C. F., Hardy, F. R. F., Turner, D. W., *Proc. Roy. Soc.* **A251,** 265 (1959).
(4) Fish, A., *Quart. Rev.* **18,** 243 (1964).
(5) Hay, J., Knox, J. H., Turner, J. M. C., "Tenth International Symposium on Combustion," p. 331, The Combustion Institute, Pittsburgh, Pa., 1965.
(6) Hiatt, R., Mill, T., Mayo, F. R., Irwin, K. C., Castleman, J. K., Gould, C. W., *J. Org. Chem.*, in press.
(7) Howard, J. A., Ingold, K. U., *Can. J. Chem.* **42,** 1044 (1964).
(8) Knox, J. H., *Chem. Commun.* **1965,** 108.
(9) Knox, J. H., Wells, C. H. J., *Trans. Faraday Soc.* **59,** 2786, 2801 (1963).
(10) Mayo, F. R., *J. Am. Chem. Soc.* **80,** 2465 (1958).
(11) Mayo, F. R., Miller, A. A., *J. Am. Chem. Soc.* **80,** 2480, 6701 (1958).
(12) Mayo, F. R., Egger, K., Irwin, K. C., *Rubber Chem. Technol.*, in press.
(13) Semenov, N. N., "Some Problems in Chemical Kinetics and Reactivity," Vol. 1, p. 92, Pergamon Press, New York, 1958.
(14) Van Sickle, D. E., Mayo, F. R., Gould, E. S., Arluck, R. M., *J. Am. Chem. Soc.* **89,** 977 (1967).

Liquid-Phase Oxidations

N. M. EMANUEL
Session Chairman

Present State and Main Trends of Research on Liquid–Phase Oxidation of Organic Compounds

N. M. EMANUEL

Institut of Chemical Physics, Academy of Sciences, Vorobyevskoye Chaussee 2-b, Moscow, U.S.S.R.

> *Oxidative chain reactions of organic compounds are current targets of theoretical and experimental study. The kinetic theory of collisions has influenced research on liquid-phase oxidation. This has led to determining rate constants for chain initiation, branching, extension, and rupture and to establishing the influence of solvent, vessel wall, and other factors in the mechanism of individual reactions. Research on liquid-phase oxidation has led to studies on free radical mechanisms and the role of peroxides in their formation.*

This symposium has been a very good, necessary, and important experiment, organized as a Faraday-type discussion. At the Faraday Discussion on oxidation in 1946, Eric Rideal said that in the field of kinetics we float over seas that are not to be found in maps. Things have changed drastically during these 20 years. Previously unknown reactions of chain generation, propagation, and termination were discovered during this period, and the over-all scheme of oxidation now contains many new elementary steps. Various efficient physical methods are now available. The ESR technique makes it possible to "see" the free radicals involved in gas- and liquid-phase oxidations. Great possibilities are opened up by the chemiluminescence technique, which, however, is not yet widely used. It also appears possible to understand the reasons for the strong effect of the dielectric constant on reaction rates. Moreover, this effect is found to obey laws that were previously considered valid only for ionic reactions—*i.e.*, the Kirkwood equation.

The chemistry of initial and later stages of oxidation can be distinguished. In the later stages, the reaction proceeds, in fact, in a different chemical system that has undergone great changes with time. Thus, at higher conversions, reaction control should be quite different from that for initial stages. This simple conclusion seems promising in that it should permit control of oxidation by changing the medium, introducing inhibitors, catalysts and other admixtures, as well as by changing temperature, pressure, and other similar factors.

At present the slower oxidative chain reactions of organic compounds, primarily hydrocarbons, are perhaps the most widespread targets of theoretical and experimental investigations of chain reactions. The significance of these investigations is enhanced by the fact that direct oxidation of hydrocarbons and other organic compounds possesses more than theoretical importance; it underlies many technological processes for producing valuable chemical products.

Production of phenol and acetone is based on liquid-phase oxidation of isopropylbenzene. Synthetic fatty acids and fatty alcohols for producing surfactants, terephthalic, adipic, and acetic acids used in producing synthetic and artificial fibers, a variety of solvents for the petroleum and coatings industries—these and other important products are obtained by liquid-phase oxidation of organic compounds. Oxidation processes comprise many parallel and sequential macroscopic and unit (or very simple) stages. The active centers in oxidative chain reactions are various free radicals, differing in structure and in reactivity, so that the "nomenclature" of these labile particles is constantly changing as oxidation processes are clarified by the appearance in the reaction zone of products which are also involved in the complex mechanism of these chemical conversions.

The kinetic theory of collisions, which has been so effective in developing the kinetics of vapor-phase reactions, has substantially influenced research on the processes of liquid-phase oxidation and in describing these processes. It has been thought that the lack of laws on which to base liquid-state theory (in contrast to the well-developed kinetic theory of gases) would in principle severely limit the development of a quantitative theory of liquid-phase reactions. At present the characteristics of the liquid state are carefully considered in discussing the mechanism of intermolecular reactions, influence of the medium on reactivity of compounds, etc.

This opened the possibility of ascertaining quantitative characteristics in the numerous individual reactions comprising the complex mechanism of oxidative chain reactions. Thus, rate constants have been determined for reactions with respect to their initiation and the branching, extension, and chain rupture, establishing specific details concerning the influence of solvents, reaction vessel surfaces, and other factors on the mechanism of individual reactions. Results of these investigations have made it possible to lay out effective plans for novel industrial oxidation processes in modern petroleum chemistry.

In developing oxidation processes a major source of free radical formation was found to be degenerate chain branching. Among the products derived from the branching were intermediate peroxides ROOH. Formation of radicals from the hydroperoxides proceeded not only by monomolecular breakdown of hydroperoxides:

$$ROOH \xrightarrow{k_3'} RO^{\cdot} + {}^{\cdot}OH,$$

but also by interaction of two saturated molecules:

$$ROOH + RH \xrightarrow{k_3'} RO^{\cdot} + R^{\cdot} + H_2O.$$

In some cases the main portion of the final products is formed by decomposition of peroxides. This is the current aspect of the classic Bach-Engler peroxide theory with respect to the chain theory of oxidation processes.

The mechanism of chain branching in the later stages of oxidation is more complex. For example, it is being demonstrated that intermolecular hydrogen bonds substantially influence the process by forming hydroperoxides and oxidation products among the molecules and by accumulating these in the reaction mixture.

In studying the reaction mechanism for chain extension it is assumed, in line with the fundamental reaction of chain extension,

$$RO_2^{\cdot} + RH \xrightarrow{k_2} ROOH + R^{\cdot}$$

where k_2 is the rate constant for the chain extension reaction, that the peroxide radical $RO_2^{\cdot}$ can often isomerize, accompanied by rupture of the C-C bond and by formation of carbonyl compounds and alcohol radicals. Analogous reactions involving radical decomposition were observed previously in vapor-phase oxidation studies.

As oxidation processes were clarified, it was observed in other chain extension reactions that RO_2 radicals reacted with oxidation products: hydroperoxides, alcohols, and ketones. The high reactivity of hydroperoxides and alcohols strongly influences the mechanism of oxidation processes. Chain rupture results from recombination of RO_2 radicals.

Research on liquid-phase oxidation processes has opened up a tremendous area for research on free radical mechanisms. Such research will be concerned, not with hypothetical particles like those studied a few years ago, but with thoroughly material active centers, amenable to investigation by experimental techniques currently in use. Problems of foremost importance are involved in correlating data on unit processes and on free radical mechanisms with identification of free radicals and their characteristic properties. Widely varied types and classes of organic compounds come under one or another of the prospective problems in constructing a general kinetic theory of unit mechanisms in oxidative chain reactions among these compounds, elucidating specific characteristics, correlating their properties with various functional groups, etc.

There seems to be special promise in oxidizing liquefied hydrocarbon gases at temperatures and pressures approximating critical levels. That such reactions are highly effective is attested, for example, by the liquid-phase oxidation of butane, one of the simplest and most efficient methods of producing acetic acid and methyl ethyl ketone.

Isomerization and decomposition of peroxide radicals—reactions which obviously do occur and which exert significant influence—depend on the nature of the reaction vessel's surface. For example, when butane is oxidized in a glass reactor, products formed by breakdown of RO_2 radicals are completely absent, whereas in a steel reactor they correspond to 20 mole % of the reacted butane. The quantity of these products rises to 35 mole % when the reactor is filled with metal packing shapes.

Formation of products containing less than four carbon atoms is not related to the catalytic activity of the metal on the decomposition of hydroperoxides. Hence, the liquid-phase oxidation of hydrocarbons involves heterogeneous catalytic reactions of isomerization and decomposition of peroxide radicals, proceeding on the reactor surface.

Results of these investigations demonstrate that changes of the reactor surface can be an effective method for directing chemical reactions. Thus, developing a theory of how heterogeneous factors influence liquid-phase chain reactions is one of the important lines of advancement in this area. Only a few years ago it was thought, almost *a priori,* that there are practically no heterogeneous factors in liquid-phase oxidation and that liquid-phase processes differ from vapor-phase processes in this respect.

Greater possibilities for discovering new kinetic phenomena and designing novel technological processes are opened up by combinations of chemical reactions depending on the utilization of free radicals and intermediate compounds formed in one of the reactions occurring in a system in order to obtain products of their interaction as active components with other components of the reaction mixture.

On the basis of this principle, a process involving oxidation of unsaturated hydrocarbons and other organic compounds, more readily oxidized than alkenes, contribute substantially to solving problems in direct single-stage production of propylene and higher alkylene oxides.

Upon oxidizing ethene, propene, or isobutene together with aldehydes, alkylated aromatic hydrocarbons, methyl ethyl ketone or other compounds, the products include, along with hydroperoxide conversion products (acids, ketones, alcohols) some alkylene oxides.

Continued investigation revealed that the principal epoxidizing agents for combined oxidation of unsaturated compounds and aldehydes are not the corresponding peracids, but the radicals of acyl peroxides.

Including mixed systems among the research on the mechanism of liquid-phase oxidation reactions aids subsequent development of the chain theory and undoubtedly contributes to practical chemistry.

An alluring field of research is the mechanism of action of oxidation inhibitors. This research will undoubtedly yield in the near future a theory for inhibition of undesirable oxidation processes. The relatively stable free radicals observed on such inhibition display extremely interesting properties. Of great interest are the effects of synergism, of inhibitor mixtures, and of mixtures of inhibitors with catalysts. A strictly quantitative and elegant description of all these phenomena may be made within the scope of the chain theory for slow oxidation.

It has always been considered that the condition of the reactor wall is less important for liquid-phase processes than for gas-phase reactions. Now there are numerous examples of marked wall effects which induce essentially new chemical results in liquid-phase oxidations. Hence, the parts played by reactor walls, by solid surfaces, and by other solid catalysts in liquid-phase oxidations should be considered as one of the most important remaining problems.

We are on the right path in attempting to establish the kinetics of oxidation reactions in flow systems. This is the scientific basis of continuous processes in chemical industry and an invaluable source of additional information on reaction mechanisms.

Research on oxidation problems is in progress in nearly every country. Many projects for directed research on oxidation processes have been objects of sustained, organized international cooperation. The quantity of information on theoretical and applied aspects in this area grows ever larger.

This symposium shows that research on oxidation processes constitutes a fertile field, tremendously rich in possibilities. This is particularly true of complex multicomponent chemical systems, where particularly great progress is expected. At present science is equipped with adequate experimental techniques for solving the problems. This symposium constitutes an important step in advancing chemical kinetics and will undoubtedly exert significant influence on future research on oxidation reactions and their practical applications.

Finally, I wish to express my sincerest thanks to Dr. Mayo and the Program Chairman, Dr. Mill for the great amount of work they did in connection with the symposium as well as for their kind hospitality.

2

Absolute Rate Constants for Hydrocarbon Autoxidation

VII. Reactivities of Peroxy Radicals Toward Hydrocarbons and Hydroperoxides

J. A. HOWARD, W. J. SCHWALM,[1] and K. U. INGOLD
Division of Applied Chemistry, National Research Council of Canada, Ottawa, Canada

> *The four propagation and three termination rate constants for the co-oxidation of cumene and Tetralin were determined at 30° and 56°C. The secondary tetralylperoxy radical is about four times as reactive in hydrogen atom abstraction as the tertiary cumylperoxy radical. The rate constant for termination by reaction of a tetralylperoxy and a cumylperoxy radical has about one quarter of the value for termination by two tetralylperoxy radicals. Rate constants for the transfer of hydroperoxidic hydrogen to peroxy radicals were measured. This reaction exhibits an exceptionally large deuterium isotope effect ($k_H/k_D \sim 17$ to 30). A possible explanation for the high value is suggested. The absolute rate constants at 30°C. for hydrogen atom abstraction from toluene, ethylbenzene, cumene, and Tetralin by cumylperoxy, tetralylperoxy, and 9,10-dihydroanthracyl-9-peroxy radicals were measured.*

In 1955 Russell showed that mixtures of cumene and small amounts of Tetralin oxidized at rates considerably below the rates of oxidation of either of the pure hydrocarbons (26). Russell suggested that this decrease in rate was caused by the fact that tetralylperoxy radicals terminate oxidation chains much more readily than cumylperoxy radicals. The mixtures, therefore, oxidize at a lower rate than pure cumene because of

[1] National Research Council summer student, 1966.

the lowered steady-state concentration of peroxy radicals. It was shown that small amounts of other reactive hydrocarbons that produce secondary peroxy radicals also retard the oxidation of cumene. The co-oxidation of many pairs of hydrocarbons has since been examined (1, 2, 3, 4, 14, 16, 19, 25, 29, 36). All subsequent work has tended to confirm Russell's explanation of the minimum in the rate vs. composition curves.

The co-oxidation of cumene (CH) and Tetralin (TH) can be represented by the following simplified reaction scheme:

$$\text{Initiator} \xrightarrow{R_i} \text{production of } C\dot{O}O \text{ or } T\dot{O}O \tag{1}$$

$$C\dot{O}O + CH \xrightarrow{k_p^{CC}} COOH + \dot{C} \xrightarrow{O_2} C\dot{O}O \tag{2}$$

$$C\dot{O}O + TH \xrightarrow{k_p^{CT}} COOH + \dot{T} \xrightarrow{O_2} T\dot{O}O \tag{3}$$

$$T\dot{O}O + TH \xrightarrow{k_p^{TT}} TOOH + \dot{T} \xrightarrow{O_2} T\dot{O}O \tag{4}$$

$$T\dot{O}O + CH \xrightarrow{k_p^{TC}} TOOH + \dot{C} \xrightarrow{O_2} C\dot{O}O \tag{5}$$

$$C\dot{O}O + C\dot{O}O \xrightarrow{2k_t^{CC}} \text{inactive products} \tag{6}$$

$$C\dot{O}O + T\dot{O}O \xrightarrow{4k_t^{CT}} \text{inactive products} \tag{7}$$

$$T\dot{O}O + T\dot{O}O \xrightarrow{2k_t^{TT}} \text{inactive products} \tag{8}$$

The rate of oxidation is given by Russell (26):

$$\frac{-d[O_2]}{dt} = \frac{[\,[CH]^2 k_p^{CC} k_p^{TC} + 2[CH][TH] k_p^{CT} k_p^{TC} + [TH]^2 k_p^{TT} k_p^{CT}\,] R_i^{1/2}}{[2[CH]^2 k_t^{CC} (k_p^{TC})^2 + 4[CH][TH] k_t^{CT} k_p^{CT} k_p^{TC} + 2[TH]^2 k_t^{TT} (k_p^{CT})^2]^{1/2}} \tag{9}$$

[In this paper the rate expressions have all been corrected for nitrogen evolution from the azo initiator, oxygen absorption by initiator radicals, and oxygen evolution in termination. It is assumed that the initiator which decomposes without starting oxidation chains does not react with oxygen (21). This correction involves the addition of $(1-e)R_i/2e$ to the measured rate, where e is the efficiency of chain initiation, found to be 0.5 at 30°C. and 0.6 at 56°C. The rate constant for Reaction 7 has been written as $4k_t^{CT}$ in order that the three termination constants may be comparable (26, footnote 27).]

The rate constants k_p^{CC}, k_t^{CC}, and k_p^{TT}, k_t^{TT} can be obtained by rotating sector studies on the pure hydrocarbons. The cross-propagation and termination constants can be calculated by substituting into Equation 9 experimental rates at different mixture compositions and solving the resultant set of equations with a computer (*4*). An analogous, but somewhat simpler approach, which also depends only on the measurement of oxidation rates at different mixture compositions has been employed by Tsepalov *et al.* (*36*). More usually, cross-rate constants have been determined by measuring the rates of oxidation of each component in two different mixtures, by following either the consumption of the reactants (*17*) or the accumulation of the products (*1, 2, 3, 4, 26*). These rates can be combined with a measurement of the over-all rate of oxidation of a mixture to yield the three cross-rate constants. However, accurate values for these rate constants are extremely difficult to obtain by any of these methods. The present paper describes a simple and elegant method for determining cross-propagation constants with an accuracy equal to that with which propagation constants for pure materials may be obtained. The method is based on the oxidation of one hydrocarbon in the presence of the hydroperoxide of a second hydrocarbon. The cross-termination rate constant is then derived by substitution of the known rate constants into Equation 9.

Experimental

Our application of the rotating sector technique to hydrocarbon oxidations has been described (*14, 15, 18*). Oxidation rates were measured at the longest convenient chain lengths and corrected for the absorption and evolution of gas in initiation and in peroxy radical–peroxy radical reactions. α,α'-Azobiscyclohexylnitrile (ACHN) was used as the photoinitiator at 30°C. and α,α'-azobis-1-propanol diacetate as the photoinitiator at 56°C. α,α'-Azobisisobutyronitrile (AIBN) was used as a thermal initiator at 30° and 56°C.

Tetralin hydroperoxide (1,2,3,4-tetrahydro-1-naphthyl hydroperoxide) and 9,10-dihydroanthracyl-9-hydroperoxide were prepared by oxidizing the two hydrocarbons and purified by recrystallization. Commercial cumene hydroperoxide was purified by successive conversions to its sodium salt until it no longer increased the rate of oxidation of cumene at 56°C. All three hydroperoxides were 100% pure by iodometric titration. They all initiated oxidations both thermally (possibly by the bimolecular reaction, $R'OOH + RH \rightarrow R'O\cdot + H_2O + R\cdot$ (*33*)) and photochemically. The experimental conditions were chosen so that the rate of the thermally initiated reaction was less than 10% of the rate of the photoreaction. The rates of chain initiation were measured with the inhibitors 2,6-di-*tert*-butyl-4-methylphenol and 2,6-di-*tert*-butyl-4-methoxyphenol. None of the hydroperoxides introduced any kinetically first-order chain termination process into the over-all reaction.

Results and Discussion

Co-oxidation of Cumene and Tetralin. The present method for determining cross-propagation constants is based on Thomas and Tolman's (34) observation that the oxidation of cumene is strongly inhibited by adding low concentrations of Tetralin hydroperoxide. These workers concluded that TOO· radicals formed in the transfer reaction:

$$COO\cdot + TOOH \xrightarrow{k_{trans}^{CTOOH}} COOH + TOO\cdot \qquad (10)$$

take over chain termination from the COO· radicals.

The rate of oxidation of 6.7M cumene in chlorobenzene at 30°C. was found to be decreased to a constant value by adding $\geq$ 0.1M Tetralin hydroperoxide. At this point, all the COO· radicals are being converted into TOO· radicals without undergoing any other reactions. Hence, the TOO· radicals are propagating and terminating the chain. The propagation constant is the cross constant k_p^{TC} (Reaction 5) and the termination constant is k_t^{TT}. The rate is given by

$$\left(\frac{-dO_2}{dt}\right)^{CH}_{TOOH} = k_p \, [CH] \left(\frac{R_i}{2k_t}\right)^{1/2} \qquad (11)$$

Tetralin hydroperoxide has little or no effect on the thermally or photochemically initiated oxidation of Tetralin, nor are the absolute rate constants for the oxidation of Tetralin (1.7M in chlorobenzene) affected by adding 0.1M [TOOH] (Table I). [Hydrogen-bonded peroxy radicals are either unimportant in this system or have the same reactivity as the peroxy radicals formed in the absence of hydroperoxide. A similar conclusion applies to propagation in cumene and cumene–COOH mixtures (see Table I).]

The measured rate constant for chain termination in the oxidation of cumene containing $\geq$ 0.1M [TOOH] (2.8 × 10^6 Mole^{-1} sec.$^{-1}$ at 30°C.) is in fairly good agreement with the value found for Tetralin (3.8 × 10^6) (14) and Tetralin + TOOH [3.6 × 10^6 Mole^{-1} sec.$^{-1}$ (Table I)]. The termination constant for Tetralin alone was considered to be the most accurate and this value was therefore combined with the measured rate of oxidation of cumene + TOOH to give k_p = 0.5 Mole^{-1} sec.$^{-1}$ at 30°C. [The small value of $k_p/\sqrt{2k_t}$ [1.8 × 10^{-4} M$^{-1/2}$ sec.$^{-1/2}$] for cumene in the presence of TOOH meant that these measurements had to be made at fairly short chain lengths ($\sim$2 to 5). It was not possible to increase the chain length by reducing R_i since thermal initiation by the TOOH became important for $R_i <$ 1 × 10^{-7} Mole sec.$^{-1}$.]

With cumene hydroperoxide the situation is more complex. Traylor and Russell (35) found that adding this hydroperoxide to cumene increased the oxidation rate of the cumene (cf. Table I). This phenomenon has been explained on the basis of the termination mechanism for the oxidation of cumene which is shown below.

$$C\dot{O}O + C\dot{O}O \rightarrow [COOOOC]_{cage} \begin{matrix} \nearrow \text{inactive products} \\ \xrightarrow{\text{direct}} \\ \searrow \text{indirect} \\ \searrow 2C\dot{O} + O_2 \end{matrix} \quad (6)$$

$$C\dot{O} \xrightarrow{O_2} C_6H_5COCH_3 + CH_3O\dot{O} \quad (12)$$

$$C\dot{O}O + CH_3O\dot{O} \rightarrow \text{inactive products} \quad (13)$$

The over-all rate constant for chain termination, $2k_t^{cc}$ ($2 \times 7.5 \times 10^3$ Mole^{-1} sec.$^{-1}$ for neat cumene at 30°C.), can be considered to be divided into two parts: a direct termination constant $2k_{t(\text{direct})}^{cc}$ which represents the fraction of the $C\dot{O}$ radicals which react with one another in the cage, and an indirect termination constant which represents the fraction of $C\dot{O}$ radicals which escape from the cage and whose products enter into Reaction 13 rather than being consumed in the propagation steps

$$C\dot{O} + CH \xrightarrow{O_2} COH + C\dot{O}O \quad (14)$$

$$CH_3O\dot{O} + CH \xrightarrow{O_2} CH_3OOH + C\dot{O}O \quad (15)$$

The addition of COOH to cumene converts the $C\dot{O}$ and $CH_3O\dot{O}$ radicals to $C\dot{O}O$ radicals by rapid chain transfer reactions (see Appendix). The indirect termination process is therefore suppressed, and with sufficient

Table I. Oxidation of Neat Cumene and 1.7M Tetralin (in and 0.1 to 0.4M Tetralin

Hydrocarbon	Hydroperoxide	Chain Length	Initiation[a]
Cumene	COOH	40	Th
Cumene	—	120	Th
Cumene	COOH	24	Ph
Cumene[c]	—	25–50	Ph
Cumene	TOOH	2–5	Ph
Tetralin	TOOH	40	Ph
Tetralin	—	40	Th
Tetralin	TOOH	10	Ph
Tetralin	—	10	Ph
Tetralin	COOH	40	Ph
Tetralin	COOH	340	Th

[a] Th = thermal initiation with AIBN. Ph = photochemical initiation with ACHN.
[b] Calculated from $k_p/(2k_t)^{1/2}$ assuming $k_p = 0.18$ Mole^{-1} sec.$^{-1}$ (see text).

COOH (0.5 to 1.6M) all termination occurs by the direct process. The rate is increased and is given by

$$\left(\frac{-d[O_2]}{dt}\right)^{CH}_{COOH} = k_p^{CC}[CH]\left(\frac{R_i}{2k_{t(direct)}^{CC}}\right)^{1/2} \quad (16)$$

The experimentally measured direct chain termination constant was found to be 5.5×10^3 Mole^{-1} sec.$^{-1}$. However, this value was not considered very accurate because there is a large correction to the measured oxidation rates for oxygen evolved in the self-reactions of $C\dot{O}O$ radicals at the relatively high photo-initiation rates required to reduce the importance of thermal initiation from the added COOH. A more accurate value of 2.9×10^3 mole^{-1} sec.$^{-1}$ was calculated from the limiting value of $k_p^{CC}/[2k_{t(direct)}^{CC}]^{-1/2}$ at high [COOH] for the AIBN thermally initiated reaction at 30°C. combined with the measured value of k_p^{CC} for neat cumene (0.18 Mole^{-1} sec.$^{-1}$).

Adding COOH to Tetralin increases its oxidation rate to a limiting value which is given

$$\left(\frac{-d[O_2]}{dt}\right)^{TH}_{COOH} = k_p^{CT}[TH]\left(\frac{R_i}{2k_{t(direct)}^{CC}}\right)^{1/2} \quad (17)$$

when all propagation and termination by $T\dot{O}O$ radicals is prevented by their conversion to $C\dot{O}O$ radicals.

$$T\dot{O}O + COOH \xrightarrow{k_{trans}^{TCOOH}} TOOH + C\dot{O}O \quad (10a)$$

A best value for the cross-propagation constant $k_p^{CT} = 1.65$ Mole^{-1} sec.$^{-1}$ was obtained at 30°C. by taking $k_{t(direct)}^{CC} = 2.9 \times 10^3$ Mole^{-1} sec.$^{-1}$. The

Chlorobenzene) in Presence of 0.5 to 1.5M Cumene Hydroperoxide Hydroperoxide at 30°C.

$k_p/(2k_t)^{1/2} \times$ 10^3 Mole$^{-1/2}$ sec.$^{-1/2}$	$\dfrac{[k_p/(2k_t)^{1/2}]_{ROOH}}{[k_p/(2k_t)^{1/2}]_0}$	k_p, Mole^{-1} sec.$^{-1}$	$k_t \times 10^{-4}$, Mole^{-1} sec.$^{-1}$
2.36	1.6	—	0.29[b]
1.50	—	—	—
2.02	1.35	0.21	0.55
1.50	—	0.18	0.75
0.18	0.12	0.43	280
2.4	1.04	—	—
2.3	—	—	—
2.3	1.00	6.2	360
2.3	—	6.4	380
21.0	9.1	1.95	0.43
21.6	9.4	—	—

[c] Absolute rate constants for oxidation of 7.17M cumene measured over a range of chain lengths, ν. At $\nu = 25 k_t^{cc} = 8.8 \times 10^3$ Mole^{-1} sec.$^{-1}$ and at $\nu = 50$ k_t^{cc} = 5.0×10^3 Mole^{-1} sec.$^{-1}$.

measured values of k_p^{CT} and $k_{t\,(direct)}^{CC}$ were 1.95 and 4.3×10^3 Mole^{-1} sec.$^{-1}$, respectively (Table I).

The simplicity of the present method for obtaining cross-propagation constants recommends its use where the hydroperoxides are both available and stable at the reaction temperature.

The cross-termination constant k_t^{CT} was obtained from the rates of oxidation of cumene-Tetralin mixtures in the region of the rate minimum (Table II). The rates, rate constants, and concentrations were substituted into Equation 9, giving $k_t^{CT} = 1.0 \times 10^6$ Mole^{-1} sec.$^{-1}$ at 30°C. The rate constants for the oxidation of the cumene-Tetralin system are summarized in Table III.

Table II. Co-oxidation of Cumene and Tetralin at 30°C.

($R_i = 2.2 \times 10^{-8}$ Mole sec.$^{-1}$)

[CH], M	[TH], M	Oxidation Rate $\times 10^7$, Mole sec.$^{-1}$	k_p, Mole^{-1} sec.$^{-1}$	$k_t \times 10^{-4}$ Mole^{-1} sec.$^{-1}$	ϕ
7.17	—	16	0.18	0.75	—
7.12	0.056	8.4	—	—	3
7.01	0.165	6.4	0.30	13	3
6.76	0.42	5.5	0.61	75	6
6.36	0.85	5.4	0.94	140	6
—	7.34	25	6.4	380	—

Table III. Collected Rate Constants (in Mole^{-1} sec.$^{-1}$) for Autoxidation of Cumene and Tetralin

	30°C.	56°C.		30°C.	56°C.
k_p^{CC}	0.18	0.9 [a]	k_t^{CC}	7.5×10^3	2.6×10^4
k_p^{CT}	1.65	4.9 [a,b]	$k_{t\,(direct)}^{CC}$	2.9×10^3	9.5×10^3
k_p^{TT}	6.4	19.0 [a,b]	k_t^{TT}	3.8×10^6	6.7×10^6 [c]
k_p^{CT}	0.5	2.7 [a]	k_t^{CT}	1.0×10^6	1.6×10^6 [c]
k_{trans}^{CTOOH}	600	1100	k_{trans}^{TCOOH}	2500	2800
$(k_{trans}^{CTOOH})_{H_2O}$	360	—	$(k_{trans}^{TCOOH})_{H_2O}$	2600	—
$(k_{trans}^{CTOOD})_{D_2O}$	12	—	$(k_{trans}^{TCOOD})_{D_2O}$	140	—
$k_{trans}^{CTOOH}/k_{trans}^{CTOOD}$	30	—	$k_{trans}^{TCOOH}/k_{trans}^{TCOOD}$	17	—

[a] Mayo, Syz, Mill, and Castleman (22) gave $k_p^{TT}/k_p^{TC} = 4.6 \pm 0.4$ and $k_p^{CC}/k_p^{CT} = 0.13 \pm 0.02$ at 60°C. which can be compared with our values of 7.0 and 0.18, respectively, at 56°C.
[b] Calculated on assumption that $E_p = 8300$ cal./mole (14).
[c] Calculated on assumption that $E_t = 4300$ cal./mole (14).

Most previous workers have tended to ignore or discount any small differences observed in the reactivities of different peroxy radicals (*cf.* 36). On the basis of our measurements of propagation constants for the oxidation of pure hydrocarbons we have recently suggested that there are significant differences in the reactivities of primary, secondary, and tertiary peroxy radicals (*15, 18, 24*). The present results leave no doubt that the secondary tetralylperoxy radical is about three or four times more reactive than the tertiary cumylperoxy radicals—that is, $k_p^{TT}/k_p^{CT} = 3.9$ and $k_p^{TC}/k_p^{CC} = 2.8$. Russell's data, which were obtained at 90°C. (*26*), give 1.7 and 2.5, respectively, for these two ratios.

In chain reactions involving three termination steps (two uncrossed and one crossed) the quantity $\phi = k_t^{CT}/(k_t^{CC} k_t^{TT})^{1/2}$ is frequently used to interrelate the cross-termination constant with the two uncrossed termination constants. For many different types of radical ϕ is found to be about 1 (or alternatively, if the statistical factor of 2 favoring the crossed termination process is ignored in the definition of the rate constants, $\phi \approx 2$). In the present reaction system ~3–6, in agreement with the value obtained by Russell at 90°C. (*26*). The crossed termination constant itself is somewhat less than half the value found for k_t. This seems reasonable since only one hydrogen atom will be available for transfer in the crossed termination, compared with the two that are available in the self-reaction of two tetralylperoxy radicals. In addition, steric hindrance to reaction should be greater for the crossed termination than for Reaction 8. The products are presumably cumyl alcohol, α-tetralone [3,4-dihydro-1(2H)naphthalenone], and oxygen (*28*).

Comparatively few ϕ values have been measured for liquid-phase co-oxidations of hydrocarbon mixtures. With the exception of the cumene-Tetralin system, the reported values are all surprisingly low even for other systems giving tertiary and secondary (or primary) peroxy radicals. For example, at 60°C. ϕ values of 0.7 (*36*) and 1.3 (*4*) have been reported for the co-oxidation of cumene and ethylbenzene [$k_t = 2.0 \times 10^7$ at 30°C. (*14*)], and a value of 1.4 (*2*) has been reported for the co-oxidation of cumene and 1-hexene [which gives mainly primary peroxy radicals with k_t probably $\approx 1.3 \times 10^8$ at 30°C. (*15*)]. The confirmation that the present work provides for a relatively large ϕ value in the cumene-Tetralin system suggests that the other systems deserve a close and careful reinvestigation.

The measured average termination constants in the co-oxidation system (Table II) directly confirm Russell's original conclusion that the minimum in the curve of oxidation rate vs. composition is caused by an increase in the rate constant of chain termination.

Hydrogen Atom Transfer from Hydroperoxides to Peroxy Radicals. The reaction of cumylperoxy radicals with Tetralin hydroperoxide (Reaction 10) can be studied at hydroperoxide concentrations below those required to reduce the oxidation rate to its limiting value. The rate of oxidation of cumene alone can be represented by:

$$\left(\frac{-d[O_2]}{dt}\right)_o^{CH} = R_o^{CH} = k_p^{CC}[CH]\left(\frac{R_i}{2k_t^{CC}}\right)^{1/2} \quad (18)$$

and in the presence of added Tetralin hydroperoxide by:

$$\left(\frac{-d[O_2]}{dt}\right)^{CH}_{TOOH} = R^{CH}_{TOOH} = (k^{CC}_p [CH] + k^{CTOOH}_{trans}[TOOH]) \tag{19}$$

$$\left(\frac{R_i}{2k^{CC}_t + 4k^{CT}_t \frac{k^{CTOOH}_{trans}[TOOH]}{k_p [CH]} + 2k^{TT}_t \left(\frac{k^{CTOOH}_{trans}[TOOH]}{k^{TC}_p [CH]}\right)^2}\right)^{1/2}$$

In the limit at high [TOOH] concentrations Equation 19 reduces to Equation 11. At low [TOOH] concentrations the measured rates can be most simply handled in the form of the ratio

$(R^{CH}_{TOOH}/R^{CH}_o)^2$ at constants R_i—that is,

$$\left(\frac{R^{CH}_{TOOH}}{R^{CH}_o}\right)^2 = \tag{20}$$

$$\frac{k^{CC}_t (k^{TC}_p)^2 [(k^{CC}_p [CH])^2 + 2k^{CC}_p k^{CTOOH}_{trans}[CH][TOOH] + (k^{CTOOH}_{trans}[TOOH])^2]}{k^{CC}_t (k^{CC}_p k^{TC}_p [CH])^2 + 2k^{CT}_t (k^{CC}_p)^2 k^{TC}_p k^{CTOOH}_{trans}[CH][TOOH] + k^{TT}_t (k^{CC}_p k^{CTOOH}_{trans}[TOOH])^2}$$

which, for neat cumene (7.17M), on substitution of the measured rate constants at 30°C. from Table II reduces to:

$$\left(\frac{R^{CH}_{TOOH}}{R^{CH}_o}\right)^2 = \tag{21}$$

$$\frac{1.71 \times 10^{-2}[1.67 + 2.58\, k^{CTOOH}_{trans}[TOOH] + (k^{CTOOH}_{trans}[TOOH])^2]}{2.85 \times 10^{-2} + 2.00\, k^{CTOOH}_{trans}[TOOH] + (k^{CTOOH}_{trans}[TOOH])^2}$$

The rate constant k^{CTOOH}_{trans} is determined from a plot of $(R^{CH}_{TOOH}/R^{CH}_o)^2$ against $k^{CTOOH}_{trans}[TOOH]$ at several different [TOOH] concentrations. In a similar way, k^{TCOOH}_{trans} can be obtained by the addition of COOH to Tetralin. The numerical expression at 30°C. is

$$\left(\frac{R^{TH}_{COOH}}{R^{TH}_o}\right)^2 = \tag{22}$$

$$\frac{33.7[117 + 21.6\, k^{TCOOH}_{trans}[COOH] + (k^{TCOOH}_{trans}[COOH])^2]}{3940 + 744\, k^{TCOOH}_{trans}[COOH] + (k^{TCOOH}_{trans}[COOH])^2}$$

The data given in Tables IV and V have been plotted in Figures 1 and 2 together with the theoretical curves for the best values of k^{CTOOH}_{trans} and k^{TCOOH}_{trans} derived from Equations 21 and 22. The rate constants at 30° and 56°C., respectively, are k^{CTOOH}_{trans} = 600 and 1100 Mole^{-1} sec.$^{-1}$, k^{TCOOH}_{trans} = 2500 and 2800 Mole^{-1} sec.$^{-1}$ (At 56°C. rate constants were obtained from both photoinitiated and AIBN thermally initiated oxidations.) and

Table IV. Effect of Tetralin Hydroperoxide on Autoxidation of Cumene at 30°C.

($R_o^{CH} = 2.30 \times 10^{-6}$ Mole sec.$^{-1}$, $R_i = 4.2 \times 10^{-8}$ Mole sec.$^{-1}$)

[TOOH] × 10^5 [TOOD] × 10^4, M		$\left(\dfrac{R_{TOOH}^{CH}}{R_o^{CH}}\right)^2$	
		With H_2O	With D_2O
3.85	0.55	0.64	—
7.7	0.32	0.43	0.68
15.4	0.20	0.28	0.59
30.8	—	—	0.36
39	0.09	—	—

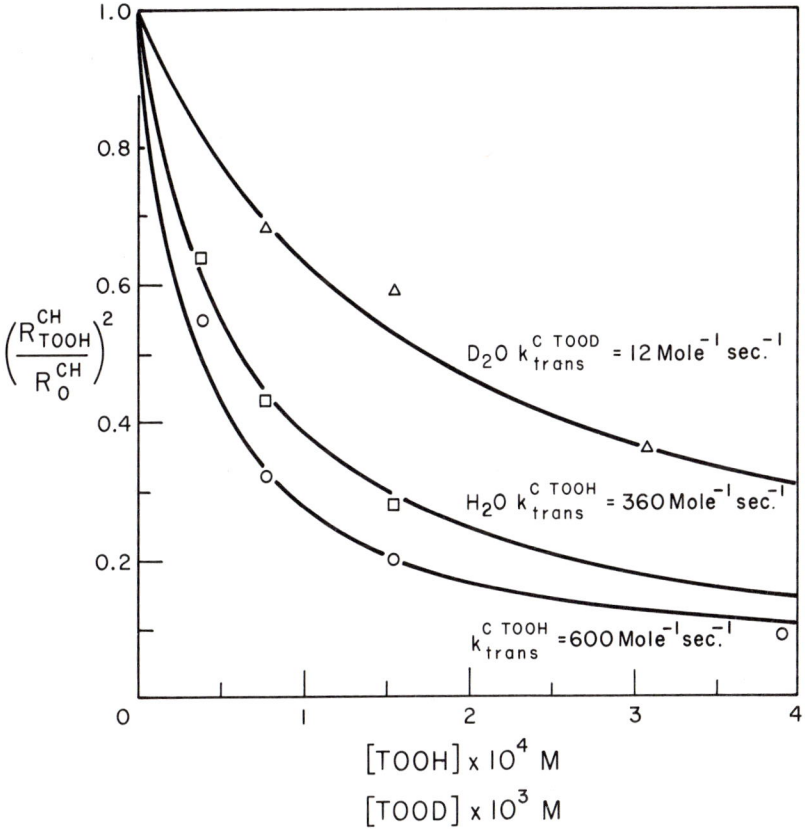

Figure 1. Effect of Tetralin hydroperoxide on autoxidation of cumene at 30°C.

○ Cumene
□ Cumene + H_2O
△ Cumene + D_2O

Table V. Effect of Cumene Hydroperoxide on Autoxidation of Tetralin at 30°C.

($R_o^{TH} = 9.84 \times 10^{-7}$ Mole sec.$^{-1}$, $R_i = 4.2 \times 10^{-8}$ Mole sec.$^{-1}$)

[COOH] × 10³ [COOD] × 10³, M	$\left(\dfrac{R_{COOH}^{TH}}{R_o^{TH}}\right)^2$	With H$_2$O	With D$_2$O
7.7	1.75	—	—
15.4	2.3	2.3	—
39	4.2	—	—
240	—	11	2.2

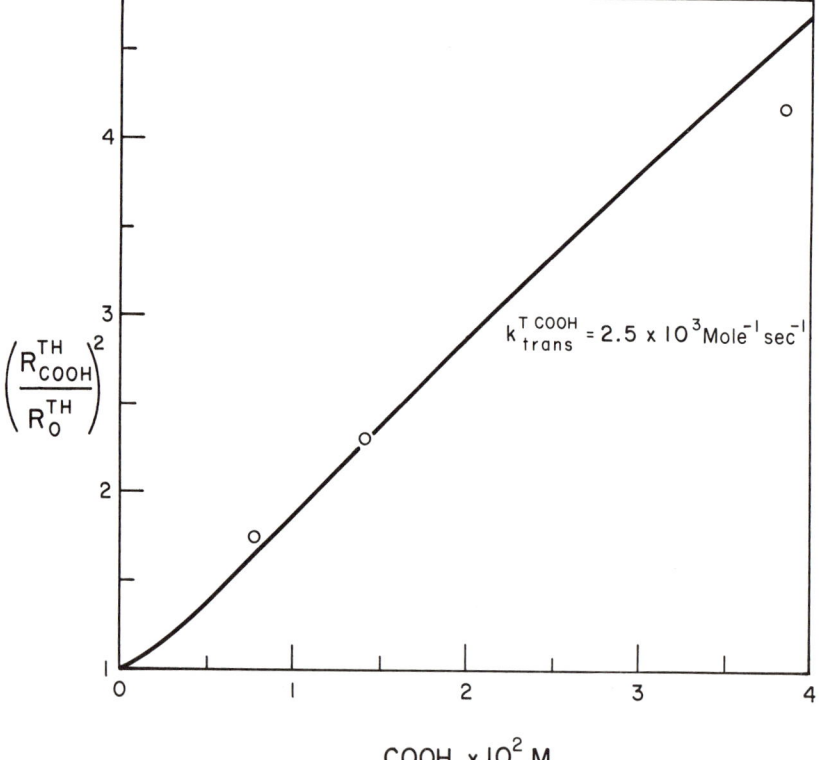

Figure 2. Effect of cumene hydroperoxide on autoxidation of Tetralin

hence K_{trans}, the equilibrium constant for Reaction 10 = 0.24 and 0.4. [k_{trans} and K_{trans} are very sensitive to the magnitudes of the rate constants used in Equations 21 and 22. For example, if k_t^{CT} had only half its measured value—i.e., if k_t^{CT} were only 5×10^5 Mole^{-1} sec.$^{-1}$—then $k_{\text{trans}}^{CTOOH} \approx 1400$, $k_{\text{trans}}^{TCOOH} \approx 1000$ Mole^{-1} sec.$^{-1}$, and $K_{\text{trans}} = 1.4$. The mean values of k_{trans}—i.e., $K_{trans} = 1.0$—at 30° and 56°C. are 1200 and 1760 Mole^{-1} sec.$^{-1}$, respectively.]

The value of 1.1×10^3 Mole^{-1} sec.$^{-1}$ found for k_{trans}^{CTOOH} in this work at 56°C. is considerably larger than the value of 12 Mole^{-1} sec.$^{-1}$ reported by Thomas and Tolman (34). The discrepancy is chiefly caused by the high values chosen by these workers for k_t^{TT} and k_t^{TC}. Using the rate constants employed in this paper, Thomas and Tolman's results give $k_{trans}^{CTOOH} \approx 300$ Mole^{-1} sec.$^{-1}$, in reasonable agreement with our own value.

The work of Hiatt, Gould, and Mayo (10) on the transfer reaction between poly(peroxystyryl) peroxy radicals and *tert*-butyl hydroperoxide at 60°C. gave $k_{transfer}/k_{propagation} \approx 4.5$. This ratio can be combined with our previously determined value for $k_{propagation}$ of 143 Mole^{-1} sec.$^{-1}$ at 60°C. (13) to give $k_{transfer} \approx 645$ Mole^{-1} sec.$^{-1}$. However, the transfer was measured at relatively high hydroperoxide concentrations, where a large fraction of the hydroperoxide would be dimerized (37) and this will certainly tend to reduce $k_{transfer}$ below the value for the free hydroperoxide.

The rate constant for exchange Reaction 10 found in this work is of about the same order of magnitude as the rate constants for a number of similar, but completely symmetric, transfer reactions of hydrogen between oxygen at this temperature (20)—that is, for reactions of the type,

$$ROH + \dot{O}R \rightleftarrows R\dot{O} + HOR$$

the rate constants are in the range 2×10^2 to 2×10^3 Mole^{-1} sec.$^{-1}$ for *o*-di-*tert*-butylphenols, *tert*-butylhydroxylamines, etc. (20). A rate constant $\geqslant 4 \times 10^5$ Mole^{-1} sec.$^{-1}$ has been estimated (31) for the transfer reaction

$$C_6H_5\dot{O} + TOOH \rightarrow C_6H_5OH + TO\dot{O}$$

Thomas and Tolman reported that Reaction 10 did not exhibit a deuterium isotope effect when the hydroperoxidic hydrogen was replaced by deuterium. This surprising result was obtained before we had shown that previous failures to observe an isotope effect in the abstraction of the phenolic hydrogen of phenols by peroxy radicals were caused by the rapid loss of deuterium from the phenol by exchange reactions with oxygenated products such as hydroperoxides (11). This problem is readily overcome by carrying out the reactions in the presence of a little heavy water (11). We have applied the same technique in the present instance. The data given in Table IV were obtained with 1 ml. of H_2O or D_2O and 13 ml. of cumene, slightly greater rates being obtained with

2 ml. of water. The results in Figure 1 and Table IV show that in cumene at 30°C. the isotope effect $(k_{trans}^{CTOOH})_{H_2O}/(k_{trans}^{CTOOD})_{D_2O}$ is about 30. The lower value of k_{trans}^{CTOOH} in the presence of water presumably arises from hydrogen bonding between TOOH and the water dissolved in the cumene. An isotope effect of about 17 was obtained for k_{trans}^{TCOOH} (Table V). This value is less accurate than the value for k_{trans}^{CTOOH} partly because it was difficult to deuterate the hydroperoxide completely at high concentrations.

The remarkably high isotope effects found in Reactions 10 and 10a are almost as embarrassing as was our discovery of an isotope effect of about 15 in the reaction with phenols referred to above. [For 2,6-di-*tert*-butyl-4-methylphenol in oxidizing styrene at 65°C. we obtained an isotope effect $k_H/k_D = 10.6$ (*11*). We have since carefully measured the isotope effect for 2,4,6-tri-*tert*-butylphenol under the same conditions and have obtained a value ≥ 15. In these cases, hydrogen bonding to the solvent is relatively unimportant (*6, 12*). Similarly, DaRooge and Mahoney (*9*) have reported that for the reaction of 2,4,6-tri-*tert*-butylphenoxy radicals with 4-phenylphenol $k_H/k_D \geq 7.5$.] Although

Table VI. Effect of *tert*-Butyl Alcohol on Hydrogen Atom Transfer from Tetralin Hydroperoxide to Cumyl Peroxy Radicals at 30°C.

[CH], M	[*tert*-BuOH], M	k_{trans}, Mole^{-1} sec.$^{-1}$
7.2	—	600
6.6	0.8	180
3.9	4.9	30
2.2	7.3	10

the isotope effect will be maximized for this almost symmetric reaction (*8*), the value of 30 is about twice the "maximum theoretical value" calculated by Melander at this temperature (*23*). The enhancement of the isotope effect above the "maximum theoretical value" might be caused by the formation of a stronger hydrogen bond between TOOD and D_2O than between TOOH and H_2O. Walling and Heaton's interesting infrared study of hydrogen bonding and complex formation by *tert*-butyl hydroperoxide in solution (*37*) may also explain the enhancement of the isotope effect. These workers report that *tert*-BuOOD formed stronger complexes with aromatic solvents than *tert*-BuOOH. Unfortunately, we could not confirm this important point for Tetralin hydroperoxide. Although the fundamental O—H stretching band maxima for TOOH in several solvents occur at about the same frequencies as those reported for *tert*-BuOOH (*37*), the two bands in CCl_4-cumene mixtures were not considered sufficiently well resolved to obtain the TOOH- and TOOD-cumene association constants with the required degree of precision. One reason for the poor resolution with TOOH is that the main band in non-bonding solvents has a low frequency shoulder which arises from a weak intramolecular hydrogen bond between the hydroperoxidic hydrogen and the aromatic ring. If TOOD does in fact form a stronger complex with cumene than TOOH, then the measured isotope effect will be enhanced. A difference of about 400 cal. per mole would double the apparent isotope effect.

The retarding effect of hydrogen bonding on the rates of hydrogen atom abstraction from ROH compounds has been observed several times (7, 12, 30). In the present reaction system, k_{trans}^{CTOOH} was decreased by adding tert-butyl alcohol. Since the tert-butyl alcohol had comparatively little effect on the rate of oxidation of cumene, k_{trans}^{CTOOH} was calculated from Equation 20 with the appropriate value of $(R_o^{CH})^2$ at each cumene concentration. [With 4.9M [tert-BuOH] and 3.9M [CH] at 30°C. $k_p^{CC}/\sqrt{2k_t^{CC}}$ was 1.24×10^{-3} M$^{-1/2}$ sec.$^{-1/2}$ compared with 1.50×10^{-3} M$^{-1/2}$ sec.$^{-1/2}$ for neat cumene. Rotating sector studies suggested that the decreased rate in tert-butyl alcohol (2-methyl-2-propanol) was due to an increase in k_t^{CC}. This was almost certainly due to increased β-scission (Reaction 12) in the alcoholic media, since the addition of COOH gave a limiting value R_{COOH}^{CH}/R_o^{CH} of 2.0 in tert-butyl alcohol, compared with 1.6 in cumene (Table I).] The results are given in Table VI. Within the limits of accuracy of the data k_{trans}^{CTOOH} extrapolates to zero as the reciprocal of the tert-butyl alcohol concentration goes to zero. This implies that the rate constant for hydrogen atom abstraction from hydrogen-bonded TOOH is small compared with the value for the unbonded compound. The hydroperoxides do not dimerize significantly in hydrocarbon solvents at concentrations below about 0.01 to 0.02M.

Hydrogen Atom Transfer from Hydrocarbons to Peroxy Radicals. The ready conversion of one chain carrier to another in hydrocarbon oxidations by the addition of a hydroperoxide is illustrated in Table VII.

Table VII. Rate Constants (in Mole^{-1} Sec.$^{-1}$) for Hydrocarbon Oxidation in Absence and Presence of 0.1 to 0.4M Tetralin Hydroperoxide at 30°C.

Hydrocarbon (R)	No TOOH[a]			0.1 to 0.4M [TOOH]		
	$k_p^{RR}/(2k_t^{RR})^{1/2} \times 10^3$	k_p^{RR}	$k_t^{RR} \times 10^{-6}$	$k_p^{TR}/(2k_t^{TT})^{1/2} \times 10^3$	k_p^{TR} [b]	$k_t^{TT} \times 10^{-6}$
Toluene	0.014	0.24	150	0.04	0.1	4.0
Ethylbenzene	0.21	1.3	20	0.18	0.5	5.3
Cumene	1.5	0.18	0.0075	0.18	0.5	2.8
Tetralin	2.3	6.4	3.8	2.3	6.2[c]	3.6

[a] Neat hydrocarbon or hydrocarbon in chlorobenzene.
[b] Calculated from $k_p^{TR}/(2k_t^{TT})^{1/2}$ with $k_t^{TT} = 3.8 \times 10^6$ Mole^{-1} sec.$^{-1}$.
[c] Best value, 6.4 Mole^{-1} sec.$^{-1}$.

Table VIII. Absolute (and Relative) Rate Constants for Hydrogen Atom Abstraction from Hydrocarbons at 30°C.

(Absolute rate constants in Mole^{-1} sec.$^{-1}$. Relative rate constants given per benzylic hydrogen in brackets after absolute values; based on toluene = 1.0.)

Hydrocarbon (R)	Peroxy Radical			
	Self (ROȮ)[a]	Cumyl (COȮ)	Tetralyl (TOȮ)	Dihydroanthracyl
Toluene	0.24 (1.0)	0.03 (1.0)	0.1 (1.0)	—
Ethylbenzene	1.3 (8.1)	0.2 (10)	0.5 (5.0)	0.6
Cumene	0.18 (2.2)	0.18 (18)	0.5 (5.0)	—
Tetralin	6.4 (20)	1.65 (40)	6.4 (64)	6.5

[a] See also Ref. 27.

In the presence of 0.1 to 0.4M Tetralin hydroperoxide the measured termination constants for toluene, ethylbenzene, and cumene are in good agreement with the value 3.8×10^6 Mole^{-1} sec.$^{-1}$ obtained with pure Tetralin. Since the variations in the measured values are within the limits of experimental error, the cross-propagation constants, k_p^{TR}, were obtained using this termination constant and the measured values of $k_p^{TR}/(2k_t^{TT})^{1/2}$.

The rate constants of α-hydrogen atom abstraction from four hydrocarbons by cumylperoxy, tetralylperoxy, and 9,10-dihdroanthracyl-9-peroxy radicals and by the normal chain-carrying peroxy radicals are compared in Table VIII. The results show that the reactivities of peroxy radicals are affected by the nature of the organic group. The relatively low propagation constant for the oxidation of pure cumene may be caused by the low reactivity of the cumylperoxy radical.

Acknowledgment

The authors are indebted to G. A. Mortimer, Monsanto Co., for a generous gift of α,α'-azobis-1-propanol diacetate.

Literature Cited

(1) Alagy, J., Clement, G., Balaceanu, J. C., *Bull. Soc. Chim. France* **1959**, 1325.
(2) *Ibid.* **1960**, 1495.
(3) *Ibid.* **1961**, 1303.
(4) *Ibid.*, p. 1792.
(5) Antonovskii, V. L., Denisov, E. T., Solntseva, L. U., *Kinetika i Kataliz* **7**, 409 (1966).
(6) Bellemy, L. J., Williams, R. L., *Proc. Roy. Soc. (London)* **254A**, 119 (1960).
(7) Berezin, I. V., Kazavskaya, N. F., Ugarova, N. N., *Zh. Fiz. Khim.* **40**, 766 (1966).
(8) Bunton, C. A., Shiner, V. J., Jr., *J. Am. Chem. Soc.* **83**, 3214 (1961).
(9) DaRooge, M. A., Mahoney, L. R., *J. Org. Chem.* **32**, 1 (1967).

(10) Hiatt, R., Gould, C. W., Mayo, F. R., *J. Org. Chem.* **29**, 3461 (1964).
(11) Howard, J. A., Ingold, K. U., *Can. J. Chem.* **40**, 1851 (1962).
(12) *Ibid.* **42**, 1044 (1964).
(13) *Ibid.* **43**, 2729 (1965).
(14) *Ibid.* **44**, 1119 (1966).
(15) *Ibid.* **45**, 785, 793 (1967).
(16) Howard, J. A., Robb, J. C., *Trans. Faraday Soc.* **59**, 1590 (1963).
(17) Ingles, T. A., Melville, H. W., *Proc. Roy. Soc. (London)* **218A**, 163 (1953).
(18) Ingold, K. U., *Pure Appl. Chem.* **15**, 49 (1967).
(19) Karpukhin, O. N., Shlyapintokh, V. Ya. Mikhailov, I. D., *Zh. Fiz. Khim.* **38**, 156 (1964).
(20) Kreilick, R. W., Weissman, S. I., *J. Am. Chem. Soc.* **88**, 2645 (1966).
(21) Mahoney, L. R., Bayma, R. W., Warnick, A., Ruof, C. H., *Anal. Chem.* **36**, 2516 (1964).
(22) Mayo, F. R., Syz, M. G., Mill, T., Castleman, J. K., ADVAN. CHEM. SER. **75**, 46 (1968).
(23) Melander, L., "Isotope Effects on Reaction Rates," p. 22, Table 2-1 Chap. 2, Ronald Press, New York, 1960.
(24) Middleton, B. S., Ingold, K. U., *Can. J. Chem.* **45**, 191 (1967).
(25) Parlant, C., de Roch, I. Sérée, Balacéanu, J. C., *Bull. Soc. Chim. France* **1964**, 3161.
(26) Russell, G. A., *J. Am. Chem. Soc.* **77**, 4583 (1955).
(27) *Ibid.* **78**, 1047, (1956).
(28) *Ibid.* **79**, 3871 (1957).
(29) Russell, G. A., Williamson, R. C., Jr., *J. Am. Chem. Soc.* **86**, 2357, 2364 (1964).
(30) Sukhanova, O. P., Buchachenko, A. L., *Zh. Fiz. Khim.* **39**, 2413 (1965).
(31) Thomas, J. R., *J. Am. Chem. Soc.* **86**, 4807 (1964).
(32) *Ibid.* **89**, 4872 (1967).
(33) Thomas, J. R., Harle, O. L., *J. Phys. Chem.* **63**, 1027 (1959).
(34) Thomas, J. R., Tolman, C. A., *J. Am. Chem. Soc.* **84**, 2079 (1962).
(35) Traylor, T. G., Russell, C. A., *J. Am. Chem. Soc.* **87**, 3698 (1965).
(36) Tsepalov, V. F., Shlyapintokh, V. Ya., Pei-huang, C., *Zh. Fiz. Khim.* **38**, 52, 351 (1964).
(37) Walling, C., Heaton, L., *J. Am. Chem. Soc.* **87**, 48 (1965).

RECEIVED October 10, 1967. Issued as N.R.C. No. 9777.

Appendix

Chain Termination in the Oxidation of Cumene. Traylor and Russell (35) assume that the acceleration in the rate of oxidation of CH which is produced by added COOH is solely caused by a chain transfer reaction between CO· radicals and COOH. This assumption implies that all $CH_3OO\cdot$ radicals enter into termination *via* Reaction 13. However, Thomas (32) has found that acetophenone is formed even in the presence of sufficient COOH to raise the oxidation rate of CH to its limiting value. (The receipt of Thomas' manuscript prior to publication stimulated the present calculations.) From this fact, and from a study of the acetophenone formed during the AIBN-induced decomposition of COOH, Thomas concludes that the accelerating effect of added COOH is primarily caused by trapping of $CH_3OO\cdot$ radicals rather than $CO\cdot$ radicals—that is, Reaction 13 is sufficiently slow that under many conditions a majority of the $CH_3OO\cdot$ radicals propagate rather than terminate oxidation chains. A similar conclusion is reached by considering a CH oxidation scheme consisting of Reactions 2, 6, 12, 13, 14 and 15. The acetophenone formed per chain terminated,

$$= \frac{k_{12}[CO\cdot]_o}{R_i} = \frac{k_{12}}{R_i} \cdot \frac{2k_{6i}[COO\cdot]_o^2}{(k_{12} + k_{14}[CH])}$$

$$= \frac{k_{6i}}{k_{6d}} \cdot \frac{2k_{6d}[COO\cdot]^2_{COOH}}{R_i} \cdot \left(\frac{R_o^{CH}}{R_{COOH}^{CH}}\right)^2 \cdot \frac{1}{1 + (k_{14}/k_{12})[CH]}$$

where $2k_{6d}$ is the rate constant for the direct chain-terminating reaction of two $COO\cdot$ radicals and $2k_{6i}$ is the rate constant for their reaction to form two $CO\cdot$ radicals. At 56°C. with 3.6M [CH] in chlorobenzene $R_{COOH}^{CH}/R_o^{CH} = 1.6$; $k_{14}/k_{12} = 0.75$ Mole^{-1}; $k_{6i}/k_{6d} = 9.5$ [at 57°C. (32)]; and with $8 \times 10^{-3}M$ [AIBN] [conditions almost identical to those of Thomas (32)] $R_i = 5.5 \times 10^{-8}$ Mole sec.$^{-1}$ ($R_o^{CH} = 2.6 \times 10^{-6}$ Mole sec.$^{-1}$). Hence, the acetophenone formed per chain terminated = $9.5 \times 1.0 \times (1/1.6)^2 \times [1/(1 + 0.75 \times 3.6)] = 1.0$. Measured values under these experimental conditions are ~0.8 at 57°C. (32) and ~1.2 at 56°C. [Antonovski *et al.* (5) have reported that at higher temperatures (110°C.) the yield of acetophenone chain-terminated decreases from 2 to 0.6 upon extensive oxidation of cumene (to 1.76M [COOH]). They suggest that acetophenone is formed not only by Reaction 12 but also by the rather unlikely reaction $2COO\cdot \rightarrow C_6H_5COCH_3 + CH_2O + COOH$. The decrease in the acetophenone chain–terminated ratio with increasing oxidation is probably mainly caused by the trapping of $CO\cdot$ radicals by the COOH formed in the reaction. A small decrease in this ratio is observable at 56°C. for much less extensive oxidation—*e.g.*, with 3.6M [CH] and $8 \times 10^{-3}M$ [AIBN] the ratio drops to ~1.0 after 40,000 seconds, at which time [COOH] ≈ 0.1M.]

Furthermore, since $k_{12}[CO\cdot]_o/R_i = 1.0$ under these conditions, the steady-state approximation gives

$$R_i = k_{12}[CO]_o = k_{15}[CH][CH_3OO\cdot]_o + 2k_{13}[COO\cdot]_o[CH_3OO\cdot]_o$$

Also,

$$R_i = 2k_{6d}[COO\cdot]_o^2 + 4k_{13}[COO\cdot]_o[CH_3OO\cdot]_o$$

Therefore,

$$\frac{k_{15}[CH][CH_3OO\cdot]_o}{R_i} = \tfrac{1}{2} + \frac{k_{6d}[COO\cdot]_o^2}{R_i} = \tfrac{1}{2} + \frac{k_{6d}[COO\cdot]^2_{COOH}}{R_i}\left(\frac{R_o^{CH}}{R_{COOH}^{CH}}\right)^2$$

$$= \tfrac{1}{2} + \tfrac{1}{2}\left(\frac{1}{1.6}\right)^2 = 0.695 =$$

the fraction of methylperoxy radicals formed which propagate the chain —i.e.,

$$\frac{k_{15}[\text{CH}]}{k_{15}[\text{CH}] + 2k_{13}[\text{COO·}]_o}$$

Similarly, 39% of the chain termination occurs by Reaction 6d ($k_{6d}[\text{COO·}]_o^2/R_i = 0.195$) and 61% by Reaction 13. The contribution of Reaction 15 to the over-all rate of oxidation is very small—i.e., $0.305 \times 5.5 \times 10^{-8} = 1.7 \times 10^{-8}$ Mole sec.$^{-1}$, whereas $R_o^{\text{CH}} = 2.6 \times 10^{-6}$ Mole sec.$^{-1}$.

[The possibility that significant amounts of 2-phenylpropyl-1-peroxy (POO·) radicals were formed via CO· abstraction from the β-position of CH was investigated using cumyl hypochlorite as the source of CO· radicals. At 56°C. with 3.6M [CH] in chlorobenzene, $(k_{14})\alpha/(k_{14})\beta = 23$ and therefore the formation of POO· radicals via Reaction 14 can be neglected under these conditions. Since COO· radicals are more selective than CO· radicals, POO· will not be formed in Reaction 2. The hypochlorite experiments gave $k_{14}/k_{12} = 0.75$ Mole^{-1} at 56°C.]

The present calculations confirm Thomas' conclusion that termination via Reaction 13 must be relatively slow. A significant fraction of the CH$_3$OO· radicals propagate rather than terminate chains under many conditions.

3

Co-Oxidation of Butadiene with Various Hydrocarbons

DALE G. HENDRY

Department of Physical Organic Chemistry, Stanford Research Institute, Menlo Park, Calif. 94025

> *Butadiene has been co-oxidized with a number of aralkyl hydrocarbons and cyclic olefins. The order of increasing reactivity toward butadieneperoxy radicals ($X-O_2-C_4H_6O_2\cdot$) is cumene, sec-butylbenzene $<$ cyclooctene $<$ cyclohexene $<$ Tetralin $<$ cycloheptene $<$ styrene $<$ cyclopentene $<$ butadiene. The organic structure of the peroxy radical appears to have little effect on its selectivity when $RO_2\cdot$ is derived from a hydrocarbon. The relative reactivity of butadiene and styrene toward the butadieneperoxy radicals has been determined as a function of temperature and solvent. Correlation of relative reactivities of hydrocarbons toward $RO_2\cdot$ radicals with those toward methyl radicals show the former to be less selective in addition and abstraction.*

The effect of structure on rate of autoxidation of hydrocarbons has been a topic of concern for some time; however, only recently has there been much effort to separate the structural effects on termination and propagation. This separation requires measuring absolute rates of termination and propagation or determining relative propagation rates. Determination of conversion of each reactant for evaluating relative propagation rates by the copolymerization equation (13) may involve measuring either the products formed or the decrease in concentration of reactants. The first method has been used in special cases by Walling and McElhill (26) for the co-oxidation of substituted benzaldehydes; by Russell (17) for the co-oxidation of cumene and Tetralin; by Mayo, Miller, and Russell (11) for the co-oxidation of some substituted olefins; and by Alagy, Clément, and Balacéanu (1) for some saturated hydrocarbons. However, Russell and Williamson (21) have studied a number of combinations of ethers and both saturated and unsaturated hydrocarbons by determining the decrease of reactants by GLPC techniques. Both methods have been used by Mayo, Syz, Mill, and Castleman (*12*), who have discussed their relative merits. Middleton and Ingold (*15*) have attempted to determine relative reactivities of hydrocarbons toward the standard peroxy radicals from rates of disappearance of a number of hydrocarbons in oxidizing mixtures where the hydrocarbon acting as the peroxy radical source was present in sufficient excess to produce the majority of the radicals.

We have used a different approach for determining the conversion of hydrocarbon. Our approach, which requires measuring the consumption of only one hydrocarbon plus the oxygen, has a distinct advantage where reactants have large differences in reactivity. At low conversions Equation 1 applies,

$$\Delta O_2 = \Delta R_1 H + \Delta R_2 H + C \tag{1}$$

where C equals the oxygen consumed by the initiator minus the oxygen evolved by the termination reaction. This approach is useful because oxygen consumption can be determined easily and accurately by gasometric methods and because consumption of only one hydrocarbon need then be determined accurately. Thus, it is not necessary to measure a very low conversion of an unreactive hydrocarbon. For our studies we have co-oxidized various hydrocarbons with butadiene since this hydrocarbon has a high reactivity, is easily separated and measured by vacuum line techniques, and the nature of its oxidation is well known (*4*, *5*).

The following reactions describe the propagation reactions for a co-oxidation of butadiene, B, with a hydrocarbon, A:

$$BOO\cdot + B \xrightarrow{k_{pbb}} BOOB\cdot$$

$$BOO\cdot + A \xrightarrow{k_{pba}} BOOH + A\cdot \text{ (or } BOOA\cdot \text{)} \quad r_b = k_{pbb}/k_{pba}$$

$$AOO\cdot + B \xrightarrow{k_{pab}} AOOB\cdot$$

$$AOO\cdot + A \xrightarrow{k_{paa}} AOOH + A\cdot \text{ (or } AOOA\cdot \text{)} \quad r_a = k_{paa}/k_{pab}$$

$$A\cdot \text{ (or } B\cdot) + O_2 \rightarrow AO_2\cdot \text{ (or } BO_2\cdot)$$

These reactions are sufficient if there is no secondary reaction (*10*) where

$$BOOA\cdot \text{ (or } AOOA\cdot) \rightarrow BO\cdot \text{ (or } AO\cdot) + AO \text{ (epoxide)}$$

Such systems have been avoided in the present work.

If there are no complicating reactions and if conversions are small, the copolymerization equation (13)

$$\frac{\Delta B}{\Delta A} = \frac{[B]}{[A]} \frac{r_b[B] + [A]}{r_a[A] + [B]} \qquad (2)$$

is applicable. It can be converted into the form (3)

$$\frac{\Delta B/\Delta A - 1}{[B]/[A]} = r_b - \frac{r_a \Delta B/\Delta A}{([B]/[A])^2} \qquad (3)$$

for plotting a large amount of data. If the relative reactivities of A and B are the same toward both $AO_2\cdot$ and $BO_2\cdot$, $r_b = 1/r_a = \alpha = [A][\Delta B]/[B][\Delta A]$. In this case, the value of α is independent of the composition of the feed. If r_b does not equal $1/r_a$, the apparent value of α approaches r_b (or $1/r_a$) as the composition of the reactants approaches pure B (or A). Thus, inspection of easily calculated α values for series of experiments will indicate the consistency of the data and the similarity of reactivity of $AO_2\cdot$ and $BO_2\cdot$.

Co-Oxidation of Aralkyl Hydrocarbons with Butadiene

Data from co-oxidations of butadiene with cumene, Tetralin, styrene, and sec-butylbenzene are included in Table I. The data for cumene show the versatility of the method. Generally 1 to 2% of the cumene reacts, but the conversions of butadiene and oxygen are usually between 20 to 25% and 30 to 70%, respectively. If the precision of measurement of reactants before and after reaction were 0.5%, the experimental error would be about 5% for the consumption of butadiene and oxygen. If the cumene were measured directly, the experimental error for consumed cumene would be from 50 to 100% (because the change in its concentration is so small); however, when the consumed oxygen and butadiene and Equation 1 are used, the error is 10% if there is no error in estimating C. If this factor is 100% wrong, the experimental error would still be only 20% for cumene. Thus, this technique can be applied to systems which are from 1/25 to 4 to 10 times as reactive as butadiene—a range greater than two powers of 10. In the co-oxidation with Tetralin, hydrocarbon consumptions were also determined by NMR analyses of the reaction mixtures. These values for $\Delta B/\Delta RH$ and α are given in parenthesis in Table I.

The r values obtained by the Fineman-Ross procedure (3) are listed in Table II. Reactivity toward the butadiene peroxy radical increases in the order cumene (0.14), sec-butylbenzene < Tetralin (1.00) < styrene (1.5) < butadiene (3.3), in reasonable agreement with previous efforts. Russell and Williamson (21) found cumene 0.10 and 0.40 as reactive as

Tetralin at 60°C., higher than reported earlier (0.06 and 0.04) at 90°C. (*17*), and styrene 1 to 3 times as reactive as Tetralin, depending on the peroxy radical. Mayo *et al.* (*12*) found cumene 0.22 and 0.13 as reactive as Tetralin and styrene 2.3 times as reactive. Alagy, Clément, and Balaceanu found cumene 0.1 as reactive as Tetralin, using a mathematical instead of a chemical analysis (*1*). Howard and Ingold found cumene 0.45, 0.48, and 0.50 times as reactive as Tetralin toward the cumyl, tetralyl, and dihydroanthracylperoxy radicals, respectively (*8*). These numbers show qualitative agreement; whether the differences are caused by the reactivity of the various peroxy radicals or by experimental error is discussed by Mayo, Syz, Mill, and Castleman (*12*).

The $r_a r_b$ products are 0.7, 0.7, and 0.9 for the co-oxidations of butadiene with cumene, Tetralin, and styrene, respectively. Hence, there is little variation among the selectivities of the peroxy radicals from these hydrocarbons and butadiene.

Effect of Temperature and Solvent on Co-Oxidation of Styrene and Butadiene

To investigate the importance of temperature for at least one pair of hydrocarbons, butadiene and styrene were co-oxidized at 30° and 80° as well as 50°C. The data for these experiments are also given in Tables I and II. A plot of the data in the Arrhenius form [$\log r_b$ (or r_s) $= \log A_b/A_s - (E_b - E_s)/4.575T$] gives a ratio of A factors for reaction of either radical with butadiene and styrene of about 100 and a difference in the energy of activation of the two reactions of 2700 ± 700 cal. per mole. The higher reactivity of butadiene compared with styrene is caused by the favorable A factor, which is partly compensated for by an unfavorable energy of activation. Similar differences are seen in the Arrhenius factors for the homopolymerization rate constants for propagation ($A_b/A_s = 27$ and $E_b - E_s = 2000$ cal.) (*24*). These large differences in A factors for the reaction of butadiene and styrene indicate the danger in attempting to interpret small difference in relative rates at one temperature.

Tables I and II include data for the co-oxidations of styrene and butadiene in chlorobenzene and *tert*-butylbenzene solutions, as well as with no added solvent. These solvents were chosen because the rate of oxidation of cyclohexene varies significantly in them at the the same rate of initiation (*6*). There is a variation in the over-all rate of oxidation under these solvent conditions, but there appears to be no significant difference in the measured r_a and r_b (Table II). If the solvent does affect the propagation reaction in autoxidation reactions, it affects the competing steps to the same degree.

Table I. Oxidations of Mixtures of Butadiene and Various

			Initial			Millimoles of Final	
Expt.	Time, Hr.	O_2	$C_4H_6{}^a$	RH	O_2	$C_4H_6{}^a$	RH^b
			RH =	Cumene			
20	20.7	1.280	0.00	6.97	0.672	0.00	6.38
22	23.6	1.044	0.886	13.45	0.516	0.598	13.26
15	23.9	1.474	1.27	13.22	0.788	1.31	12.98
19	20.2	0.987	1.61	12.93	0.318	1.18	12.76
14	24.3	1.452	0.96	5.88	0.953	0.73	5.79
18	18.6	1.180	2.33	12.45	0.378	1.76	12.29
17	20.0	0.957	3.20	11.90	0.065	2.59	11.80
11	24.4	1.357	2.29	5.56	0.394	1.75	5.46
21	12.0	1.291	4.44	11.19	0.357	3.72	11.05
32	1.97	1.382	17.29	3.05	0.929	16.85	^{c}
120	4.04	1.627	10.99	0	1.200	13.32	0
			RH =	Tetralin			
26	4.00	1.014	0.00	14.33	0.554	0.00	13.87
28	4.39	1.021	0.893	13.77	0.559	0.836	13.37
142	6.80	1.2753	2.270	13.01	0.851	1.936	12.50
4	7.67	1.3477	3.004	13.74	0.4492	2.616	13.23
24	3.92	1.197	3.10	12.08	0.690	2.90	11.80
27	4.70	1.496	5.40	10.74	0.671	4.92	10.43
149	2.75	1.4609	9.159	9.293	0.9454	8.799	9.141
23	4.14	1.613	8.15	8.05	0.800	7.54	7.97
136	3.66	1.1196	12.275	7.867	0.2646	12.568	7.719
			RH =	sec-Butylbenzene			
39	15.25	1.554	0.00	11.98	1.382	0.00	11.82
48	18.08	1.197	0.630	12.09	0.934	0.484	11.99
41	15.88	1.043	1.443	11.52	0.637	1.16	11.45
38	14.80	0.978	2.69	11.26	0.351	2.23	11.17
52	6.92	0.576	3.51	10.45	0.181	3.21	10.39
47	3.00	1.290	13.64	4.54	0.693	13.23	^{c}
			RH =	Styrene at 50°			
55	2.05	1.095	3.29	14.11	0.394	3.09	13.63
54	2.37	1.270	6.25	11.83	0.471	5.80	11.49
61	2.08	1.083	9.26	9.48	0.367	8.82	9.27
51	2.00	1.238	9.43	9.48	0.575	9.06	9.25
		RH = Styrene	(2.0M)	in Chlorobenzene	Solvent		
59	4.32	0.999	0.00	4.00	0.611	0.00	3.62
57	8.29	1.061	1.00	3.00	0.448	0.79	2.62

Hydrocarbons at 50°C. in Presence of 0.10M ABN

Reactants

\multicolumn{3}{c}{Consumed}		ΔB		R_0,
O_2	C_4H_6	RH^b	$(B/RH)_{av}$	$\overline{\Delta RH}$	α	Mole/l.-hr.

RH = Cumene

O_2	C_4H_6	RH^b	$(B/RH)_{av}$	$\Delta B/\Delta RH$	α	R_0
0.608	0.00	0.589	—	—	—	0.0294
0.528	0.332	0.190	0.055	1.70	31.5	0.0111
0.687	0.423	0.248	0.085	1.80	20.0	0.0142
0.669	0.485	0.168	0.121	2.89	23.8	0.0166
0.472	0.371	0.092	0.145	4.03	27.8	0.0199
0.802	0.632	0.157	0.165	4.03	24.4	0.0216
0.892	0.775	0.101	0.245	7.67	31.5	0.0226
0.953	0.853	0.102	0.365	8.36	22.9	0.0390
0.934	0.773	0.140	0.366	5.52	15.1	0.0388
0.453	0.45	^{c}	5.60	—	—	0.113
0.427	0.41	0	—	—	—	0.104

RH = Tetralin

O_2	C_4H_6	RH^b	$(B/RH)_{av}$	$\Delta B/\Delta RH$	α	R_0
0.460	0.000	0.46	—	—	—	0.0575
0.462	—	0.40	0.0638	—	—	0.0525
0.424	0.334	0.52	0.165	0.647 (0.737)	3.92 (4.48)	0.0617
0.8985	0.388	0.51	0.223	0.76 (0.88)	3.41 (3.95)	0.0530
0.506	0.220	0.280	0.250	0.778	3.10	0.0605
0.825	0.510	0.31	0.487	1.65	3.39	0.0877
0.516	0.360	1.52	0.974	2.37 (3.59)	2.43 (3.69)	0.0865
0.813	0.655	0.155	0.983	4.22	4.37	0.105
0.855	0.707	0.148	1.53	4.78 (4.64)	3.13 (3.03)	0.1044

RH = sec-Butylbenzene

O_2	C_4H_6	RH^b	$(B/RH)_{av}$	$\Delta B/\Delta RH$	α	R_0
0.172	0.00	0.158	—	—	—	0.00564
0.263	0.168	0.10	0.0462	1.66	35.9	0.00729
0.406	0.326	0.068	0.113	4.80	42.5	0.0128
0.627	0.519	0.094	0.219	5.52	25.2	0.0206
0.393	0.322	0.065	0.322	4.95	15.9	0.0284
0.597	0.48		3.01	—	—	0.0955

RH = Styrene at 50°

O_2	C_4H_6	RH^b	$(B/RH)_{av}$	$\Delta B/\Delta RH$	α	R_0
0.695	0.229	0.47	0.230	0.486	2.11	0.171
0.799	0.455	0.34	0.508	1.33	2.49	0.168
0.716	0.50	0.21	0.964	2.33	2.42	0.172
0.663	0.43	0.23	0.987	1.87	1.89	0.161

RH = Styrene (2.0M) in Chlorobenzene Solvent

O_2	C_4H_6	RH^b	$(B/RH)_{av}$	$\Delta B/\Delta RH$	α	R_0
0.388	0.00	0.385	—	—	—	0.0449
0.613	0.23	0.38	0.318	0.612	1.92	0.0369

Table I.

		Initial			Final	Millimoles of	
Expt.	Time, Hr.	O_2	C_4H_6 [a]	RH	O_2	C_4H_6 [a]	RH [b]

RH = Styrene (2.0M) in Chlorobenzene Solvent

| 56 | 5.06 | 1.026 | 2.01 | 2.00 | 0.591 | 1.76 | 1.84 |
| 58 | 3.26 | 1.057 | 3.86 | 0.00 | 0.721 | — | 0.00 |

RH = Styrene (2.0M) in tert-Butylbenzene Solvent

64	5.50	1.214	0.00	4.00	0.875	0.00	3.67
62	5.60	1.106	1.07	3.00	0.806	0.99	2.81
60	5.22	1.18	2.00	2.00	0.898	1.85	1.90
63	4.52	1.177	4.05	0.00	0.886	3.80	0.00

RH = Styrene at 30° [d]

69	3.00	1.185	0.00	17.20	0.412	0.00	16.43
73	2.90	1.164	6.84	12.41	0.626	6.628	12.10
71	3.10	1.118	9.77	9.75	0.633	9.57	9.49
74	2.50	1.261	9.75	9.61	0.915	9.48	9.54
79	2.72	1.395	13.68	6.81	1.003	13.40	6.72

RH = Styrene at 80° [e]

76	0.765	1.240	0.00	16.33	0.547	0.00	15.43
77	0.756	1.269	4.66	12.75	0.608	4.32	12.43
67	0.750	1.254	5.93	12.92	0.540	5.50	12.59
68	0.720	1.068	6.39	11.51	0.465	6.06	11.28
78	0.800	1.121	9.14	9.045	0.364	8.62	8.85

RH = Cyclopentene

115	6.26	1.200	0.00	12.28	0.449	0.00	11.53
116	8.80	1.217	1.17	9.79	0.294	0.99	9.05
114	7.12	1.337	1.98	9.65	0.636	1.80	9.16
112	4.94	1.188	4.98	5.40	0.682	4.68	5.20

RH = Cyclohexene

36	4.13	1.123	0.00	19.08	0.584	0.00	—
49	3.00	1.114	2.94	15.90	0.736	2.86	15.61
42	5.00	1.613	3.37	16.19	0.917	3.00	15.88
46	5.33	1.310	5.00	14.68	0.520	4.56	14.41
30	2.78	1.037	10.19	10.13	0.497	9.77	10.03

RH = Cycloheptene

108	7.37	1.392	0.00	16.61	1.030	0.00	16.25
109	8.74	1.396	4.45	13.22	0.472	4.06	12.75
110	5.80	1.098	8.51	10.13	0.227	7.90	9.87
106	3.13	1.609	9.22	9.39	1.107	8.85	9.26

Continued

Reactants

Consumed				$\dfrac{\Delta B}{\Delta RH}$		R_0,
O_2	C_4H_6	RH^b	$(B/RH)_{av}$		α	Mole/l.-hr.

RH = Styrene (2.0M) in Chlorobenzene Solvent

0.435	0.27	0.16	0.980	1.69	1.72	0.0429
0.336	—	0.00	—	—	—	0.0521

RH = Styrene (2.0M) in *tert*-Butylbenzene Solvent

0.339	0.00	0.334	—	—	—	0.0328
0.300	0.104	0.191	0.355	0.545	1.54	0.0268
0.284	0.174	0.105	0.991	1.66	1.68	0.0272
0.291	0.294	0.000	—	—	—	0.0321

RH = Styrene at 30°[d]

0.773	0.00	0.773	—	—	—	0.129
0.538	0.227	0.311	0.549	0.730	1.33	0.0905
0.485	0.22	0.26	1.00	0.846	0.85	0.0783
0.346	0.22	0.13	1.00	1.69	1.68	0.0693
0.392	0.30	0.09	2.01	3.33	1.66	0.0717

RH = Styrene at 80°[e]

0.693	0.00	—	—	—	—	0.459
0.660	0.34	0.32	0.349	1.06	2.95	0.438
0.714	0.383	0.331	0.437	1.16	2.65	0.438
0.603	0.368	0.235	0.546	1.57	2.87	0.410
0.757	0.58	0.19	0.992	3.04	3.06	0.476

RH = Cyclopentene

0.751	0.00	0.75	—	—	—	0.112
0.923	0.18	0.74	0.135	0.237	1.75	0.105
0.701	0.18	0.51	0.201	0.345	1.72	0.0928
0.506	0.31	0.20	0.869	1.60	1.84	0.1067

RH = Cyclohexene

0.539	0.00	—	—	—	—	0.0653
0.377	0.086	0.289	0.184	0.297	1.61	0.0628
0.696	0.386	0.306	0.198	1.26	6.37	0.0696
0.790	0.512	0.273	0.328	1.88	5.73	0.0741
0.540	0.44	0.10	0.993	4.40	4.45	0.0967

RH = Cycloheptene

0.362	0.00	0.365	—	—	—	0.0246
0.924	0.445	0.471	0.332	0.946	2.85	0.0529
0.871	0.610	0.256	0.823	2.38	2.89	0.0748
0.502	0.37	0.13	0.970	2.85	2.94	0.0807

Table I.

Expt.	Time, Hr.	Initial O_2	Initial $C_4H_6{}^a$	Initial RH	Final O_2	Final $C_4H_6{}^a$	Final RH^b
						Millimoles of	

RH = Cyclooctene

93	18.83	1.260	0.00	14.61	0.961	0.00	14.33
102	15.15	1.123	1.538	13.60	0.565	1.320	13.31
100	10.50	1.064	1.517	13.58	0.687	1.358	13.40
104	10.10	1.195	2.340	13.05	0.721	2.096	12.87
101	10.05	1.211	3.228	12.40	0.616	2.886	12.21
103	8.03	1.049	4.00	11.87	0.489	3.64	11.73

^{a} Calculated to be in liquid phase assuming vapor density to equal (mole fraction) × (vapor density of pure butadiene—13.06 grams/liter).
^{b} Consumption of RH calculated from Equation 1 using $C = 1/2R_i tV$ where $R_i = 8.7 \times 10^{-4}$ mole/liter/hr. for ABN experiments at 50°. $R_i = 7.4 \times 10^{-4}$ mole/liter/hr. for ABC experiments at 80°. Time t is that indicated in table and liquid volume V is 2.0 ml. except in Expts. 11, 14, 20, 112, 114, 115, 116 where it was 1.0 ml. $C = 0$ for DBPO experiments.

Co-Oxidations of Cyclic Olefins

The rates of oxidation of either pure cyclic olefins (23) or their 2.0M solutions in chlorobenzene (6) indicate that the order of decreasing $k_p/k_t^{1/2}$ is $C_5 > C_6 > C_7 > C_8$. Thus, if one assumes from the similarities of structures that the termination rate constants are the same for each olefin, the relative reactivities in chain propagation would be the same as the relative rates at the same rate of initiation. This order differs from the observed orders of reactivity of these olefins toward the bromine atom (19) and the methyl free radical (7, 9, 14). We have now co-oxidized butadiene with these olefins and obtained relative propagation rates. The experimental data are given in Table I and the resulting values of r_b and r_a in Table II. Thus, the relative reactivity of the cyclic olefins toward the butadiene peroxy radical is $C_5 > C_7 > C_6 > C_8$, more similar to those observed for bromine and methyl. However, the systems are complicated by the fact that propagation can involve either addition or hydrogen abstraction. In Table III the composite propagation rates are separated into relative addition and abstraction rates according to the data of Van Sickle, Mayo, and Arluck (23). Included in Table III are the data for the other hydrocarbons from Table II as well as some data for bromine and methyl radicals. With bromine atoms no addition occurs because of insufficient concentration of bromine to trap the Br-olefin intermediates.

Continued

Reactants

Consumed							
O_2	C_4H_6	RH^b	$(B/RH)_{av}$	$\dfrac{\Delta B}{\Delta RH}$	α		R_0, Mole/l.-hr.

RH = Cyclooctene

0.299	0.00	0.285	—	—	—		0.00793
0.588	0.253	0.290	0.0981	0.871	8.90		0.0184
0.377	0.184	0.183	0.1064	1.007	9.45		0.0179
0.474	0.283	0.181	0.171	1.56	9.13		0.0235
0.595	0.393	0.192	0.248	2.04	8.23		0.0296
0.560	0.412	0.139	0.326	2.97	9.07		0.0348

[c] Conversion of RH was too small to be calculated accurately.
[d] Initiator 0.01M DBPO.
[e] Initiator 0.01M ABC.

Table II. Autoxidation Parameters for Various Hydrocarbons

Hydrocarbon	r_b (C_4H_6)	r_a (HC)	$r_a r_b$	$\left(\dfrac{k_p}{k_t^{1/2}}\right)$
Butadiene[a]				
in tert-butylbenzene	—	—	—	0.73
in chlorobenzene	—	—	—	1.19
Cumene	23 ± 3	0.03 ± 0.01	0.7	0.20
Sec-Butylbenzene	25 ± 10	—	—	0.030
Tetralin	3.3 ± 0.5	0.22 ± 0.5	0.7	0.38
Styrene (50°C.)				—
in tert-butylbenzene	2.2 ± 0.2	0.42 ± 0.05	0.9	0.70
in chlorobenzene				1.02
Styrene (30°C.)	1.5 ± 0.4	0.7 ± 0.2	1.0	0.44
Styrene (80°C.)	2.8 ± 0.3	0.30 ± 0.05	0.8	2.91
Cyclopentene	1.8 ± 0.5	0.58 ± 0.07	1.0	0.46
Cyclohexene	4 ± 1	0.10 ± 0.03	0.41	0.32
Cycloheptene	3.0 ± 0.1	0.38 ± 0.02	1.1	0.14
Cyclooctene	8 ± 1	0.10 ± 0.02	0.8	0.040

[a] Since oxidation of neat butadiene is not homogeneous, experiments in tert-butylbenzene are used to evaluate $k_p/k_t^{1/2}$.

It is clear from Table III that addition of methyl radicals occurs 1.5 to 3 times as often as abstraction, while for the peroxy radical addition occurs only 0.1 to 2 times as often. Despite this large variation in the proportions reacting by addition, the relative rates of addition of these two radicals to cyclic olefins are about the same. In fact, the reactivity of cyclohexene relative to cyclopentene in these systems agrees with that observed (25) for the trichloromethyl radical (0.14 to 0.30). However, for the relative reactivity toward abstraction for the three radicals the agreement is not as good, although the qualitative changes in going from C_5 to C_8 are the same. Presumably the reactivity pattern for addition is determined largely by the olefins, while for abstraction the reactivity pattern is also determined by the attacking radical.

Table III. Relative Reactivities of Hydrocarbons Toward Various Radicals[a]

	ROO· (50°C.)			CH_3· (65°C.) (7, 9, 14)			Br· (40°C.) (19)
	k_{total}	k_{addn}[b]	k_{abst}[b]	k_{total}	k_{abst}	k_{addn}	k_{abst}
Butadiene	1.00	1.00	—	1.00	1.00	—	
Styrene	0.45	0.45	—	0.39	0.39	—	
Cyclopentene	0.56	0.07	0.49	0.0045	0.0029	0.0016	1.00
		(1.00)	(1.00)		(1.00)	(1.00)	
Cycloheptene	0.33	0.07	0.26	0.0033	0.0021	0.0012	1.41
		(1.0)	(0.54)		(0.74)	(0.69)	
Tetralin	0.30	—	0.30	0.018	—	0.008	
Cyclohexene	0.25	0.05	0.20	0.00075	0.00045	0.00030	0.20
		(0.33)	(0.46)		(0.16)	(0.15)	
Cyclooctene	0.12	0.084	0.036	0.0046	0.0033	0.0013	0.75
		(1.2)	(0.08)		(1.15)		
Cumene	0.043	—	0.043	0.0055	—	0.0055	
sec-Butyl-benzene	0.04	—	0.040	—	—	—	

[a] All reactivities are molecular.
[b] Total relative reactivities of cyclic olefins assigned broken to abstraction and addition on basis of Ref. 23.

Correlation of Reactivities Toward Peroxy and Methyl Radicals

In Figure 1 the logarithms of the relative reactivities for the peroxy radicals from Table III are plotted against the logarithms of the corresponding relative reactivities of the methyl radicals. Molecular reactivities in Table III have been corrected for the number of reactive positions; addition reactivities have been divided by 2 except for styrene, and abstraction reactivities have been divided by 4 except for cumene. The correlation of the addition data is very good; only the cyclohexene point shows significant deviation, less than a factor of 2. For a range of methyl reactivities of 2000 there is a range of only 50 in reactivities toward the peroxy radical. One might have expected the peroxy radical to show a greater selectivity than the methyl radical since its addition (and also abstraction) reactions are less exothermic and less rapid than for the methyl radical. Szwarc and Binks (22) have shown that the reactivities of a number of compounds toward the methyl radical correlate reasonably

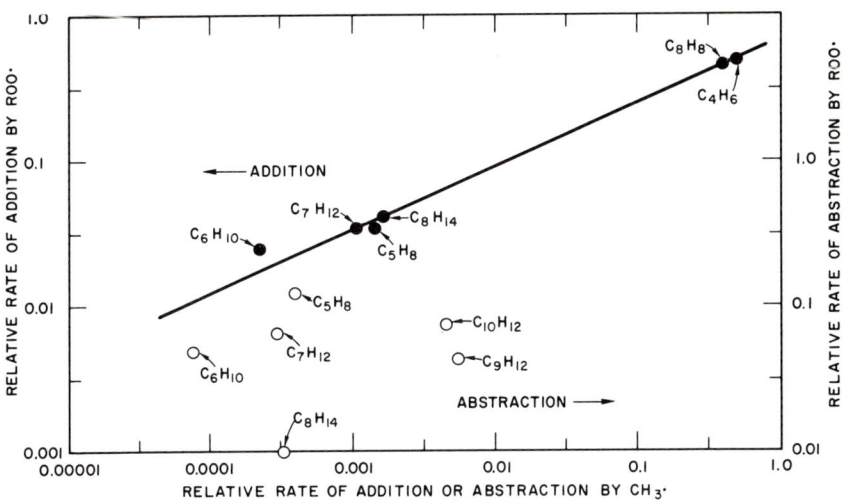

Figure 1. Correlation of reactions of methyl and peroxy radicals

well with the reactivities toward ethyl, 1-propyl, trichloromethyl, and styrenyl radicals and that the correlations are consistent with the expected relationship between selectivity and exothermicity or reactivity. However, in our case $RO_2\cdot$ is less selective than $CH_3\cdot$, even though it is less reactive.

The transition states of free radical reactions generally show evidence of polar character wherein electron transfer to or from the radical has occurred (20). Thus, the electron affinity or ionization potential of the radical involved should affect the reaction. The much higher electron affinity (16) of $RO_2\cdot$ than $CH_3\cdot$ radicals no doubt alters the transition state so that the reactivities toward it show less selectivity. The results of Szwarc and Binks (22) center around the fact that only carbon radicals were used for the correlation, and thus the electron affinity does not vary sufficiently to show in the correlation any deviation from the expected reactivity-selectivity relationship.

The reactivities in abstraction toward the $RO_2\cdot$ radicals correlate poorly with the methyl data. For example, toward $RO_2\cdot$ radicals the cyclic olefins and Tetralin are as reactive as cumene (within a factor of about 2), while toward $CH_3\cdot$, Tetralin and cumene are at least 10 times more reactive than the cyclic olefins. This apparent lack of selectivity in abstraction by the peroxy radical is no doubt related to the factors controlling the selectivity in addition but to some degree is a function of the reactants. The peroxy radical appears to show a somewhat more typical selectivity (8, 18) toward a broad selection of aralkyl hydrocarbons, although the changes in exothermicity are more significant in this series.

Experimental

Chemicals. Research grade butadiene (>99.8 mole % pure) (Phillips Petroleum Co.) was distilled from −80° to −195°C. traps to remove butadiene dimer and water. All hydrocarbons and solvents were obtained from commercial sources and fractionally distilled through a 4-foot column packed with metal helices; only fractions of sufficient purity were employed after passage through silica gel. Purity was always greater than 99.5% as judged by GLPC using a polar column packing such as Carbowax on Chromosorb-W. In cases where impurities were detected, GLPC showed that they did not react significantly during oxidation.

Azobis-(2-methylpropionitrile) (ABN) was recrystallized from methanol. Azobis-1-cyanocyclohexane (ABC) was furnished by D. E. Van Sickle, and di-*tert*-butylperoxalate (DBPO) was prepared by a literature procedure (2). Oxygen (99.9%) was passed through a −80°C. trap to remove trace amounts of water.

Procedure. The reaction tubes were about 12 ml. in volume and had break tips which allowed the contents of the tube after reaction to be transferred to the vacuum line without contact with the atmosphere. The oxidation catalyst was weighed into the tube in the crystalline form, and the hydrocarbon to be oxidized with butadiene was added with a syringe. At this point the reaction tube was joined to the vacuum line by a ground-glass joint, and the contents were degassed to remove air dissolved in the hydrocarbon. Butadiene was measured in a calibrated bulb as a gas and transferred to the reaction tube by opening the necessary valves and freezing in the tube. Oxygen was measured in a gas buret and then allowed to expand into the reaction tube and partially condense at −195°C.; the tube was sealed off, and the oxygen that was not sealed in the reaction tube was pumped (using a Toepler pump) back into the gas buret and measured. The difference between the two readings from the gas buret equals the oxygen transferred and sealed in the reaction tube.

Reactions were carried out at $30.0° \pm 0.05°$, $50.0° \pm 0.05°$, or $80.0° \pm 0.10°$C. for the indicated times. The reaction tube was sealed by deKhotinsky wax to a device on the vacuum line suitable for breaking the break tip in the line, the contents were cooled to −195°C., and the tip was broken. The noncondensable gases (unreacted oxygen and nitrogen from the thermal decomposition of the initiator) and the unreacted butadiene were isolated and measured. The butadiene was not always completely separated from the reaction mixture, and it was necessary to determine the purity of the recovered butadiene by GLPC. A portion of the second hydrocarbon was distilled from the reaction mixture and analyzed by GLPC to determine whether all the butadiene had been removed. With the more volatile hydrocarbons it was necessary to correct for the incomplete separations.

Because of the high volatility of cyclopentene, the above procedure was not attempted, and after the reaction the butadiene and cyclopentene were distilled into a 2-liter bulb to which was added a measured amount of fluorotrichloromethane as an internal standard for GLPC analysis. The amount of unreacted butadiene was obtained by comparison of the area ratio of butadiene and halomethane in the analysis with the ratio in a known mixture of nearly identical composition. Attempts to measure the cyclopentene in the same manner were not as satisfactory, possibly because of the difficulty of distilling cyclopentene quantitatively from the reaction products.

Acknowledgment

The author is grateful to David A. Jones and Dennis Schuetzle for their assistance in performing the experiments and to Frank R. Mayo for his helpful discussions.

Literature Cited

(1) Alagy, J., Clément, G., Balacéanu, J. C., *Bull. Soc. Chim. France* **1959**, 1325; **1960**, 145; **1961**, 1303, 1792.
(2) Bartlett, P. D., Benzing, E. P., Pincock, R. E., *J. Am. Chem. Soc.* **82**, 1762 (1960).
(3) Fineman, M., Ross, S. D., *J. Polymer Sci.* **5**, 259 (1950).
(4) Handy, C. T., Rothrock, H. S., *J. Am. Chem. Soc.* **80**, 5306 (1958).
(5) Hendry, D. G., Mayo, F. R., Schuetzle, D., unpublished data.
(6) Hendry, D. G., Russell, G. A., *J. Am. Chem. Soc.* **86**, 2386 (1964).
(7) Herk, L., Stefani, A., Szwarc, M., *J. Am. Chem. Soc.* **83**, 3005 (1961).
(8) Howard, J. A., Schwalm, W. J., Ingold, K. U., ADVAN. CHEM. SER. **75**, 6 (1968).
(9) Leavitt, F., Levy, M., Szwarc, M., Stannett, V., *J. Am. Chem. Soc.* **77**, 5493 (1955).
(10) Mayo, F. R., Miller, A. A., *J. Am. Chem. Soc.* **80**, 2465, 2480, 2493, 2497 (1958).
(11) Mayo, F. R., Miller, A. A., Russell, G. A., *J. Am. Chem. Soc.* **80**, 2500 (1958).
(12) Mayo, F. R., Syz, M. G., Mill, T., Castleman, J. K., ADVAN. CHEM. SER. **75**, 38 (1968).
(13) Mayo, F. R., Walling, C., *Chem. Rev.* **46**, 191 (1950).
(14) Meyer, J. A., Stennett, V., Szwarc, N., *J. Am. Chem. Soc.* **83**, 25 (1961).
(15) Middleton, B. S., Ingold, K. U., *Can. J. Chem.* **45**, 191 (1967).
(16) Pritchard, H. O., *Chem. Revs.* **52**, 579 (1953).
(17) Russell, G. A., *J. Am. Chem. Soc.* **77**, 4583 (1955).
(18) *Ibid.*, **78**, 1047 (1956).
(19) Russell, G. A., Desmond, K. M., *J. Am. Chem. Soc.* **85**, 3139 (1963).
(20) Russell, G. A., Williamson, R. C., *J. Am. Chem. Soc.* **86**, 2357 (1964).
(21) *Ibid.*, p. 2364.
(22) Szwarc, M., Binks, J. H., "Kekulé Symposium," p. 262, Butterworths Scientific Publications, London, 1958.
(23) Van Sickle, D. E., Mayo, F. R., Arluck, R. M., *J. Am. Chem. Soc.* **87**, 4824 (1965).
(24) Walling, C., "Free Radicals in Solution," p. 95, Wiley, New York, 1957.
(25) *Ibid.*, p. 254.
(26) Walling, C., McElhill, E. A., *J. Am. Chem. Soc.* **73**, 2927 (1951).

RECEIVED November 2, 1967. Research performed under the SRI general oxidation program, supported by 19 petroleum and chemical companies.

4

Co-Oxidations of Hydrocarbons

FRANK R. MAYO, MARTIN G. SYZ, THEODORE MILL, and JANE K. CASTLEMAN

Stanford Research Institute, Menlo Park, Calif. 94025

Reactivity ratios for all the combinations of butadiene, styrene, Tetralin, and cumene give consistent sets of reactivities for these hydrocarbons in the approximate ratios 30:14:5.5:1 at 50°C. These ratios are nearly independent of the alkylperoxy radical involved. Co-oxidations of Tetralin-Decalin mixtures show that steric effects can affect relative reactivities of hydrocarbons by a factor up to 2. Polar effects of similar magnitude may arise when hydrocarbons are co-oxidized with other organic compounds. Many of the previously published reactivity ratios appear to be subject to considerable experimental errors. Large abnormalities in oxidation rates of hydrocarbon mixtures are expected with only a few hydrocarbons in which reaction is confined to tertiary carbon-hydrogen bonds. Several measures of relative reactivities of hydrocarbons in oxidations are compared.

The original objectives of this work were to determine how much the relative reactivity of two hydrocarbons toward alkylperoxy radicals, $RO_2\cdot$, depends on the substituent R—, and whether there are any important abnormalities in co-oxidations of hydrocarbons other than the retardation effect first described by Russell (30). Two papers by Russell and Williamson (31, 32) have since answered the first objective qualitatively, but their work is unsatisfactory quantitatively. The several papers by Howard, Ingold, and co-workers (20, 21, 23, 24, 29) which appeared since this manuscript was first prepared have culminated (24) in a new and excellent method for a quantitative treatment of the first objective. The present paper has therefore been modified to compare, experimentally and theoretically, the different methods of measuring relative reactivities of hydrocarbons in autoxidations. It shows that large deviations from linear rate relations are unusual in oxidations of mixtures of hydrocarbons.

By using the differential form of the copolymer composition equation (26, 28) the products of oxidation of mixtures at low conversions permit comparison of rates of chain propagation in autoxidations of various compounds, essentially free from effects of chain initiation, chain termination, and over-all rates.

The copolymer composition equation was first applied to co-oxidations in mixtures of aldehydes (25, 39) and later to numerous pairs of hydrocarbons and their derivatives (1, 2, 3, 4, 8, 27, 31, 32, 33). For oxidations of mixtures of A and B, attack by a peroxy radical first gives (by addition or hydrogen abstraction) A· and B· radicals; in the presence of sufficient oxygen all these are then converted to $AO_2\cdot$ and $BO_2\cdot$ peroxy radicals. From the relative rates of reaction, $\Delta[A]/\Delta[B]$, of A and B at two or more average feeds $[A]/[B]$, in long kinetic chains, the copolymer composition equation

$$\frac{-\Delta[A]}{-\Delta[B]} = \frac{r_a[A]/[B] + 1}{r_b[B]/[A] + 1} \tag{1}$$

then permits solution for the two reactivity ratios:

$$r_a = \frac{k_{aa}(\text{for } AO_2\cdot + A)}{k_{ab}(\text{for } AO_2\cdot + B)} \qquad r_b = \frac{k_{bb}(\text{for } BO_2\cdot + B)}{k_{ba}(\text{for } BO_2\cdot + A)} \tag{2}$$

When the relative reactivity of A and B toward $AO_2\cdot$ is the same as toward $BO_2\cdot$, then $r_a = 1/r_b$ and $r_a r_b = 1$. Thus, r values permit us to determine how change in R— affects the selectivity of $RO_2\cdot$ radicals (our first objective).

In general, if $r_a r_b = 1$ and if the chain termination constants for oxidations of A alone and B alone are the same ("ideal" reactivities), then the total rates of oxidation of mixtures at constant rate of initiation are a linear function (for "ideal" solutions) of the volume % of B in the A—B feed. Russell (30, 31, 32) and Alagy and co-workers (1, 2, 3, 4, 8, 33) have shown for several systems that when B has a higher termination constant than A and when B is sufficiently reactive, B can reduce (sometimes fourfold) the rate of oxidation to less than the "ideal" rate. A secondary objective was to see if there were any other important abnormalities (particularly rates much greater than ideal) in co-oxidations of hydrocarbons.

Previous and Contemporary Work

Since the third objective of this paper is to compare, theoretically and experimentally, different methods of measuring reactivities of hydrocarbons in autoxidation, the most pertinent work is reviewed below.

Liquid-Phase Co-Oxidations at Institut Francais du Pétrole (IFP).
From 1959 to 1961, Alagy, Clément, and Balaceanu published four papers (1, 2, 3, 4) on the liquid-phase co-oxidations of 43 combinations of 17 hydrocarbons; in 1964 a fifth paper (8) on three combinations of styrene appeared. These authors measured initial rates of oxygen absorption by undiluted hydrocarbons and their mixtures with 0.06M ABN [2,2'-azobis-(2-methylpropionitrile)] at 60°C. and used deviations of these rates from linear functions of composition to compare rate constants for chain termination in autoxidations of single hydrocarbons. Since this work has been well reviewed and brought up to date in this volume (33), details of their methods are not discussed here. In our opinion, their method is not capable of the fine discrimination their papers suggest (4, 33); neither is their revised scheme as poor as suggested by the criticism of Howard and Ingold (21). In any event, the IFP conclusions about rate constants have now (33) been brought into agreement with the absolute measurements of Howard and Ingold.

More important at present are their r values obtained from rate data by solving the copolymerization rate equation for three unknowns, r_a, r_b, and the cross-termination factor, ϕ. The results are unsatisfactory; consideration in our discussion of possible reasons brings out some weaknesses in the IFP estimates of termination constants.

Liquid-Phase Co-Oxidations by Russell and Williamson. Russell and Williamson (31, 32) reported numerous co-oxidations of hydrocarbons and some ethers at 1M total concentrations in benzene at 60°C., with 0.1M ABN as initiator. Oxidations were carried to conversions up to 50%; the oxygenated products were then removed by treatment with silica gel, and the remaining original hydrocarbons were determined by gas chromatography. This method depends on the reaction of enough of both A and B so that significant values of $\Delta[A]/\Delta[B]$ can be obtained for use in Equation 1. A 1% error in determining recovered A or B at 10% conversion of that substrate leads to a 10% error in $\Delta[A]$, and so on. Equation 1 always leads to larger errors in r than in $\Delta[A]$ or $\Delta[B]$. The consistency of the results is determined by the spread in values of r_a and r_b by the intersection method (26, 28).

This chromatographic method is relatively fast and simple. The results range from good to poor. From stated errors in 20 combinations of hydrocarbons (and ethers) on which Russell and Williamson report some results, 25 of the 40 r values seem to be reliable within about 50%; 15 may have greater errors or are not reported. Fifteen r values seem consistent to better than 15%, but five of these are based on at least one conversion below 10%, so that there is some doubt about their reliability. Of eight of their values that we have checked, six differ from ours by a factor of about 3 or more.

The variable quality of Russell and Williamson's results has three origins:

(1) Most of the combinations were tested only in 20 to 80, 50 to 50, and 80 to 20 starting ratios; feeds should have been adjusted to give such ratios in the products.

(2) Some substrates (particularly cumene) were so unreactive that the necessary substantial conversions of the unreactive component were not attained.

(3) There is no evidence that inconsistent or unreasonable results were checked.

We have made special efforts to overcome all these difficulties in our own work.

Their work adequately supports their qualitative conclusion "that peroxy radicals have about the same reactivity irrespective of the nature of the alkyl substituent" if the term "about" admits a factor of 2 or 3 and if we overlook the factor of 50 for Tetralin-Decalin. Their work leaves unanswered the question of whether these factors are real or experimental.

Co-Oxidations of Substituted Styrenes. Dulog, Kern, and co-workers (9) have studied the rates and products of co-oxidation of styrene and several substituted styrenes. Oxidations were carried out to ⩽5% conversion with undiluted monomers and $0.302M$ ABN at 50°C. The polyperoxides were isolated and analyzed by elementary analysis and infrared absorption. Cleavage of the styrenes to benzaldehydes and formaldehydes was neglected. This neglect will introduce a minor error if the two styrenes cleave to different extents. Russell and Williamson (31) report 16 and 22% cleavage for styrene and p-nitrostyrene with 1 atm. of oxygen at 60°C. (the extent of cleavage would be less at 50°C.).

Results of Dulog, Kern et al., and a few others (27, 31) are summarized in Table I. From the r_a values, toward the unsubstituted styrene-peroxy radical, the most reactive styrene—p-methoxystyrene—is three times as reactive as the least reactive, m-nitrostyrene. Thus, the best electron donor shows the highest reactivity (lowest enthalpy of activation) toward the electron-accepting styrene-peroxy radical. The deviations from unity of the $r_a r_b$ products measure the effects of substitution on the selectivity of the peroxy radicals. These products depart more from unity with increasing differences in the electron-donating and accepting properties of the nuclear substituents. The same trends appear within the other styrene combinations.

The consistency of these results suggests that while nuclear substitution in styrenes affects their reactivity toward peroxy radicals by a maximum factor of only about 3 (r_a and r_b values), the effect on the selectivity of the peroxy radicals ($r_a r_b$ products) is nearly as great.

Absolute and Relative Rate Constants of Ingold et al. For some time, Ingold and co-workers (20, 21, 23, 24, 29) have been reporting

Table I. Co-Oxidations of Substituted Styrenes[a]

Monomer A	r_a	Monomer B	r_b	$r_a r_b$
Styrene	0.38 (0.55[b])	p-Methoxystyrene	2.10 (1.2[b])	0.80 (0.66[b])
	0.705	p-Iodostyrene	0.995	0.70
	0.71	p-Methylstyrene	1.51	1.07
	0.88	p-Bromostyrene	0.97	0.86
	0.88	p-Fluorostyrene	0.92	0.81
	0.89	p-Chlorostyrene	0.98	0.87
	1.10	p-Cyanostyrene	0.50	0.55
	1.20 (1.5[b])	p-Nitrostyrene	0.42 (0.22[b])	0.50 (0.33[b])
	1.24	m-Nitrostyrene	0.32	0.40
p-Methoxystyrene	(2.3[b])	p-Nitrostyrene	(0.17[b])	(0.39[b])
Styrene	(0.48,[c] 0.9[b])	α-Methylstyrene	(2.1,[c] 1.2[b])	(1.0,[c] 1.1[b])
	9.2 (7.9[c])	Methyl methacrylate	0.09 (0.078[c])	0.83 (0.62)
	0.37	2,3-Dimethylbutadiene	2.75	1.02
	0.32	Methyl sorbate	2.45	0.79
	0.33	Methyl eleostearate	29.2	0.96
α-Methylstyrene	(15[c])	Methyl methacrylate	(0.041[c])	(0.62[c])

[a] Results of Dulog et al. (9) at 50°C. except as noted.
[b] (31).
[c] (27).

absolute rate constants for chain propagation and chain termination in oxidations of hydrocarbons. One of the earlier papers (21) shows that primary alkylperoxy radicals are more reactive in chain propagation than tertiary alkylperoxy radicals.

Early this year, Middleton and Ingold (29) reported relative rates of chain propagation of a primary, secondary, and tertiary peroxy radical (from allylbenzene, Tetralin, and α-methylstyrene, respectively) with a series of nine hydrocarbons. By using a large excess of one of the first three hydrocarbons, they dealt almost entirely with one chain carrier in each co-oxidation. Relative reactivities were determined by a GLPC method like that of Russell and Williamson (31, 32). They concluded that the average relative reactivities of the primary, secondary, and tertiary peroxy radicals toward the nine hydrocarbons were 5.2/2.2/1.0 but that the relative reactivities of the nine hydrocarbons were about the same toward each type of radical. These results are acceptable as semiquantitative. However, despite numerous replicate analyses, their experimental method suffers from the same limitations as that of Russell and Williamson, and they give no *primary* data—only the calculated results.

A subsequent paper (*23*) gives propagation and termination constants for numerous additional hydrocarbons and deals mostly with relative reactivities of active hydrogen atoms and with effects of structure on termination constants. A comparison of relative reactivities of hydrocarbons toward alkylperoxy, *tert*-butoxy, and phenyl radicals uses a different alkyl in each alkylperoxy radical in spite of the differences in reactivity among different alkylperoxy (*29*) radicals.

This difficulty has now been overcome. Howard, Schwalm, and Ingold (*24*) show that the rate constant for reaction of any alkylperoxy radical with any hydrocarbon can be determined (by the sector method) by carrying out the autoxidation of the hydrocarbon in the presence of $>0.1M$ hydroperoxide corresponding to the chosen radical. All the absolute propagation and termination constants for the co-oxidation of cumene and Tetralin were thus determined. Our Tetralin-cumene work suggests that their results agree well with the best we have been able to get.

Experimental Procedure and Results

Except as otherwise noted, all oxidations were carried out in the presence of $0.06M$ ABN at 60°C. The quantities of materials, reaction times, and analytical results are given in Tables II to V. The oxygen reacting has been corrected in the tables and discussion for that absorbed in chain initiation and evolved in chain termination: each ABN decomposed is assumed to evolve 1 N_2 and to absorb 0.60 O_2 net. The rate constant of decomposition of ABN is taken to be 9.94×10^{-6} per second and the initiation efficiency to be 60% (*37*).

The r values were determined by the method of Fineman and Ross (*10, 38*). When $(\rho - 1)/R$ is plotted against ρ/R^2, a straight line should be obtained with intercept r_a and slope $-r_b$. Here, ρ is defined as $\Delta A/\Delta B$ and R as the average $[A]/[B]$ in the feed.

Styrene-Tetralin (Syz). PROCEDURE. Co-oxidations of styrene (Eastman Kodak Co., distilled and passed through alumina before each use) and Tetralin (Matheson, Coleman, and Bell, practical grade, washed with sulfuric acid and distilled, then passed through alumina before use) were carried out in sealed tubes of ~110-ml. capacity, fitted with two capillary sidearms and a break seal. The vessel was connected to the vacuum line through one sidearm, and through the other a weighed amount of a solution of ABN in Tetralin and styrene was injected. This sidearm was then sealed off, and the reaction mixture was degassed by freezing under vacuum; a measured amount of oxygen (average pressure about 300 mm. at 25°C.) was then added through the other sidearm, which was then sealed off. The reaction vessel was then shaken in a water bath at 60° ± 0.2°C. for 10 hours. After the vessel was sealed to the vacuum line and cooled in liquid nitrogen, the seal was broken, and the total noncondensable gases were measured volumetrically. Oxygen was then removed through a Cu—CuO furnace at 280°C., and CO was oxidized to CO_2. The latter was condensed in a trap at −196°C., and the noncondensables (essentially nitrogen) were measured. The CO_2 was then determined volumetrically by volatilization from any other products (H_2O) at −78°C.

DETERMINATION OF HYDROCARBONS. To determine oxidized styrene in co-oxidation products, known amounts of the neat oxidation mixtures were mixed with small amounts of the internal standard, hexamethylcyclotrisiloxane (single band at 9.8 τ). NMR spectra were obtained and the bands at 0.2 τ (Ph—$\overline{\text{CHO}}$) and 5.8 τ (—CPhH—$\overline{CH_2}$—OO—) were electronically integrated several times and compared with the area of the standard. The proportion of benzaldehyde ranged from 13% (Run 93) to 19% (Run 84) of the total oxidized styrene. An oxidation of styrene without Tetralin in the same manner (2-hour reaction time, 13.5% conversion) gave 21% benzaldehyde. The sum of the benzaldehyde (and assumed formaldehyde) and styrene polyperoxide thus determined was 101.0 ± 3.2% of the corrected oxygen uptake. Our average standard deviation for determining ΔS in a reaction mixture was ±1.5%. The amount of Tetralin reacted (ΔT) was then taken as the difference between the (corrected) oxygen reacted (ΔO_2) and the styrene reacted (ΔS).

RESULTS. The data in Table II, when plotted in Figure 1, give $r_s = 2.3 \pm 0.3$; $r_t = 0.43 \pm 0.03$; $r_s r_t = 1.0$. For this system, Chevriau, Naffa, and Balaceanu (8) have reported $r_s = 4.2$; $r_t = 0.85$; $r_s r_t = 3.6$. Russell and Williamson (32) have reported $r_s = 2.8$; $r_t = 1.2$; $r_s r_t = 3.3$.

Styrene-Cumene (Syz). The experimental procedure for starting co-oxidations of styrene and cumene (Phillips 98%, washed with sulfuric acid, dried with calcium hydride, fractionated, and then passed through alumina before each use) was essentially the same as that used for styrene and Tetralin. However, the retarding effect of styrene on the oxidation of cumene led to the use of 0.12M ABN and to a different analytical procedure. Conversions were so low that styrene peroxide units and benzaldehyde could not be determined satisfactorily by NMR in very dilute solution in cumene. We therefore took amount of cumene reacted to be the hydroperoxide that could be titrated by the Wibaut (40) method. This is an adequate approximation since in oxidizing cumene alone, 97.7% of the reacting oxygen (corrected for ABN decomposition) appeared as titratable hydroperoxide (Run 122 in Table III). Other experiments in Table III are corrected for this blank. In the oxidation of styrene alone (Run 26), only 1.06% of the reacting oxygen could be titrated. However, cumene hydroperoxide is unstable in the presence of oxidizing styrene at 60°C. (16). To correct for this complication, two sets of styrene-cumene mixtures with mole ratios of 0.0271 and 0.192 were

Table II. Co-oxidations of Styrene (S) and Tetralin (T) for 10 Hours at 60°C.

Run Number (Quantities in mmoles)

	88	93	90	95	80	82	84
S_o	0.652	0.796	0.820	1.054	1.985	2.499	3.558
T_o	2.669	2.804	2.539	2.614	1.983	1.202	0.708
ΔO_2	0.386	0.507	0.443	0.447	0.654	0.755	1.074
ΔS NMR	0.129	0.195	0.188	0.209	0.460	0.599	0.959
ΔT diff	0.257	0.312	0.255	0.238	0.194	0.156	0.115
S_{av}	0.587	0.699	0.726	0.944	1.755	2.200	3.079
T_{av}	2.541	2.648	2.412	2.498	1.886	1.124	0.651
$R = S_{av}/T_{av}$	0.231	0.240	0.300	0.377	0.941	1.957	4.729
$\rho = \Delta S/\Delta T$	0.501	0.628	0.737	0.880	2.371	3.839	8.339
$(\rho - 1)/R$	−2.16	−1.40	−0.87	−0.32	1.45	1.45	1.55
ρ/R^2	9.75	9.01	8.18	6.19	2.67	1.00	0.37

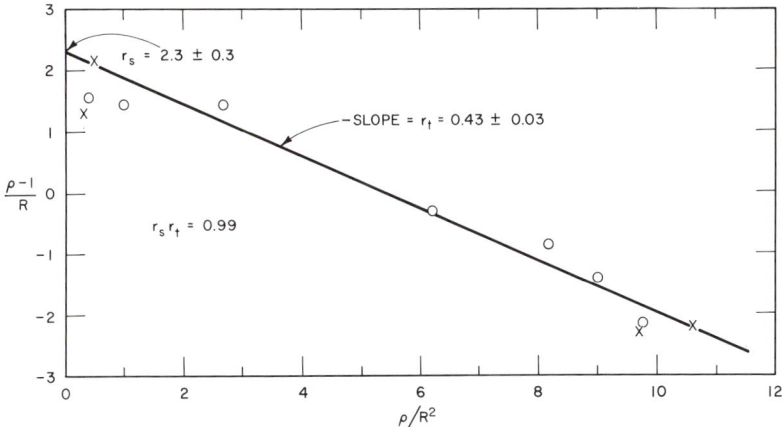

Figure 1. Finemann-Ross plot of styrene-Tetralin co-oxidations in Table II. The symbol × indicates the results if styrene consumption were 1.5% greater or less than shown in Table II

oxidized for 2.0, 4.0, and 8.25 hours, and the calculated values of $(\rho -1)/R$ and of ρ/R^2 (open circles in Figure 2) were extrapolated to zero time (solid circles in Figure 2). The two points thus determined corresponded to $r_s = 20.5$ and $r_c = 0.053$.

Since these reactivity ratios indicated a larger difference in relative reactivities of styrene and cumene than were indicated by previous work [13 and 0.1 (8), 6.5 and −0.2 (32)], we checked our conclusion in two ways.

1. An experiment like 137 was carried out in which the cumene was replaced with an equal volume of chlorobenzene, and 0.0186M cumene

Table III. Co-oxidations of Styrene (S) and

	Run No.					
	122	26	132	135	134	
Time, hr.	15	2	8.25	4	2	0
S_o	0	7.225	0.202	0.320	0.287	
C_o	4.968	0	7.444	11.987	10.549	
O_o	1.466	1.766	1.550	1.486	1.397	
ΔO_2 [b]	0.991	0.962	0.446	0.365	0.195	
ΔC tit [b]	0.968		0.320	0.253	0.131	
ΔS diff [c]		0.970 [c]	0.126	0.112	0.064	
C_{av}			7.284	11.861	10.480	
S_{av}			0.139	0.264	0.255	
$R = S_{av}/C_{av}$			0.0191	0.022	0.0244	
$\rho = \Delta S/\Delta C$			0.395	0.442	0.488	
$(\rho - 1)/R$			−31.5	−25.2	−21.0	−18.7 [a]
ρ/R^2			1090	902	826	740 [a]

[a] In the left column, ΔC is by titration (after corrections) and ΔS by difference; in the right column, ΔS is by GLPC and ΔC by difference.
[b] Corrected for blank Expt. 122.

hydroperoxide was added so that the average concentration of this material would correspond to that in Run 137. Titrations before and after 4 hours of heating at 60°C. showed that 6.1% of the titratable peroxide disappeared during oxidation of the styrene. Applying this correction to Run 137 (instead of extrapolating) gave $(\rho - 1)/R = 12.3$ and $\rho/R^2 = 135$, indicated by a solid square in Figure 2, in satisfactory agreement with the extrapolated value.

2. In Run 65, the cumyl hydroperoxide concentration was corrected as in Run 1, and the reacted styrene was determined by GLPC. For the latter determination, 10 ml. of cyclohexane were added to the reaction mixture, from which the oxygenated products were then removed by treatment with 3.5 grams of aluminum oxide. $\Delta S + \Delta C$ then equaled 1.042 ΔO_2, a good check.

Table III and Figure 2 show the alternate results obtained by determining each reactant directly and by difference in Run 65. It is difficult to escape the conclusion that $r_s = 20 \pm 2$; $r_c = 0.052 \pm 0.003$; $r_s r_c = 1.0$.

Tetralin-Cumene (Mill and Castleman). Co-oxidations of Tetralin and cumene were carried out by shaking the mixtures indicated in Table IV with oxygen at a total pressure of 1 atm. at 60°C. Oxygen absorption was measured and corrected for ABN decomposed. Oxidation products were determined by reducing the hydroperoxides with a slight excess of triphenylphosphine (18) and analyzing by GLPC for the ratio of (tetralol + tetralone) to (cumyl alcohol + acetophenone). In most runs acetophenone was absent or present only in traces; no products other than these were noted in the gas chromatograms. Although some nonterminating

Cumene (C) with 0.12M ABN at 60°C.

(Quantities in mmoles)

136	137	138		65[a]	
8.25	4	2	0	4	
1.681	1.804	1.694		2.111	
9.782	10.492	9.862		10.980	
1.540	1.545	1.417		1.945	
0.966	0.626	0.327		0.517	
0.247	0.156	0.075		0.123	0.101
0.719	0.470	0.252		0.394	0.416
9.659	10.414	9.825		10.919	10.929
1.320	1.569	1.568		1.914	1.903
0.137	0.151	0.162		0.175	0.174
2.82	3.02	3.35		3.20	4.11
13.8	13.4	14.6	14[d]	12.6	17.9
157	137	128	118[d]	105	136

[c] $\Delta S = \Delta O_2 - \Delta C$. In Expt. 26, ΔS was determined directly by NMR; 1.06% of the reacted O_2 could be titrated as hydroperoxide.
[d] Extrapolated.

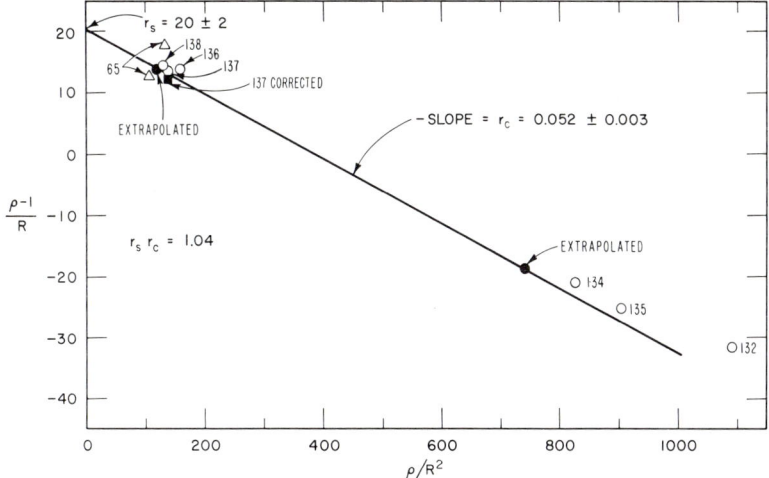

Figure 2. *Finemann-Ross plot of styrene-cumene co-oxidations in Table III*

interactions of cumylperoxy radicals evolve oxygen in the oxidation of cumene alone (7, 13, 35), the strong retardation by Tetralin suggests that most radical interactions in the co-oxidation with Tetralin involve $TO_2\cdot$ radicals and are terminating. The presence of tetralone and near-absence of acetophenone in the products support this conclusion. Hence the correction at the beginning of this section was still applied in this system.

Table IV summarizes results of co-oxidations of seven mixtures of cumene and Tetralin containing 50 to 90 mole % cumene. Data are plotted in Figure 3. The scatter in the plot is somewhat greater than expected from the analytical procedure. Run 12 with 70% cumene shows

Table IV. Co-Oxidations of Tetralin (T) and Cumene (C) at 60°C.

	Run Number (Quantities in mmoles)						
	100	7	12	40	98	58	24
Time, hr.	9.1	11.75	14.25	11.10	11.5	16.5	16.0
C_o	100.87	100.23	139.75	221.38	210.54	212.25	180.22
T_o	101.17	100.07	58.63	57.16	37.46	37.84	20.91
$-\Delta O_2$	18.31	19.01	14.39	13.83	11.14	14.72	18.87
Correction[a]	0.17	0.24	0.28	0.31	0.29	0.39	0.31
$-\Delta O_2$ cor.	18.48	19.25	14.67	14.14	11.43	15.11	19.18
ΔC[b]	3.34	3.42	4.14	5.96	5.78	8.20	11.4
ΔT[b]	15.14	15.83	10.53	8.18	5.65	6.91	7.78
$R = T_{av}/C_{av}$	0.944	0.935	0.388	0.243	0.167	0.165	0.0977
$\rho = \Delta T/\Delta C$	4.53	4.74	2.54	1.37	0.979	0.842	0.683
$(\rho - 1)/R$	3.74	4.00	3.96	1.53	−0.126	−0.956	−3.24
ρ/R^2	5.08	5.42	16.88	23.2	35.2	30.8	71.5

[a] Corrected for evolution of one N_2 and net absorption of 0.6 O_2 for each ABN decomposed.
[b] Values derived from measured ρ and total O_2 consumption.

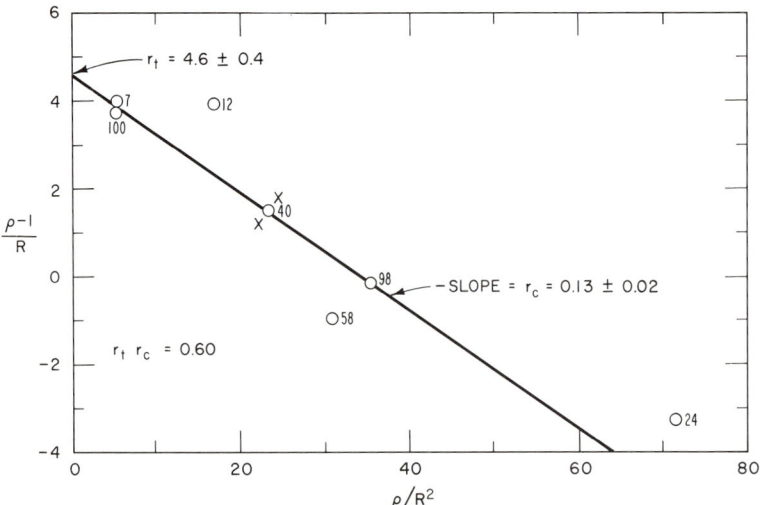

Figure 3. Finemann-Ross plot of Tetralin-cumene co-oxidations in Table IV. The symbol × indicates result if $\Delta T/\Delta C$ is 5% greater or less than shown in Table IV

the poorest agreement and has been neglected in plotting the best line, which corresponds to $r_t = 4.6 \pm 0.4$; $r_c = 0.13 \pm 0.02$; $r_t r_c = 0.60$. The additional points for Run 40 show the effect of a 5% change in $\Delta T/\Delta C$ on the results. For the same systems at 90°C., Russell (*30*) gave $r_t = 16$; $r_c = 0.044 \pm 0.02$; $r_t r_c = 0.70$. Russell and Williamson (*32*) later gave for 60°C. $r_t = 2.5$; $r_c = 0$. Alagy *et al.* gave only $r_t = 1.0$. The absolute propagation constants of Howard, Schwalm, and Ingold (*24*) permit the calculations, $r_t = 4.5$, $r_c = 0.18$, $r_t r_c = 0.83$ at 56.6°C., in good agreement with our own, and 8.3, 0.11, and 0.90 at 30°C. Since Tetralin has four times as many benzyl hydrogen atoms as cumene, the reactivities per hydrogen atom toward peroxy radicals are the same within a factor of 2.

Tetralin–*cis*–Decalin (Syz). Co-oxidations of this system were undertaken because the internally consistent results of Russell and Williamson (*32*) ($r_t = 1.89 \pm 0.01$; $r_d = 0.01 \pm 0.002$; $r_t r_d = 0.019$) indicated an extraordinary difference in selectivity of the two alkylperoxy radicals. Oxidations of *cis*-Decalin (7% *trans* by GLPC, from K and K Laboratories, passed over aluminum before use) and Tetralin were carried out in sealed tubes and analyzed as described for styrene-Tetralin.

Neither NMR nor titration of the hydroperoxide distinguishes between the two reaction products in this system. Because appreciable amounts of CO and CO_2 are generated from secondary oxidations of Decalin, the reacted Decalin could not be determined by difference from the oxidized Tetralin and the consumed oxygen. (In an oxidation of *cis*-Decalin alone to 2.9% conversion in 19.5 hours, 14.5% of the oxygen consumed appeared as CO and CO_2.) Since the conversion of Decalin never exceeded 3%, an accuracy of at least 10% in the determination of reacted Decalin requires that the error in the GLPC analysis for Decalin be $< 0.3\%$. This precision could be obtained by replicate analyses and integration of the peak heights by a Wilkens 471 digital integrator. A Carbowax 20 M column was used at 130°C., and acetophenone served as internal standard.

Results are summarized in Table V and Figure 4. On the Fineman-Ross plot the limits of the experimental error shown were calculated for runs 20 and 29 taking the errors as $+1.0\%$ for ΔT and -15% for ΔD, and as -1.0% for ΔT and $+15\%$ for ΔD.

Discussion

Background of Present Research. We have been critical enough of our own work so that we can afford to criticize other work. Our studies on co-oxidations began in 1962 with the idea that co-oxidations of styrene with hydrocarbons which gave mainly hydroperoxides which could be easily titrated would provide an economical route for accumulating data

Table V. Co-Oxidations of *cis*-Decalin (D) and Tetralin (T) at 60°C.

	Run Number (Quantities in μmoles)				
	7	29	11	17	20
Time, hr.	21.5	19.0	~21	22.2	20.5
T_o	2425	2262	1739	632	518
D_o	3341	3163	4175	4445	5405
ΔT^a	705	547	803	235	155
ΔD^a	85	69	191	142	118
T_{av}	2073	1989	1337	514	441
D_{av}	3299	3128	4080	4374	5346
$R = (T/D)_{av}$	0.629	0.637	0.327	0.117	0.0824
R^2	13.2	12.4	12.9	14.1	15.9
$\rho = \Delta T/\Delta D$	8.29	7.92	4.21	1.65	1.31
$(\rho - 1)/R$	11.6	10.9	9.82	5.55	3.69
ρ/R^2	21.0	19.6	39.4	158	193

^{a} ΔD and ΔT determined by GLPC.

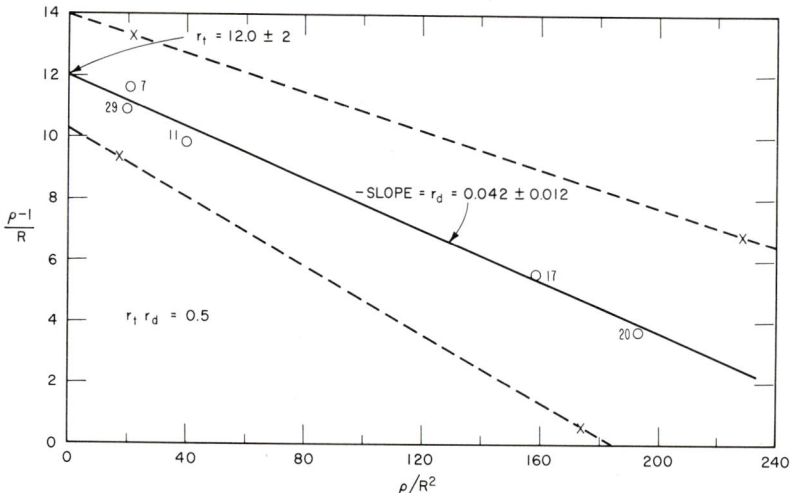

Figure 4. Finemann-Ross plots of cis-Decalin-Tetralin co-oxidations in Table V. The symbol × indicates probable limits of experimental error for Runs 20 and 29

on relative reactivity. We soon obtained fairly consistent results on styrene-Tetralin (r_s = 5.4, r_t = 0.54, $r_s r_t$ = 2.9) and styrene-cumene (r_s = 100, r_c = 0.5, $r_s r_c$ = 50), but they did not seem reasonable nor were they consistent with the Williamson thesis (*31, 32*) or with the work of Alagy *et al.* (*8, 33*). We then found that difficulty arose from the accelerated decomposition of titratable peroxide in systems containing oxidizing styrene (*16*). Seeing the difficulties (as well as advantages) in the Russell and Williamson approach, and finding that monomeric styrene interfered with determining hydroperoxides by gas chromatography and that styrene polyperoxide was not cleanly reduced by triphenylphosphine, we turned to determining reacted styrene (as polyperoxide and benzaldehyde) by NMR absorption. Total oxygen absorption was measured, and the amount of Tetralin or cumene reacting was then determined by difference. This method gave consistent results with cumene ($r_s = 3.3$, $r_c = 0.14$, $r_s r_c = 0.5$) but rather poor results with Tetralin ($r_s = 3$, $r_t = 0.6$, $r_s r_t = 1.8$). However, neither set of results seemed reasonable; the difference in reactivity between styrene and cumene was improbably low, and the $r_s r_t$ product for Tetralin was too high. The results given above for co-oxidations of Tetralin, cumene, and styrene are internally consistent, reasonable, and also (as shown below) consistent with a simple general pattern. The Tetralin-cumene results are based on analyses for both kinds of products and are otherwise satisfactory.

Reactivity Ratios and Relative Rates of Oxidation. Table VI summarizes reactivity ratios for all combinations of butadiene, styrene, Tetralin, and cumene reported here and by Hendry (*12*) for butadiene. The underscored r values in the table are calculated from the base point in the same column (1.00) and the reciprocals (because of the definition of r in Equation 2) of the relative reactivities of the hydrocarbons toward an average $RO_2 \cdot$ radical, B:S:T:C = 1.0:2.2:4.5:5:30 (all underscored values in the butadiene column). The calculated and experimental values always agree within 50%, and in seven out of 12 systems, within 15%. Thus, among these four hydrocarbons a change of R— in $RO_2 \cdot$ affects relative reactivities by less (often much less) than 50%.

Most of the largest differences between observed and calculated r values in Table VI occur for cumene combinations. Such discrepancies are inevitable if cumene tends to alternate with butadiene and Tetralin but not with styrene. (This distinction could be partly experimental error, which tends to be greatest in combinations of the least reactive hydrocarbon—cumene—with the other hydrocarbons.) These alternation tendencies are measured by the products of the reactivity ratios, the last number in each group of three in Table VI, 1 corresponding to no effect and 0 to inability of one or both peroxy radicals to react with the hydrocarbon from which it is derived. If we take cumene to be 1/40 as reactive as butadiene (instead of 1/30, as in the table), agreement is better for the two styrene-cumene reactivity ratios and poorer for the other cumene ratios. If instead we take cumene to be 1/20 as reactive as butadiene, agreement is much better for the butadiene-cumene and Tetralin-cumene ratios but worse for the styrene-cumene ratios. These considerations illustrate how the R— in R—O_2· can affect selectivity in reactions of R—O_2· up to 50% (when the product of the ratios is 0.5) and how this effect is readily measured by the ratio products.

Among hydrocarbons, the differences between the experimental and calculated values of r must be almost entirely a steric effect; polar effects are small and must be transmitted through two oxygen atoms. The largest authentic steric effect known to us is that in the Tetralin-Decalin system, where the $r_a r_b$ product is 0.5. This product means that if we take any steric effects in the reactions of tetralylperoxy radicals with Tetralin or Decalin as base points, reaction of decalylperoxy with Decalin is retarded sterically by 50% more than it is retarded with Tetralin. This result is plausible when we consider the structures of Decalin and the derived radical.

Table VI. Reactivity Ratios in Co-Oxidations of Butadiene, Styrene, Tetralin, and Cumene at 60°C.[a]

B \ A	Butadiene, r_b	Styrene, r_s	Tetralin, r_t	Cumene, r_c
Butadiene	1.00	0.45 ± 0.05 / 0.45 / 0.9	0.24 ± 0.03 / 0.22 / 0.84	0.03 ± 0.01 / 0.033 / 0.7
Styrene	2.2 ± 0.2 / 2.2 / 0.9	1.00	0.43 ± 0.03 / 0.49 / 1.0	0.052 ± 0.003 / 0.073 / 1.0
Tetralin	3.5 ± 0.5 / 4.5 / 0.84	2.3 ± 0.3 / 2.0 / 1.0	1.00	0.13 ± 0.02 / 0.15 / 0.6
Cumene	23 ± 3 / 30 / 0.7	20 ± 2 / 13.6 / 1.0	4.6 ± 0.4 / 6.7 / 0.6	1.00

[a] First number in each set is reactivity ratio for peroxide radical of hydrocarbon at top of column. Second number ___ is monomer reactivity ratio calculated from base point in column (1.00) and reciprocals of the average reactivities of hydrocarbons toward RO_2·: B:S:T:C = 1.0:2.2:4.5:30. The third number is the $r_a r_b$ product (each appearing in two places in the table).

For co-oxidation of nuclear-substituted styrenes, Table I shows a consistently decreasing $r_a r_b$ product as the polarity difference increases in pairs of substituted styrenes, the minimum value being about 0.4. This effect cannot be steric and must be polar. The polar effects could be caused by transmission of substituent effects through the O—O link, to some tendency of different styrenes to associate in solution—i.e., actual complex formation—or to a nonrandom distribution of substituted styrenes in their solutions. At present, these polar effects together account for deviations from average reactivity by a factor as large as 2.5. Somewhat larger deviations seem possible for more-substituted styrenes and for oxygen-substituted hydrocarbons.

From the considerations above, in co-oxidations of hydrocarbons all $r_a r_b$ products less than 0.5 or greater than 1.0 (where no explanation seems necessary or plausible) should be considered suspect unless unusual care has been taken to confirm them. Most previously published results are therefore suspect.

Over-all Rates of Co-Oxidations. We now discuss rates of co-oxidation of hydrocarbons at constant rates of initiation, in the presence of sufficient oxygen, and in terms of the "ideal" rate of co-oxidation. Published rate curves for co-oxidations range between nearly ideal and sharply depressed rates. As shown previously (1, 2, 3, 4, 8, 30, 31, 32, 33) a depressed rate results when hydrocarbon B, with a high termination constant and adequate reactivity, is added to hydrocarbon A with a low termination constant. It now appears from the work of Alagy et al. (1-4, 8, 33) and work with isobutane (5) and tert-butyl hydroperoxide (17) that few hydrocarbons have termination constants low enough to produce striking effects—e.g., isobutane, 2,3-dimethylbutane, and cumene, but not sec-butylbenzene. However, no examples have appeared of a mixture of hydrocarbons oxidizing much faster than the ideal rate, nor are any likely to be found. While considerations of termination constants allow for an increased reaction rate when a reactive but slow-terminating hydrocarbon is added to a fast-terminating hydrocarbon, they do not permit a rate much greater than "ideal." The only other apparent way to obtain a rate much higher than ideal is through a strong alternating effect in chain propagation ($r_a r_b \ll 1$); our conclusions on reactivity ratios rule out this route.

Hence, in co-oxidations of hydrocarbons at constant rate of initiation in the presence of sufficient oxygen, it will be easy to depress oxidation rates of the relatively few hydrocarbons with slow termination constants, but otherwise oxidation rates will not differ much from a linear function of composition. When rates of chain initiation are not known or controlled, other relations may appear.

While conventional studies of over-all rates of co-oxidation may (4, 33) [with an occasional exception (33)] or may not (21) be capable of detecting small difference in termination constants, the discussion below shows that there are enough unknowns in co-oxidation rate studies that the careful rate measurements at the IFP are inadequate for measuring reactivity ratios (r values). Thus, from 12 pairs of r values determined by rate studies ($1, 2, 3, 4$) on hydrocarbon mixtures, only three $r_a r_b$ products fall between our acceptable limits 0.5–1.0; the others range from 0.18 (styrene-cyclohexene) to 3.57 (styrene-Tetralin). Of 12 ϕ values given, nine indicate a less than statistical tendency toward crossed termination. Values of ϕ greater than 1 are common in copolymerization ($26, 28$) and easy to understand in oxidation; ϕ values less than 1 are not. Further, in the most recent summary (33), these calculations on hydrocarbon pairs are neglected, and instead data are presented on 19 combinations containing one or two ethers. Here, only six combinations give $r_a r_b$ products between 0.5 and 1 (although four more are close) and only three (none of five cumene combinations!) indicate a more than statistical tendency toward crossed termination. These improbable results show that the usual copolymerization rate equation must neglect one or more significant effects. Tsepalov et al. (36) obtained similar results (with the same weaknesses) from their studies of rates of co-oxidations of cumene and ethylbenzene at 60°C., $r_c = 0.44$, $r_e = 0.97$, $r_a r_b = 0.43$, $\phi = 0.7$.

Traylor and Russell (35) have recently suggested that chain termination in the oxidation of cumene arises through interaction of cumylperoxy radicals to give cumyloxy radicals and that cleavage of some of the latter gives rise to methyl (and methylperoxy and methoxy) radicals which are responsible for most of the chain termination. Hendry has since shown that this conclusion is essentially correct at conventional oxidation rates (13). Thus, the fraction of cumyloxy radicals that cleave (and therefore the rate of chain termination) will depend on the reactivity of the hydrocarbon mixture and on the concentration of hydroperoxide that has been formed. The rate equation for co-oxidation ($1, 30$) does not allow for this complication and may fail for hydrocarbons which are no more reactive than cumene and in which the tertiary hydrogen atom is the principal point of attack. This limitation may not apply to co-oxidations involving only primary and secondary alkylperoxy radicals; as far as we know (17), essentially all their interactions are terminating in the liquid phase below 100°C., and there are fewer complications from alkoxy radicals or cleavage.

Several workers have pointed out the easy exchange of alkylperoxy radicals with hydroperoxides ($16, 24, 34, 35$) and how it may affect rates of co-oxidation, at least of cumene. Howard et al. (24) show how even $0.1M$ hydroperoxide affects both oxidation rate and reactivity ratios. When $r_a r_b = 1$, this exchange will make no difference in the apparent radical reactivities. Russell and Williamson have necessarily used high conversions in most of their co-oxidations. Alagy et al. have not specified their conversions.

Hendry and Russell (*15*) have shown how polyarylmethanes and polyarylethylenes may retard autoxidations of other hydrocarbons at 60° to 90°C.; because the carbon radicals involved react relatively slowly and incompletely with oxygen, the free arylmethyl radicals contribute to a fast, crossed, termination reaction. A similar effect has been reported for benzyl radicals above 300°C. (*11*). These effects can usually be overcome by sufficient oxygen pressure; hence, our restriction at the beginning of this section.

In co-oxidations of undiluted hydrocarbons over a range of compositions, the change in reaction medium may affect over-all rates of oxidation by a factor of 2 or 3 (*14, 19, 37*). Further, while many autoxidations are autoaccelerating, others are autoretarding (*37*).

In view of these difficulties, as well as that of obtaining hydrocarbons of reproducible purity, we consider that the published over-all rate studies on mixtures of hydrocarbons are useful empirical observations, but that most quantitative treatments, including calculation of ϕ values, are not yet really useful. We therefore endorse the statement by Allen and Plesch (*6*) on ϕ values in copolymerization: "It would be . . . appropriate to point out that most of the simplifying assumptions, plausible enough in their context, made in polymer kinetics come home to roost in the ϕ factor." Therefore we omit our own rate data; they are consistent with those already published (*1, 2, 3, 4, 8, 30, 31, 32, 33*).

From the $r_a r_b$ and ϕ values from co-oxidations of styrene, Chevriau and co-workers (*8*) conclude that polymeric peroxy radicals behave differently from simple peroxy radicals. This statement seems to be unjustified. We find no evidence for it in our co-oxidations of styrene and butadiene.

Over-all and Relative Rates of Oxidation of Hydrocarbons. Table VII lists rate constants or relative rates of oxidation as reported by Ingold *et al.* (*23*), Sajus (*33*), or Hendry (*12, 13*) for those hydrocarbons for which comparisons are possible.

Let us now consider what we mean by the reactivity of a hydrocarbon in autoxidation. One measure is how fast it oxidizes by itself at unit concentration and unit rate of initiation. (Rates of thermal oxidation at unknown rates of initiation are not useful enough to be considered.) The first two columns of figures in Table VII give such comparisons in terms of $k_p/(2k_t)^{1/2}$ at 30° and 60°C. and determine the order in which hydrocarbons are listed in the table. The results of Ingold and Sajus agree fairly well; the orders of reactivity are identical except for a trivial difference with the xylenes. The stated quotients at 60°C. are uniformly 2 to 3 times as large as at 30°C. for *sec*-butylbenzene and more reactive hydrocarbons but 3 to 6 times as large for less reactive hydrocarbons.

The difference suggests that at least one group was encountering uncorrected difficulties with short chain lengths.

A second measure of reactivity is the relative rates of attack of a hydrocarbon by the same alkylperoxy radical. The three middle columns of figures in Table VII give such relative rates toward the butadiene peroxy and tetralylperoxy radicals. The 50° and 30°C. results are consistent. The 65°C. results are not as good; reasons were considered above, as well as the extensive similar summary by Russell and Williamson (32).

Table VII. Comparison of Absolute and Relative Rates of Oxidation of Hydrocarbons at 30° to 65°C. (Moles/liter/sec.)

Hydrocarbons	$10^5 k_p/(2k_t)^{1/2}$		Relative k_p (Cumene = 1)			k_p	$k_t \times 10^{-6}$
	30° [a]	60° [b]	$BO_2\cdot$ [c] 50°	$TO_2\cdot$ [d] 30°	$TO_2\cdot$ [e] 65°	30° [a]	30° [a]
α-Methylstyrene	1300				40	10	0.3
Styrene	890	2000	10.4			41	21
Me$_2$C=CMe$_2$	320	1000				2.6	.32
Cyclopentene	280		12.7			8.8	3.1
Tetralin	230	616	7.0	8.3	9.3	6.3	3.8
Cyclohexene	230	510	5.7		20	6.1	2.8
Cumene	150	356	1.00	1.0	1.0	0.18	0.0075
						0.16 [f]	0.0095 [f]
						0.80 [f]	0.025 [f]
Allylbenzene	49				7.5	10	220
Ethylbenzene	21	53		0.65	1.5	1.30	20
sec-Butylbenzene	18	45	0.9			0.08	0.09
Cyclohexylbenzene	15	60				0.06	0.08
n-Butylbenzene	8.1	39				0.56	25
1-Octene (1-hexene)	6.2	37				1.0	130
p-Xylene	4.9	16.1				0.84	150
o-Xylene	3.3	16.3			0.8	0.42	77
Mesitylene		13.9			0.8		
m-Xylene	2.8	11.8			0.48		150
Toluene	1.4	5.3	0.34			0.24	150

[a] (23).
[b] (33).
[c] Butadiene peroxy radicals (12).
[d,e] Tetralin peroxy radicals, (24) or (29), respectively.
[f] (13) at 30° and 60°C., respectively.

A third measure of reactivity is the rate constant for chain propagation of individual hydrocarbons. Together, the chain propagation and chain termination constants in the last two columns in Table VII give us the only good basis for understanding oxidation rates of single hydrocarbons. However, for the listed propagation constants, the peroxy radical changes every time that the hydrocarbon changes, and the differences in absolute reactivity among peroxy radicals are great enough that absolute propagation constants are a poor measure of relative reactivities of hydrocarbons toward the same radicals. Thus tetralylperoxy is about four times as reactive as cumylperoxy toward the same hydrocarbon (24), and primary alkylperoxy radicals are still more reactive (*see* review of Ingold work).

Usually when we consider relative reactivity, we want to know about relative reactivity in competing reactions. This paper has shown that for hydrocarbons, this relative reactivity (not absolute reactivity) is nearly independent of the R in the $RO_2 \cdot$ radical. The largest authentic effect of R found was a factor of 2 and was associated with the most steric hindrance. [The recent claim of Howard, Schwalm, and Ingold (24) that R affects the selectivity of $RO_2 \cdot$ rests essentially on a single rate constant, cumylperoxy and toluene in their Table VIII, which is inconsistent with their other data by other criteria.] These considerations mean that the choice of hydrocarbon used as a standard for comparing other hydrocarbons is mostly a matter of convenience. For those with access to gasometric equipment, the method of Hendry (12) using one volatile hydrocarbon is most convenient. Although one determination of relative reactivity can be sufficient, to avoid small steric effects most of the attack on the mixture should be by the standard alkylperoxy radical. The method of Howard, Schwalm, and Ingold (24), using simple rates of oxidations in the presence of a common hydroperoxide, seems equally good, providing that trace metal catalysis is carefully avoided.

A rather surprising result coming from Table VII is that the rates of oxidation of single hydrocarbons, $(k_p/(2k_t)^{1/2}$, are a fairly good measure of the relative reactivities of the hydrocarbons, considerably better than the individual values of k_p. One reason for the agreement might be the limited number of data on relative reactivity. However, a major reason seems to be that high k_p's and high k_t's tend to go together and roughly to compensate each other. Similarly, in Table VII, cumene has the lowest termination constant and nearly the lowest propagation constant, so that the over-all rate and relative reactivity data agree well. Of the 38 sets of constants listed (23) at 30°C., the k_p's differ by a factor of 24700, the k_t's by a factor of 77800 ($k_t^{1/2}$ by only 280), but the quotients $k_p/k_t^{1/2}$ by only 7100.

The relations above seem to apply to ethers (31, 32, 33) as well as hydrocarbons. Oxidations of alcohols (33) and a few hydrocarbons (22) utilize as chain carriers $HO_2 \cdot$ radicals which have high termination constants. We are now investigating the behavior of some alcohols, ketones, and esters in autoxidations.

Literature Cited

(1) Alagy, J., Clément, G., Balaceanu, J. C., *Bull. Soc. Chim. France* **1959**, 1325.
(2) *Ibid.*, **1960**, 1495.
(3) *Ibid.*, **1961**, 1303.
(4) *Ibid.*, p. 1792.
(5) Allara, D. L., Mill, T., Hendry, D. G., Mayo, F. R., ADVAN. CHEM. SER. **76**, 000 (1968).
(6) Allen, P. E. M., Plesch, P. H., in "The Chemistry of Cationic Polymerization," P. H. Plesch, ed., p. 128, Macmillan, New York, 1963.
(7) Blanchard, H. S., *J. Am. Chem. Soc.* **81**, 4548 (1959).
(8) Chevriau, C., Naffa, P., Balaceanu, J. C., *Bull. Soc. Chim. France* **1964**, 3002.
(9) Dulog, L., Szita, J., Kern, W., *Fette, Seifen, Anstrichmittel* **65**, 108 (1963), and citations by Dulog of theses by H. Schmidt and L. Maempler at University of Mainz.
(10) Fineman, M. A., Ross, S. D., *J. Polymer Sci.* **5**, 259 (1950).
(11) Giammaria, J. J., Norris, H. D., *Ind. Eng. Chem., Prod. Res. Develop.* **1**, 16 (1962).
(12) Hendry, D. G., ADVAN. CHEM. SER. **75**, 24 (1968).
(13) Hendry, D. G., *J. Am. Chem. Soc.* **89**, 5433 (1967).
(14) Hendry, D. G., Russell, G. A., *Ibid.*, **86**, 2368 (1964).
(15) *Ibid.*, p. 2371.
(16) Hiatt, R., Gould, C. W., Mayo, F. R., *J. Org. Chem.* **29**, 3461 (1964).
(17) Hiatt, R., Mill, T., Mayo, F. R., Irwin, K. C., Castleman, J. K., Gould, C. W., *J. Org. Chem.*, in press.
(18) Horner, L., Jurgleit, W., *Ann. Chem.* **591**, 138 (1955).
(19) Howard, J. A., Ingold, K. U., *Can. J. Chem.* **42**, 1044 (1964).
(20) *Ibid.*, **44**, 1113 (1966).
(21) *Ibid.*, p. 1119.
(22) *Ibid.*, **45**, 785 (1967).
(23) *Ibid.*, p. 793.
(24) Howard, J. A., Schwalm, W. J., Ingold, K. U., ADVAN. CHEM. SER. **75**, 6 (1968).
(25) Ingles, T. A., Melville, H. W., *Proc. Roy. Soc.* **A218**, 163 (1953).
(26) Mayo, F. R., *Ber. Bunsenges. Physik. Chem.* **70**, 233 (1966).
(27) Mayo, F. R., Miller, A. A., Russell, G. A., *J. Am. Chem. Soc.* **80**, 2500, 6701 (1958).
(28) Mayo, F. R., Walling, C., *Chem. Revs.* **46**, 191 (1950).
(29) Middleton, B. S., Ingold, K. U., *Can. J. Chem.* **45**, 191 (1967).
(30) Russell, G. A., *J. Am. Chem. Soc.* **77**, 4583 (1955).
(31) Russell, G. A., Williamson, R. C., Jr., *Ibid.*, **86**, 2357 (1964).
(32) *Ibid.*, p. 2364.
(33) Sajus, L., ADVAN. CHEM. SER. **75**, 58 (1968).
(34) Thomas, J. R., Tolman, C. A., *J. Am. Chem. Soc.* **84**, 2079 (1962).
(35) Traylor, T. G., Russell, C. A., *Ibid.*, **87**, 3698 (1965).
(36) Tsepalov, V. F., Schlyapintokh, V. Ya., P'ei-huang, Chou, *Russ. J. Phys. Chem.* **38**, 26, 184 (1964).
(37) Van Sickle, D. E., Mayo, F. R., Arluck, R. M., *J. Am. Chem. Soc.* **87**, 4832 (1965).
(38) Walling, C., "Free Radicals in Solutions," p. 111, Wiley, New York, 1957.
(39) Walling, C., McElhill, E. A., *J. Am. Chem. Soc.* **73**, 2927 (1951).
(40) Wibaut, J. P., van der Leeuven, H. B., van der Wall, B., *Rec. Trav. Chim.* **73**, 1033 (1954).

RECEIVED November 2, 1967.

5

Kinetic Data on the Radical Oxidation of Petrochemical Compounds

L. SAJUS

Institut Francais du Pétrole, 92-Rueil-Malmaison, France

> *The results obtained by liquid-phase oxidation or co-oxidation of various hydrocarbons are reviewed, and new results are reported for new kinds of compounds such as alkylaromatics, alcohols, and ethers, which were also systematically studied by co-oxidation. Gathering all kinetic data and discussing them in connection with data on absolute termination constants, obtained by other groups through physical measurements, enables us to estimate the termination and propagation rate constants for about 40 compounds and to present characteristic values for some new classes of compounds. Examples demonstrate that co-oxidation studies make it possible to explain the behavior of complex compounds reacting by different kinds of bonds, and more particularly the behavior of polymers oxidized in solution.*

The role of liquid-phase radical oxidation in petrochemical synthesis and in controlling the oxygen stability of organic compounds has led industrial chemical research teams to undertake fundamental research in these fields. The problem of chemical synthesis consists of optimizing and controlling reactions which are often highly complex as shown by the following facts:

(1) Some oxidations are done on an isolated compound, while many are done directly on a mixture of compounds—*e.g.*, a hydrocarbon cut for acid synthesis, or olefin wax or olefin aldehyde mixtures for epoxidation.

(2) In any case, the initial reagents must be oxidized in the presence of the reaction products, and it is always best if the concentration of products can be increased because it corresponds to a better conversion rate per run—*e.g.*, cyclohexane oxidation must be done in the presence of cyclohexanol and cyclohexanone—and recent advances have increased the amount of conversion per run from 8 to 15% without any loss of selectivity.

(3) Low concentration products such as initial impurities, parasite reaction products, or catalytic and adjuvant agents, are all capable of greatly changing the reaction products and their rates of formation.

Mastering such phenomena first requires a thorough understanding of the elementary steps involved in the intimate reaction phenomena as well as of the rules governing the combinations of these mechanisms. This paper describes one of the experimental approaches used by the oxidation group of the Institut Français du Pétrole in rapidly determining the kinetic constants by kinetic studies of oxidation and co-oxidation. Part of the experimental work was previously presented (1, 2, 3, 4, 5, 8); more recently we have studied systematically new classes of compounds, and I wish to review, bring up to date, and discuss the whole set of kinetic data. I will also take this opportunity to rebut some criticisms of our previous conclusions about the correctness of the co-oxidation method to estimate kinetic constants. In any case the method using co-oxidation must be examined critically because it is important to know the validity of the results as well as the extent to which co-oxidations are a simple function of the oxidation characteristics of pure compounds since this is the process necessarily used for elementary rate constants. In addition, the results obtained will be discussed from the point of view of structural influences on the kinetic constants.

Oxidation Kinetics and Measurement of Relative Oxidizabilities

Oxidation consists of a radical scheme with three fundamental stages: initiation, propagation, and termination (7). Self-initiation is a process which varies from one substrate to another and is difficult to study conjointly with oxidation. Thus, it is best if the other kinetic stages can be disassociated from the initiation step, by using an added initiator, AIBN (azobisisobutyronitrile) or ACN (azobiscyclohexanenitrile), which produces a known quantity of radicals. The rates of initiation V_i produced in this manner must be faster than competing thermal initiation. This requires first a fairly low temperature so that the spontaneous auto-initiation processes are slow. Temperatures ranging from 50°–90°C. were chosen for the different materials studied, but for the sake of comparison all the results presented are extrapolated to 60°C. Second, we were obliged to limit the conversion so that the degenerate branching processes (35) caused by the accumulation of peroxides would remain negligible.

Under such conditions, the effects of the rate of initiation, V_i, the reagent concentration, and the oxygen pressure can be determined. A zero order was always obtained with respect to the oxygen at pressures varying from 0.2–1.5 atm. At low concentration, all of the reagents studied produce hydroperoxide as the main product. Propagation involves the peroxy radical, with other possible propagation (*10*) reactions remaining negligible. The termination reactions are necessarily peroxy radical recombination reactions because of the zero order with respect to oxygen.

For the rate of initiation, a half-order was usually obtained, and a first order with respect to the reagent as with tetrahydrofuran (THF) (*13*). For a few reagents, these relations do not apply—e.g., for *m*-xylene the rate of oxidation is between half and first order with respect to V_i and less than first order with respect to xylene (*11*). (In every case V_{O_2} represents the rate of consumption of O_2 corrected for N_2 evolution, and efficiency of initiation is assumed to be 0.7).

The difference in kinetic order should be compared with the difference in chain length. For THF the chain length is sufficient, $44 < \lambda < 48$, depending on the conditions, for the general kinetics to be satisfied:

$$V = V_i^{1/2} [RH] k_p/(2 k_t)^{1/2} \qquad (1)$$

The same relation applies to Tetralin, cumene, alcohols, and all compounds oxidizing with a considerable chain length. For *m*-xylene the chain is short ($\lambda \simeq 2$). In this case, the simplifying hypotheses for long chains which neglect the initiator radicals (I·) and their peroxy (IO$_2$·) forms are no longer valid (*11*). In particular the rate of oxygen consumption $V_{O_2} = \dfrac{dO_2}{dt}$ is no longer a valid measure of the rate of propagation (V_p) of the oxidation because an appreciable amount of oxygen is used for the initial formation of IO$_2$· and some is evolved when the termination occurs.

The apparent order of V_p with respect to hydrocarbon depends somewhat on the solvent, as shown in Figure 1. In *tert*-butylbenzene, which is inert chemically and clearly resembles the xylenes, V_p is first order in hydrocarbon, but deviations occur in *o*-dichlorobenzene. This principle affects generally the choice of solvents (*21*). Thus, even in radical oxidation, where solvent effects are much weaker than in ionic reactions, proper choice of solvent is essential if kinetic laws are to be observed over a wide range of reagent concentrations.

A fairly systematic study was made of various families of compounds. By measuring V_{O_2} under conditions for which V_i is known, the ratio V_{O_2}/V_i can be used to see if we are dealing with long chains $\left(\dfrac{V_{O_2}}{V_i} > 10\right)$ or short chains $\left(\dfrac{V_{O_2}}{V_i} < 10\right)$.

By applying Formula 1 for V_{O_2} or V_p, depending on the case, it is possible to determine the values of $k_p/(2 k_t)^{1/2}$ ($= V_p/V_i^{0.5}$ [RH]) which measures the apparent oxidizability, and V_p/V_i ($= \lambda =$ the chain

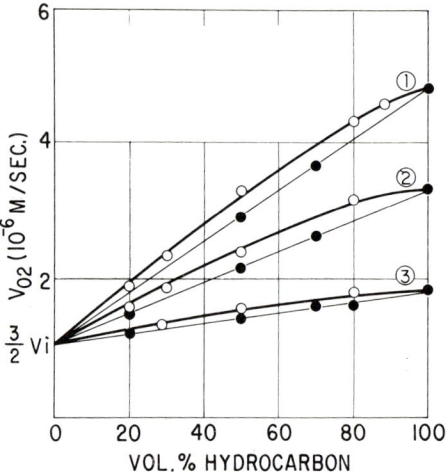

Figure 1. Order in reagent concentration for short chain oxidation; $T = 60°C$.

Reagents:
(1) Ethylbenzene
(2) sec-Butylbenzene
(3) m-Xylene

Solvents:
● o-Dichlorobenzene
○ tert-Butylbenzene

length). These two quantities are given for saturated (4), olefinic (8, 27), and aromatic (11) hydrocarbons, and for alcohols (28, 29) and ethers (13) in Table I (Columns 2 and 3). The conditions are uniform. $T = 60°C$. The initiation is caused by 0.06 mole AIBN which corresponds to an actual initiation rate of 8.82×10^{-7} mole/sec. (39). The discussion below shows how the termination constants may be estimated and the propagation constants compared.

Determining Termination Constants by Co-Oxidation

The principle of co-oxidation was set forth in 1951 (38) and has since been developed and elucidated in an increasingly quantitative way (5, 12, 31, 32). We return to this basic principle only to describe the planned scheme, the theories used, and the parameters retained for curve plotting (22). The two reagents are represented by AH and BH. Symbols a and b refer to the peroxidic radicals $AO_2\cdot$ and $BO_2\cdot$, while symbols A and B refer to molecules AH and BH. The sign k_{aB} designates the reaction constant of radical $AO_2\cdot$ on reagent BH; while k_{aa}, k_{ab}, and k_{bb} are the termination constants of the radicals. Consequently, when the rate

Table I. Kinetic Data for Oxidation of Pure Compounds with 0.06 M AIBN at 60°C.[a]

Compound	Experimental		Estimated from Table III	
	$\dfrac{10^4 \, k_p}{(2 \, k_t)^{0.5}}$	λ	$2 \, k_t \, (10^{-6})$	(k_p)
Decalin	0.9	2	20	0.40
Methylcyclopentane	1.6	3	20	0.71
Methylcyclohexane	0.95	2–3	20	0.43
Cyclohexane	0.25	0.3	20	0.112
Pinane	17.2	14	2	2.45
2,3-Dimethylbutane	6.1	7	0.04	0.12
2,3-Dimethyl-2-butene	100	100	2	14.2
Cyclohexene	51	71	2	10
1-Butene	11.0		20	4.9
2-Butene	12.4		2	1.75
1-Hexene	3.7	6–7	20	1.60
Methyl Methacrylate	20	19	0.04	0.4
Toluene	0.53	2	300	0.92
o-Xylene	1.63	3	300	2.8
m-Xylene	1.18	2–3	300	2.05
p-Xylene	1.61	3	300	2.8
1,3,5-Trimethylbenzene	1.39	2–3	300	2.4
1,2,4-Trimethylbenzene	2.40	4	300	4.2
Ethylbenzene	5.30	6	20	2.4
n-Propylbenzene	2.22	3–4	20	1.0
n-Butylbenzene	3.92	4–5	20	1.75
Cyclopentylbenzene	29	24	20	13
Cyclohexylbenzene	6.0	4–5	20	8.7
s-Butylbenzene	4.48	4–5	20	1.98
Tetralin	61.6	53	20	27.5
Cumene	35.6	31	0.04	0.72
p-Cymene	11.0	10	20	4.9
Styrene	200	180	20	90
Cyclohexanol	4.51	6	50	3.2
2-Butanol	2.9	3–4	50	2.05
Cyclohexene-3-ol	97	120	50	68
Methylphenylcarbinol	14	14	50	10
Methylisobutylcarbinol	2.46	2–3	50	1.74
Dibutyl ether	6.7	5–6	2	0.95
Diisopropyl ether	14.1	13	2	2.0
Di-α-methylbenzyl ether	20	18	2	2.8
Tetrahydrofuran	23.9	31	2	3.4
2-Methyltetrahydrofuran	86	110	2	12
Dibenzyl ether	133	110	2	18

[a] Rate constants in moles/sec.

initiator is known, the scheme contains seven unknown kinetic constants as shown below.

For initiation

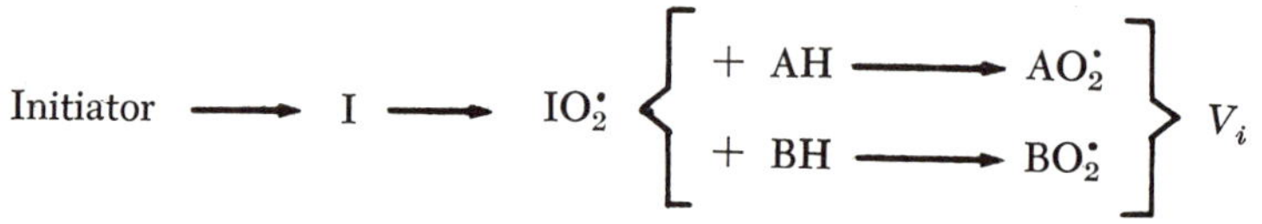

$$\text{Initiator} \longrightarrow I \longrightarrow IO_2^{\cdot} \left\{ \begin{array}{l} + AH \longrightarrow AO_2^{\cdot} \\ + BH \longrightarrow BO_2^{\cdot} \end{array} \right\} V_i$$

For propagation

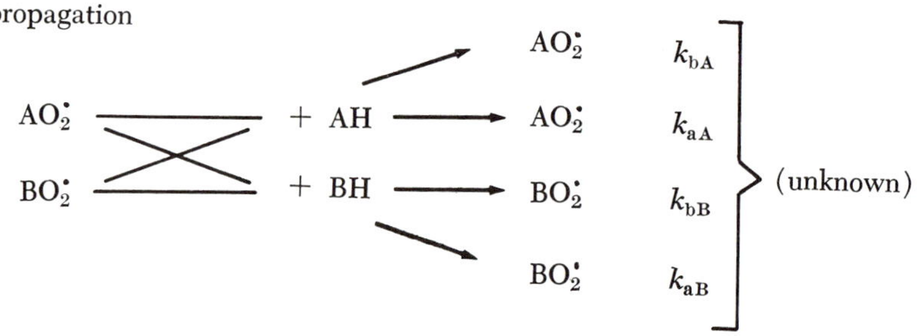

$$\begin{array}{c} AO_2^{\cdot} \\ BO_2^{\cdot} \end{array} \diagtimes \begin{array}{l} + AH \longrightarrow \\ + BH \longrightarrow \end{array} \begin{array}{l} AO_2^{\cdot} \\ AO_2^{\cdot} \\ BO_2^{\cdot} \\ BO_2^{\cdot} \end{array} \left. \begin{array}{l} k_{bA} \\ k_{aA} \\ k_{bB} \\ k_{aB} \end{array} \right\} \text{(unknown)}$$

For termination

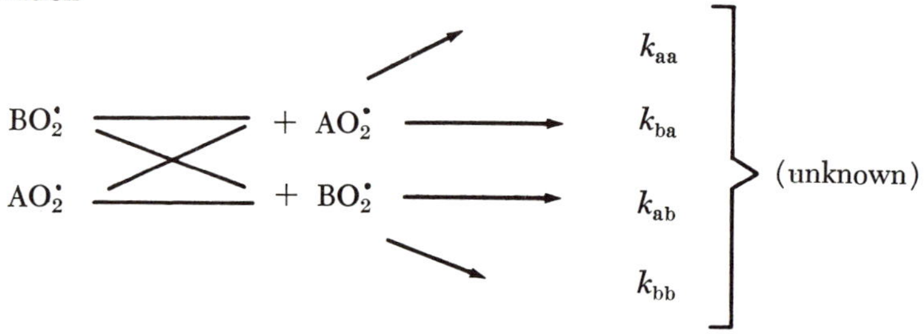

$$\begin{array}{c} BO_2^{\cdot} \\ AO_2^{\cdot} \end{array} \diagtimes \begin{array}{l} + AO_2^{\cdot} \longrightarrow \\ + BO_2^{\cdot} \longrightarrow \end{array} \left. \begin{array}{l} k_{aa} \\ k_{ba} \\ k_{ab} \\ k_{bb} \end{array} \right\} \text{(unknown)}$$

For long chains, the co-oxidation rate equation has the following form:

$$V_p = V_i^{1/2} \frac{k_{aA}k_{bA}[AH]^2 + 2 k_{aB}k_{bA}[AH][BH] + k_{bB}k_{aB}[BH]^2}{(k_{aa}k_{bA}^2[AH]^2 + 2 k_{ab}k_{bA}k_{aB}[AH][BH] + k_{bb}k_{aB}^2[BH]^2)^{1/2}}$$

For short chains this equation is unusable (*11*). The problem can be dealt with by ignoring the intervention of the $IO_2^{\cdot}$ radicals of the initiator because oxidation kinetics show that these radicals have little effect. A simplified and sufficient co-oxidation equation will thus be the same as the preceding equation provided that corrected values of V_p are used.

Curves can be interpreted in two ways to provide different kinds of data. First, the relative values of rupture constants can be determined. The tangents at the ends of a curve can be used to estimate constants k_{aa}

and k_{bb}. At A the expression of the slope is:

$$\frac{dV_p}{d(R_BH)} = \frac{V_A°}{(R_AH)°} \frac{k_{aB}}{k_{aA}} \left(2 - \frac{k_{ab}}{k_{aa}} \frac{k_{aA}}{k_{bA}}\right) - \frac{V_A°}{(R_BH)°} \qquad (2)$$

Considering the hypotheses according to which the reactivities of two radicals differ little with respect to the same molecule ($k_{aA} = k_{bA}$) and the cross-constant k_{ab} is between termination constants k_{aa} and k_{bb}, it can be seen that the slope is a function of ratio k_{ab}/k_{aa} and can be used to estimate its value. In this way, it is possible to estimate whether k_{bb} is larger or smaller than k_{aa}.

As examples of this, Figures 2, 3, and 4 show a group of co-oxidation curves (1, 2, 3, 4, 28, 29). The linearity in Figure 2 is a dependable equivalence test of the radicals in a termination constant in that it provides a check of whether a group of compounds has the same rupture constant. A co-oxidation curve which begins with a slope less than that of the ideal dilution line indicates that the compound, even though present in a low concentration, is retarding oxidation and thus has a higher termination constant (Figures 3 and 4). All alcohols, when compared with cumene, give about the same inhibition (Figure 3), and in mixture with Tetralin, which has a higher termination constant than cumene, the same alcohols offer an inhibition that is not as strong but still is similar for each one (Figure 4).

We thus concluded that alcohols on the whole have a similar termination constant which is greater than that of cyclohexene, with the termination constant of Tetralin being greater than that of cumene.

As Mayo notes (23), some co-oxidation curves are more significant than others. For example, by our method cyclohexane gives straight lines with cumene, cyclohexane, and Tetralin, co-reagents of increasing termination constant. However, this relation is consistent with Relation 2; cyclohexane is so unreactive and the ratio k_{aB}/k_{aA} is so small that the change in slope can be only slight and therefore less significant.

Some criticism (23, 24, 34) of our previously presented results does not consider all the kinetic data obtained recently in the systematic study of such families as alkylaromatics, ethers, alcohols. Taken together, the cross tests permitted classification of many of the compounds studied by substituent as well as by relative k_t values. This arrangement is shown in Table II. Once this was done, all that remained was to standardize the absolute value of this classification and to check its accuracy. For this purpose, we used several values of k_t available from the literature for the compounds studied. Table III lists these values corrected to 60°C., where the authors gave an activation energy measurement for the recombination reaction, but otherwise listed without correction. Considering the slight variation of these constants with temperature they are fair approximations at 60°C. These values were determined by one or several of the following methods: rotating sector (R), chemiluminescence (C), electronic paramagnetic resonance (E), thermocouple method (T). When several measurements were made, the differences were fairly large, sometimes by as much as a factor of 5, and we took an average value (underlined figure). Nevertheless, the values are sufficiently different—ranging from 3×10^4 to 5×10^8 depending on the nature of the compounds—to be considered significant.

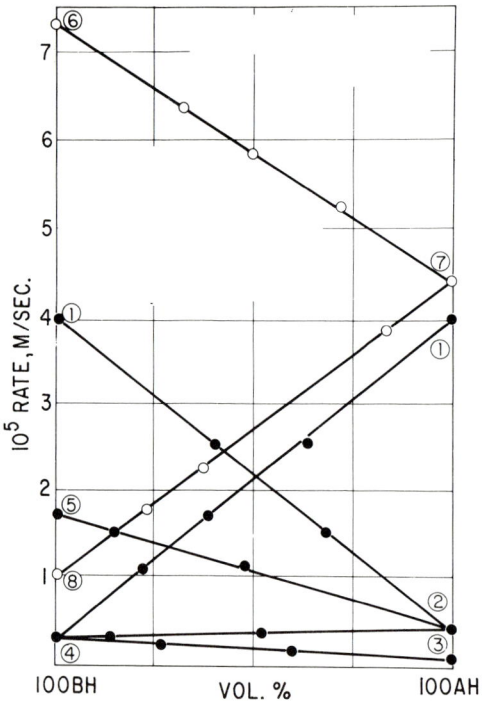

Figure 2. Examples of linear co-oxidation. Initiator, AIBN = 0.06 mole/liter; T = 60°C.

Legend:
(1) Tetralin
(2) Ethylbenzene
(3) Dicyclobenzyle
(4) Cyclohexylbenzene
(5) Cyclopentylbenzene
(6) Dimethyl-2,3-butene-2
(7) Cyclohexene
(8) Pinane

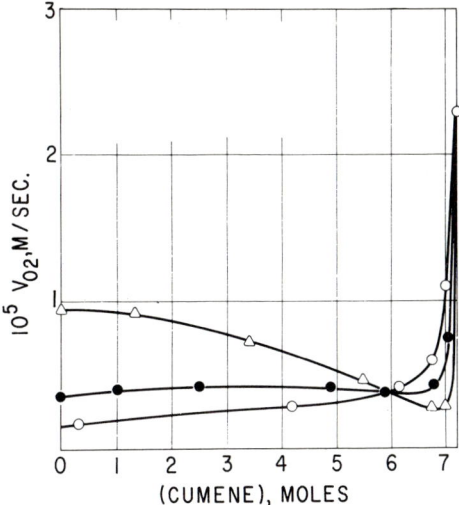

Figure 3. Co-oxidation of different alcohols with cumene. [AIBN] = 0.06 mole/liter; $T = 60°C$.

Legend:
△ Methylphenylcarbinol
● Cyclohexanol
○ Methylisobutylcarbinol

These values can thus be used to standardize the comparative scale established. By considering the published values in Table III—5×10^4 for cumene, 2×10^6 for cyclohexene, 2×10^7 for Tetralin, 3×10^7 for ethylbenzene, 4×10^7 for styrene, 3×10^8 for p-xylene, and 5×10^7 for cyclohexanol—we can attribute a numerical value to each group. The order of the values found agrees with the qualitative classification furnished by co-oxidation. Co-oxidation does not and should not suggest any difference in termination constants for ethylbenzene and Tetralin. The direct measurement values—2×10^7 and 3×10^7—are different but close, and it has not been proved that the difference is significant. Since we can use either value, 2×10^7 was chosen for this group of compounds. Styrene is also not significantly different from ethylbenzene, and its co-oxidation should thus be estimated at 2×10^7, though a value of 4×10^7 was obtained by direct measurement (*19*).

Howard and Ingold (*18*) have noted that in the co-oxidation of allylbenzene and Tetralin the experimental curve was only slightly rounded, though reported values of k_t for these co-reagents differ by a factor of 58. This argument can be rebutted. First, our qualitative analysis

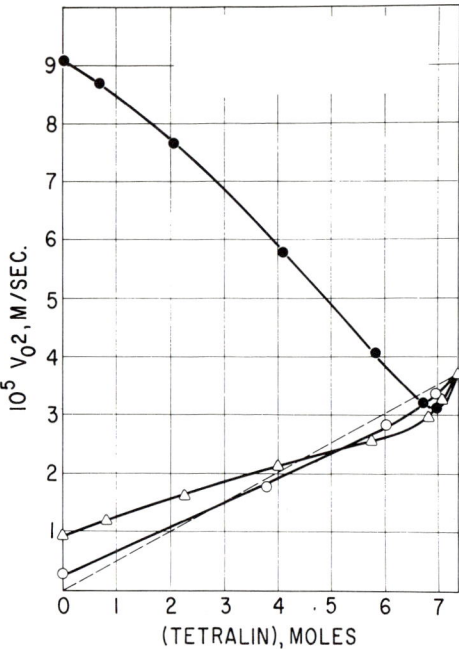

Figure 4. Co-oxidation of different alcohols with Tetralin

Legend:
△ Methylphenylcarbinol
● Cyclohexene-2-ol-1
○ 2-Butanol

is based on commonly accepted hypotheses; exceptions, especially if some complex chemical feature is occurring, are possible. Second, the ratio of 58 was provided by measurements made at 30°C. on pure compounds, and co-oxidation is carried out at 65°C. on reagents diluted to 1M by o-dichlorobenzene. Without discussing the reliability of the experimental data, it will not be surprising that such a change in experimental conditions reduces the ratio by a factor of 10 or even 5—values which would be consistent with the slope of the curve.

An over-all comparison reveals the consistency of the general results, thus bringing out the limitations of the separating power of such a method. It does not discriminate for k_t values differing by a factor of only 2 but often shows the difference when the factor is about 5. The separating power of the kinetic method is of the same order as the current average difference shown by direct measurements, meaning that this method remains of definite interest.

Table II. Classification of Compounds on the Basis of Termination Constants[a]

Cumene; 2,3-dimethylbutane.

Cyclohexene; 2,3-dimethylbutene-2; 2-butene; dibutyl ether; diisopropyl ether; tetrahydrofuran; 2-methyl THF.

Ethylbenzene; Tetralin; n-propylbenzene; cyclopentylbenzene; cyclohexylbenzene; p-cymene; styrene.

Cyclohexanol; 2-butanol; methylphenylcarbinol; methylisobutylcarbinol.

Toluene; o-xylene; m-xylene; p-xylene; 1,3,5-trimethylbenzene; 1,2,4-trimethylbenzene.

[a] In each group, compounds having a similar termination constant are grouped together. Groups are presented in order of increasing constant. We list only compounds for which extensive co-oxidation results have been obtained. For the others only some co-oxidation experiments presented in the complete experimental report, enabled us to estimate that their k_t value is similar to some of the compounds listed here, and assumed k_t values have been reported only in Table I.

Table III. Absolute Rate Constants for Interaction of Peroxy Radicals[a]

Compound	Method	$2 k_t$	$T°C.$
Cumene (26)	R	3.3×10^4	65
Cumene (36)	C	3.3×10^4	20
Cumene (16)	R	5.3×10^4	60
Cumene (33)	E	7×10^4	57
Cumene		5×10^4	60
Cyclohexene (6)	R	1.0×10^6	15
Cyclohexene (17)	R	5.6×10^6	30
Cyclohexene (30)	T	1.6×10^6	40
Cyclohexene (19)	T	2×10^6	56
Cyclohexene		2×10^6	60
Tetralin (9)		2.3×10^7	25
Tetralin (17)	R	1.4×10^7	60
Tetralin		2×10^7	60
Ethylbenzene (17)	R	3.2×10^7	30
Ethylbenzene (34)	R	2×10^7	70
Ethylbenzene (36)	C	4×10^7	40
Ethylbenzene		3×10^7	60
Cyclohexanol (25)		4.8×10^7	60
Toluene (18)	R	3.0×10^8	30
p-Xylene (18)	R	3.0×10^8	30
Octene (25)		4.2×10^7	60
Octene (18)	R	2.6×10^8	30
s-Butylbenzene (18)	R	1.8×10^5	30
2,3-Dimethylbutene (18)	R	0.64×10^6	30
Styrene (15)	R	4.2×10^7	30

[a] Values of $2k_t$ in moles per second measured by physical method for compounds studied by co-oxidation.

These values determined for k_t are listed in Column 4 of Table I. From these values it is possible to determine the constants k_p belonging to each reagent (Column 5) as well as the constant k_p' per atom of reactive hydrogen. The value of k_p', which is deduced from k_p by compensating for the statistic effect caused by the number of hydrogen atoms, is thus very significant in discussing reactivity in function of structure.

This comparison of reactivity can be completed by analyzing co-oxidation curves further, as described in the next section; this resulting in directly determining the relative reactivity toward abstraction by some free radicals.

Determining Relative Reactivities by Co-Oxidation

The mathematical processing of co-oxidation curves can be oriented to determine, independently of any hypothesis about the recombination constants, the three relative parameters:

$$r_a = \frac{k_{aA}}{k_{aB}} \qquad r_b = \frac{k_{bB}}{k_{bA}} \qquad \phi = \frac{k_{ab}}{2(k_{aa} \times k_{bb})^{1/2}}$$

by arranging the co-oxidation rate equation in the form:

$$V_p = V_i^{1/2} \frac{r_A [R_AH]^2 + 2 [R_AH] [R_BH] + r_b [R_BH]^2}{(r_A^2 \delta_A^2 [R_AH]^2 + 4\phi r_A r_B \delta_A \delta_B [R_AH] [R_BH] + r_B^2 \delta_B^2 [R_BH]^2)^{1/2}}$$

in which δ_A and δ_B which designate the reverse of the oxidizabilities of pure compounds,

$$\delta_A = \frac{(k_{aa})^{1/2}}{k_{aA}} \qquad \delta_B = \frac{(k_{bb})^{1/2}}{k_{bB}}$$

are determined in a nonambiguous manner on the pure substrates.

A mathematical program developed and checked for this purpose (35) has been used to determine the three parameters r_A, r_B, and ϕ which determine the shapes of the curves. Table IV uses as examples the values obtained (6) for r_A, r_B, and ϕ in a homogeneous family such as ethers which are treated in co-oxidation with cumene, Tetralin, cyclohexene, and among themselves. The reactivity in the propagation stage depends much more on the substrate attacked than on the attacking radical. This hypothesis, moreover, is confirmed by the values of the product of $r_A \times r_B$, values which are and have to be quite close to 1 (18, 19).

A value of r_A greater than 1 shows that AH is more reactive toward $AO_2\cdot$ than BH, and the reverse is true for r_B. In a comparative series with the same AH compound, the increasing values of r_B or decreasing values of r_A indicate the presence of BH compounds which are more and more reactive toward $RO_2\cdot$.

5. SAJUS Petrochemical Compounds

Table IV. Kinetic Parameter Values for Co-Oxidation Curves of Ethers[a]

AH	BH	r_A	r_B	2ϕ	$r_A \times r_B$
Cumene	$(n\text{-}C_4H_9)_2O$	0.17	2.3	0.21	0.4
	$(i\text{-}C_3H_7)_2O$	0.11	2.2	0.55	0.24
	THF	0.11	5.8	0.07	0.64
	Me-2-THF	0.11	17.5	0.5	1.9
	$[C_6H_5(CH_3)CH]_2O$	0.5	9.5	0.87	4.7
Cyclohexene	$(n\text{-}C_4H_9)_2O$	3.2	0.41	1.3	1.3
	$(i\text{-}C_3H_7)_2O$	4.3	0.72	0.14	3.1
	THF	2.7	0.32	1.3	0.86
	Me-2-THF	0.98	1.0	1.4	0.98
	$(C_6H_5CH_2)_2O$	0.26	1.34	6.3	0.34
	$[C_6H_5(CH_3)CH]_2O$	5	0.09	1.1	0.45
Tetralin	$(i\text{-}C_3H_7)_2O$	2.2	0.67	0.4	1.5
	THF	5.8	1.2	0.52	7
	$(C_6H_5CH_2)_2O$	0.7	2.4	2.6	1.7
$(n\text{-}C_4H_9)_2O$	$(i\text{-}C_3H_7)_2O$	0.65	1.6	1.7	1
THF	Me-2-THF	0.56	1.6	0.8	0.9
$(C_6H_5CH_2)_2O$	$[C_6H_5(CH_3)CH]_2O$	4.3	0.45	0.29	1.9
$(C_6H_5CH_2)_2O$	Me-2-THF	2.5	0.48	1.2	1.2
$(i\text{-}C_3H_7)_2O$	$[C_6H_5(CH_3)CH]_2O$	0.84	0.61	2.1	0.5

[a] Temp. = 60°C.; initiator [AIBN] = 0.06 mole per liter.

The results are consistent and can be used to define an order of oxidizability in the family considered as well as in reference compounds.

Cumene < $(n\text{-}C_4H_9)_2O$ < $(i\text{-}C_3H_7)_2O$ ≃ $(C_6H_5(CH_3)CH)_2O$ < THF < Cyclohexene ~ Me-2-THF < Tetralin < $(C_6H_5CH_2)_2O$

This classification of reactivity can be compared with the values of k_p obtained from the apparent oxidizabilities and from the termination constants (Table I). For ethers, the order of reactivity above is the same as the order of k_p. Similarly, the rates of oxidation of alkylbenzenes (11, 12) lead to the following classification.

toluene and m-xylene < 1,3,5 trimethylbenzene < o-xylene and p-xylene < n-propylbenzene < ethylbenzene and n-butylbenzene < ≳ cumene < p-cymene < Tetralin

This agrees with the order of k_p's in Table I.

This consistency of results in the realm of homogeneous families of compounds continues to a reasonable degree for all the compounds.

Whereas several ambiguous cases do indeed still exist, it can be emphasized that they do not cast any uncertainty on the consistency of the majority of data.

Special Applications

The consistency of results can be emphasized further by analyzing special problems, some of which might be useful in expanding our analysis. An initial check lies in the existence of the phenomenon of entrainment. The phenomenon of co-oxidation implies that diluting an AH reagent with a BH co-reagent of smaller termination constant will bring about a faster reaction of the AH reagent than its disappearance rate in a pure state. This phenomenon was checked by titrating the reagent consumed. Figure 5 shows that in the co-oxidation of methylphenylcarbinol and Tetralin (28, 29), for which a theoretical curve showing a maximum in alcohol oxidation rate can be calculated, analyses of the reaction products near the estimated maximum confirm this conclusion. Further, the analysis of co-oxidation is also successful in the case of vinylic compounds producing polyperoxide. The agreement was shown previously for the co-oxidation of styrene and cyclohexene (8).

The case of a pure compound in which two kinds of positions are attacked in oxidation, is often difficult to interpret. Analysis based on the composition of the parallel oxidation mechanisms sometimes enables the behavior of such a reagent to be understood. This is the case for p-cymene (11), which has a k_t value of 2×10^7 and k_p value of 1.23. Comparable molecules have different kinetic characteristics:

	k_t	k_p
Toluene	1.5×10^8	0.72
p-Xylene	1.5×10^8	2.0
Cumene	4×10^4	0.72

From the relative roles of the primary and tertiary radicals which are statistically formed on p-cymene, the termination constant of p-cymene should lie between that of cumene and that of toluene or xylene. Experimentally, this interpretation can be easily checked (12) by comparing the oxidation of p-cymene diluted by benzene with aromatic hydrocarbon mixtures having equal concentrations of methyl and isopropyl groups in comparable mixtures (Table V).

The similarity in rates for comparable solutions clearly demonstrates that the oxidation of p-cymene can actually be considered to be a co-oxidation of independent functional groups. This conclusion, moreover, fits in with the conclusions about substituent effects in oxidation (11),

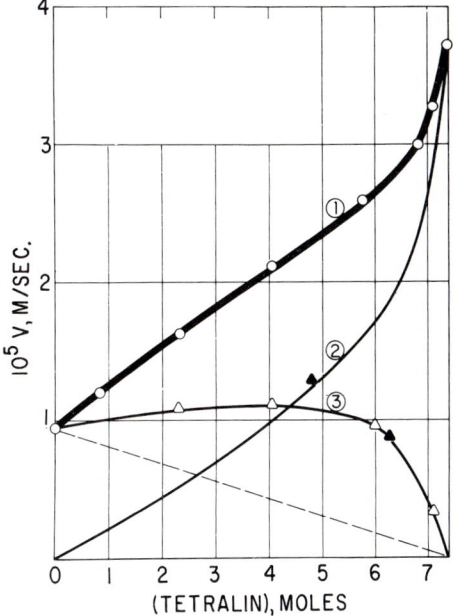

Figure 5. Experimental evidence of rate of entrainment by Tetralin in methylphenylcarbinol oxidation; $T = 60°C$.

Legend:
(1) Total oxidation rate
(2) Calculated Tetralin rate
(3) Calculated alcohol rate
▲ Measured alcohol rate

Table V. Cymene Oxidation[a]

	V_p, Mole Per Sec.
p-Cymene (4.07M) + benzene	$0.37 \cdot 10^{-5}$
Cumene (4.07M) + toluene (4.07M)	$0.38 \cdot 10^{-5}$
p-Cymene (5M) + benzene	$0.46 \cdot 10^{-5}$
Cumene (5M) + p-xylene (2.5M)	$0.43 \cdot 10^{-5}$
p-Cymene (5.4M) + benzene	$0.50 \cdot 10^{-5}$
Cumene (5.4M) + trimethyl-1,3,5-benzene (1.8M)	$0.52 \cdot 10^{-5}$

[a] Oxidation rates V_p of equivalent mixtures at 60°C. with 0.06 mole per liter AIBN.

drawn from the similarity in the reactivity of various xylenes. The electronic effects transmitted by the substituents have relatively little effect on the reactivities of C—H bonds.

A similar case of multifunctional reagent is found in the oxidation of polyethers (14). A check can be made to see that during oxidation a polyether, which includes ether reactional functions and alcohol reactional functions corresponding to the end of the chain, has a termination constant defined by the respective concentrations of both types of groups.

These relations are illustrated by Figure 6; various monodispersed polyethers with decreasing molecular weights, and hence increasing alcohol/ether ratios, behave like a polyether + 1-butanol mixture with the same alcohol content. The result obtained with monofunctional ethers and alcohols can be used to deal with the case of polyethers in a relatively quantitative manner, assuming an independence of the functional groups.

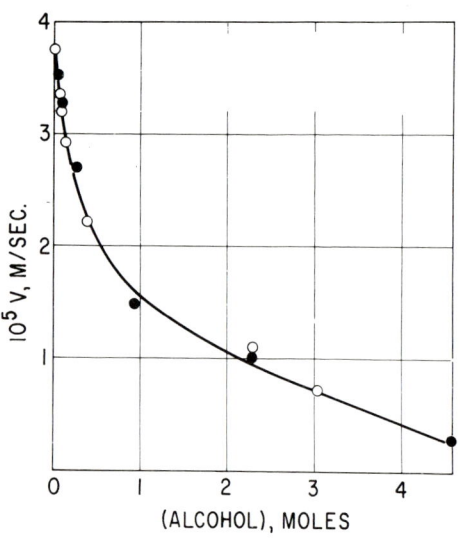

Figure 6. Experimental evidence of relative efficiency in termination of different groups in a macromolecular compound. [Ether linkage] = 4.56 moles/liter; [ACN] = 0.03 mole/liter; $T = 94°C$.

Legend:
  ○ Polyethylene oxide of different molecular weight
  ● Polyethylene oxide (MW = 25,000) and 1-butanol

Kinetic Characteristics and Structure

The over-all values found for k_t and k_p can be used to illustrate various rules linking the order of magnitude of these constants to the structure. Such a discussion was recently presented by Howard and Ingold (18) using data obtained by direct measurements and concerning

Table VI. Termination Constants for Some Peroxy Radical Families at 60°C. (moles/liter/sec.)

Trisalkyl or α-aryl-sec-alkyl radical (1, 2, 3, 4, 5)	$2 \times 10^7 < 2 k_t < 10^8$
Primary benzylic radical (11, 12)	$1 \times 10^8 < 2 k_t < 5 \times 10^8$
α-Aryl-alkyl-radical (11, 12)	$10^7 < 2 k_t < 5 \times 10^7$
α-Hydroxy-sec-alkyl-radical (28, 29)	$5 \times 10^7 < 2 k_t < 2 \times 10^8$
Primary or secondary allylic radical (27)	$2 \times 10^6 < 2 k_t < 2 \times 10^7$
α-Alkoxy-alkyl-radical (13, 14)	$2 \times 10^6 < 2 k_t < 10^7$
Trisalkyl or αaryl-sec-alkyl radical (1, 2, 3, 4, 5)	$4 \times 10^4 < 2 k_t < 10^5$

only hydrocarbon radicals. Our results enable us to support their conclusions, to extend the discussion to new classes of compounds, and to add some new ideas. It is possible to link the termination constants, k_t of the compounds studied fairly directly to a certain number of structures because the families of isostructural compounds have similar k_t constants (Table VI).

Previously, it was mentioned that the steric characteristics of the carbon atom (primary, secondary, tertiary) should be considered; it becomes clear with the new classes of compounds studied that the electronic characteristic of the substituent groups—aryl, hydroxyl, alkoxyl—must also be taken into account. Thus, whereas trisalkylperoxy radicals have weak recombination constants, dialkylhydroxy or dialkylalkoxy have much higher recombination constants, occasionally similar to values observed for secondary alkyl radicals.

In structures with various reactional positions, consideration must be given to a composite action of the different recombinations possible. The values of k_t are then intermediate between those of structures having only one reaction function. This is the case, for example, of pinane compared with cyclohexane and dimethylbutane, or else of p-cymene compared with cumene and toluene.

Previous discussions (20, 39) on the propagation rate, k_p, point out the effect caused by the resonance energy of the radical formed. Our results support this view and enable us to complete the arrangement by families according to the groups adjacent to the attacked function—alkyl, benzyl, alkoxy, allyl, hydroxyl. The steric effect does not reveal itself in any important way—e.g., α-methylbenzylic ether has a k_p which is close to that of benzylic ether, and the tertiary carbons in the former product are generally attacked at rates comparable with that of a less-encumbered carbon.

Conclusions

The kinetic method used by the Institut Francaise du Pétrole oxidation group enabled us to arrive at a consistent picture of the elementary reactivity and recombination phenomena peculiar to many organic structures. In the future it will enable us to complete our understanding of other interesting structures, and when physical methods have given us more numerous and more accurate values for the termination constants, the over-all results can be analyzed again to give a more accurate picture of the facts.

The simultaneous oxidations of various desired or undesired compounds are usually necessary in our research, and in order to understand the phenomena better, it is essential that they be dealt with as complex co-oxidations. The success obtained by using simple co-oxidation for analyzing elementary processes proves that we have a method, which when applied to actual complex transformations, makes possible a more quantitative understanding of industrial operations.

Literature Cited

(1) Alagy, J., Clement, G., Balaceanu, J. C., *Bull, Soc. Chim. France* **1959**, 1325.
(2) *Ibid.* **1960,** 1495.
(3) *Ibid.* **1961,** 1303.
(4) *Ibid.*, p. 1792.
(5) Alagy, J., Clement, G., Balaceanu, J. C., *Compt. Rend.* **247,** 2137 (1958).
(6) Bateman, L., Gee, G., *Proc. Roy. Soc. (London)* **195A,** 391 (1948).
(7) Bolland, J. L., *Quart. Rev. (London)* 3, 1 (1949).
(8) Chevriau, C., Naffa, P., Balaceanu, J. C., *Bull. Soc. Chim. France* **1964,** 3002.
(9) Dewar, M. J. C., Bamford, C. H., *Proc. Roy. Soc. (London)* **198A,** 252 (1949).
(10) Emanuel, N. M., Proceedings of the 7th World Petroleum Congress, Elsevier, London, 1967.
(11) Gadelle, C., Clement, G., *Bull. Soc. Chim. France* **1967,** 1175.
(12) Gadelle, C., Ph.D. Thesis, Paris, 1967.
(13) Grosborne, P., Seree de Roch, I., *Bull. Soc. Chim. France* **1967,** 2260.
(14) Grosborne, P., Ph.D. Thesis, Paris, 1967.
(15) Howard, J. A., Ingold, K. U., *Can. J. Chem.* **43,** 2729 (1965).
(16) *Ibid.* **44,** 1113 (1966).
(17) *Ibid.*, p. 1119.
(18) *Ibid.* **45,** 793 (1967).
(19) Howard, J. A., Robb, J. C., *Trans. Faraday Soc.* **59,** 1590 (1963).
(20) Jungers, J. C., Sajus, L., de Aguirre, I., Decroocq, D., "Analyse Cinetique de la Transformation Chimique," p. 398, Ed. Technip, Paris, 1967.
(21) Jungers, J. C., Sajus, L., de Aguirre, I., Decroocq, D., *Rev. Inst. Franc. Petrole* **21,** 109 (1966).
(22) Jungers, J. C. *et al.*, "Cinetique Chimique Appliquee," Ed. Technip, Paris, 1958.
(23) Mayo, F. R., private communication.
(24) Mayo, F. R., Syz, M. G., Mill, T., Castleman, J. K., "Preprints of International Oxidation Symposium," p. 777, 1967.
(25) McCarthy, R. L., MacLachlan, A., *Trans. Faraday Soc.* **57,** 1107 (1961).
(26) Melville, H. W., Rochards, S., *J. Chem. Soc.* **1954,** 944.

(27) Menguy, P., Chauvel, A., Clement, G., Balaceanu, J. C., *Bull. Soc. Chim. France* **1963**, 2643.
(28) Parlant, C., Seree de Roch, I., Balaceanu, J. C., *Bull. Soc. Chim.* **1963**, 2452.
(29) *Ibid.* **1964**, 3161.
(30) Robb, J. C., Shahiw, M., *J. Inst. Petrol.* **44**, 283 (1958).
(31) Russell, G. A., *J. Am. Chem. Soc.* **77**, 4583 (1955).
(32) *Ibid.* **78**, 1041 (1956).
(33) Thomas, J. R., *J. Am. Chem. Soc.* **85**, 591 (1963).
(34) Tsepalov, V. Ya., Schlyapintokh, V. Ya., *Dokl. Akad. Nauk. SSSR* **124**, 883 (1959).
(35) Van Tiggelen, A., Sajus, L., Balaceanu, J. C., *Rev. Inst. Franc. Petrole* **17**, 1533 (1962).
(36) Vichutinskii, A. A., *Dokl. Akad. Nauk. SSSR* **157**, 150 (1964).
(37) Vignes, J., *I. F. P. Rept.* **10,**565 (1964).
(38) Walling, C., McElhill, E. A., *J. Am. Chem. Soc.* **73**, 2927 (1951).
(39) Walling, C., "Free Radicals in Solution," Wiley, New York, 1957.

RECEIVED October 9, 1967.

6

Liquid-Phase Oxidation of High Molecular Weight 1-Alkenes

CHARLES J. NORTON and DENNIS E. DRAYER

Denver Research Center, Marathon Oil Co., Littleton, Colo.

> *The liquid-phase oxidation of the high molecular weight 1-alkenes—pure 1-hexadecene, and a mixture of commercial C_{15} to C_{18} alpha-olefins—has been studied over a range of conversions, temperatures, and times. The product mixtures were analyzed for alkenyl hydroperoxides, polymeric dialkyl peroxides, and decomposition products. With excess and constant oxygen the rate of olefin disappearance is pseudo-first-order. Alkenyl hydroperoxides are primary products but thermally decompose extensively above 100°C. Two types of polymeric dialkyl peroxides are postulated: Type I arising as primary products from a previously proposed branching addition mechanism and Type II arising as secondary products from the addition of alkenoxy and hydroxy radicals to olefin, followed by further reaction with oxygen and olefin.*

The literature on liquid-phase olefin oxidation has been well reviewed (1, 2, 3, 5, 6, 8, 12, 14, 15, 16, 17, 18, 19, 20). Recent attention has been focused on the effects of structure and reaction conditions on the proportions of alkenyl hydroperoxy radical reaction by the abstraction and addition mechanisms at lower temperatures and conversions. The lower molecular weight cyclic and acyclic olefins have been extensively studied by Van Sickle and co-workers (17, 18, 19, 20). These studies have recently been extended to include higher molecular weight alkenes (16).

Pure and commercial mixtures of 1-alkenes in the C_{15} to C_{18} range were studied at high temperatures and high conversion levels. The reaction variables were studied to make desired reproducible mixtures containing substantial concentrations of alkenyl hydroperoxides and polymeric dialkyl peroxides.

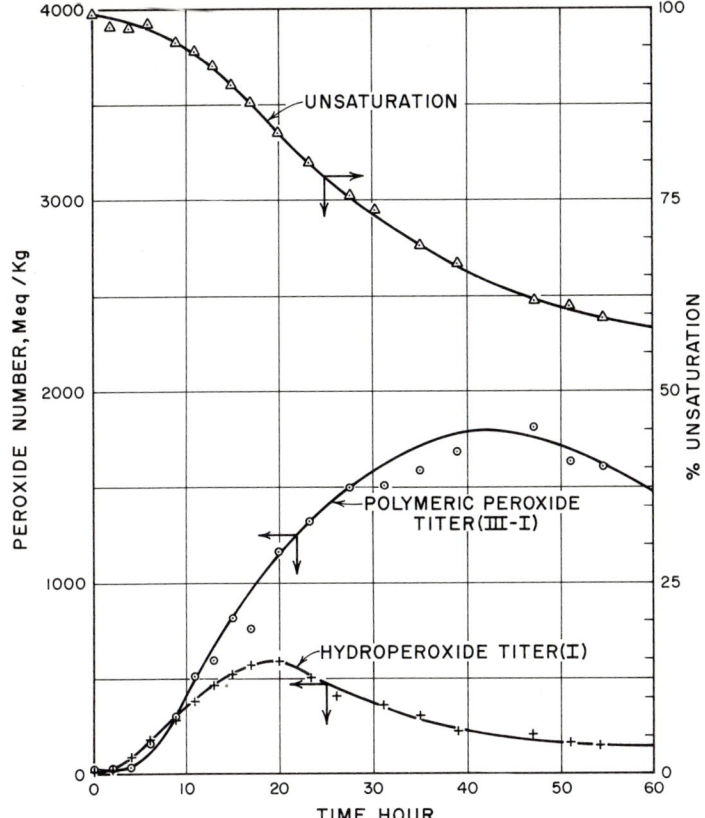

Figure 1. Autoxidation of 1-hexadecene at 110°C. in glass apparatus

Experimental

Reagents. The 1-hexadecene, 99% purity, was obtained from the Humphrey Chemical Co. The mixed alpha-olefins in the C_{15} to C_{18} range were furnished by the California Chemical Co. These contain 90.5 weight % actual alpha-olefins. Pressurized air was used for most of the experiments.

Analytical. The feeds and recovered oils were analyzed for 1-alkene using an F & M Model 720 gas-liquid chromatograph equipped with an SE-30 gum-rubber column with helium carrier gas and 99% purity Humphrey Chemical Co. 1-alkenes for reference peaks.

Unsaturation in the olefinic feeds and recovered oils was determined by mercuric acetate complex formation, which yields titratable acetic acid (21). The addition of peroxide concentrate to the blank 1-hexadecene and unaerated alpha-olefin feed did not interfere with the olefin analysis.

Table I. Comparison of Calculated and Experimental

Expt. No.	Temp., °C.	°F.	Oxygen Pressure, P.S.I.	Exhaust Air Rate, SCFM	Reaction Time, Hr.
1	129.4	265	9	1	0
					1
					2
					3
					4
					5
2	118.3	245	9	1	0
					1
					2
					3
					4
					5
					6
					7
					8
					9
					10
3	118.3	245	3	1	0
					1
					2
					3
					4
					5
					6
					7
					8
					9
					10
4	118.3	245	3	0.2	0
					1
					2
					3
					4
					5
					6
					7
					8
					9
					10

Data in Oxidation of C_{15}-C_{18} Alpha-Olefins

Compositions, Weight %						
Olefin		Hydroperoxides		Polymeric Peroxides		
Exptl.	Calcd.	Exptl.	Calcd.	Exptl.	Calcd.	
98.3	98.3	0.36	0.36	2.50	2.50	
96.0	92.6	3.41	3.96	5.76	3.98	
89.6	87.3	5.84	5.29	9.44	6.96	
83.0	82.2	6.92	5.65	12.40	9.99	
77.8	77.4	6.56	5.60	15.54	12.59	
73.2	73.0	6.28	5.40	17.51	14.63	
98.3	98.3	0.34	0.34	2.52	2.52	
95.0	95.8	0.96	2.33	2.87	2.96	
94.0	93.4	2.25	3.69	3.73	3.85	
93.5	91.1	4.03	4.60	6.11	5.12	
90.9	88.8	5.63	5.19	6.72	6.60	
88.0	86.5	7.02	5.56	11.76	8.20	
84.3	84.4	8.03	5.78	13.16	9.86	
79.6	82.2	8.85	5.88	15.46	11.50	
78.9	80.2	8.88	5.91	20.37	13.10	
71.9	78.1	8.70	5.89	23.67	14.64	
69.9	76.2	8.45	5.83	25.22	16.14	
98.3	98.3	0.36	0.36	2.50	2.50	
95.4	95.8	1.04	2.35	2.80	2.94	
92.5	93.4	2.34	3.70	4.03	3.84	
89.8	91.1	3.82	4.60	4.89	5.12	
88.7	88.8	5.17	5.20	7.50	6.60	
88.1	86.5	6.56	5.56	10.20	8.21	
83.1	84.4	7.66	5.78	11.00	9.87	
80.8	82.2	8.48	5.88	15.38	11.51	
77.0	80.2	8.93	5.91	17.78	13.11	
74.5	78.1	8.68	5.89	20.18	14.65	
71.6	76.2	8.23	5.83	23.49	16.15	
98.3	98.3	0.34	0.34	2.52	2.52	
94.9	95.8	0.68	2.33	2.90	2.96	
94.8	93.4	1.68	3.69	3.33	3.85	
92.4	91.1	2.68	4.60	4.73	5.13	
91.5	88.8	3.65	5.19	6.75	6.61	
90.8	86.5	4.32	5.56	6.99	8.21	
90.0	84.4	5.12	5.78	8.53	9.87	
85.8	82.2	5.47	5.88	10.39	11.51	
85.7	80.2	5.72	5.91	11.05	13.11	
84.0	78.1	5.98	5.89	12.09	14.65	
83.7	76.2	6.24	5.83	14.17	16.15	

Table I.

Expt. No.	Temp., °C.	°F.	Oxygen Pressure, P.S.I.	Exhaust Air Rate, SCFM	Reaction Time, Hr.
5	107.2	225	3	1	0
					2
					4
					6
					8
					10
					12
6	107.2	225	9	1	0
					2
					4
					6
					8
					10
					12
7	107.2	225	9	0.2	0
					2
					4
					6
					8
					10
					11
8	107.2	225	15	1	0
					2
					4
					6
					8
					10
					11

A Perkin-Elmer Model 21 infrared spectrophotometer was used to detect and to estimate the hydroxylic and carbonyl functions in the oxidized product mixtures. The organic hydroperoxide and peroxide functional groups in the product mixtures were determined by an iodine liberation and titration procedure (*11*). In order to get reproducible results, it is necessary to pretreat the olefins with about 10 weight % activated silica or alumina for several hours with agitation to remove adventitious peroxides and impurities. Sodium bisulfite solution rapidly destroys hydroperoxides but does not destroy peroxides completely. The hydroperoxides and peroxides decomposed extensively during attempted distillation at about 1 mm. of Hg partial vacuum. We had some success in concentration by liquid chromatography over silica gel: the unconverted olefins are eluted with *n*-hexanes, and a hydroperoxide-peroxide

Continued

Compositions, Weight %

Olefin		Hydroperoxides		Polymeric Peroxides	
Exptl.	*Calcd.*	*Exptl.*	*Calcd.*	*Exptl.*	*Calcd.*
98.3	98.3	0.32	0.32	1.62	1.62
96.7	96.3	0.48	2.00	2.57	1.92
95.0	94.3	1.01	3.25	3.39	2.59
94.9	92.4	2.00	4.16	4.38	3.60
93.0	90.5	3.35	4.83	5.37	4.75
91.3	88.7	4.63	5.30	7.62	6.06
89.5	86.9	5.98	5.62	9.74	7.46
98.3	98.3	0.32	0.32	2.54	2.54
93.4	96.3	0.82	2.00	3.04	2.82
94.0	94.3	1.72	3.25	4.19	3.48
93.9	92.4	3.34	4.16	6.25	4.48
91.5	90.5	5.14	4.83	9.26	5.62
88.8	88.7	6.37	5.30	12.32	6.92
84.5	86.9	7.40	5.62	14.57	8.30
98.3	98.3	0.36	0.36	2.54	2.54
94.6	96.3	0.70	2.03	2.87	2.82
94.8	94.3	1.59	3.27	3.87	3.48
93.2	92.4	3.02	4.18	5.30	4.48
91.5	90.5	4.46	4.84	6.51	5.62
87.5	88.7	5.85	5.31	9.79	6.92
88.9	87.8	5.98	5.49	10.27	7.60
98.3	98.3	0.32	0.32	2.54	2.54
97.0	96.3	0.82	2.00	3.02	2.82
95.0	94.3	2.05	3.25	3.64	3.48
96.2	92.4	3.70	4.16	5.29	4.48
87.0	90.5	5.45	4.83	8.23	5.62
88.6	88.7	7.02	5.30	10.97	6.92
86.6	87.8	7.62	5.49	12.66	7.60

concentrate is eluted with acetone. Thin-layer chromatography over silica gel gave discrete zones that reacted with starch-iodide reagent, indicating several possible types of peroxidic products. We have not isolated or characterized the polymeric dialkyl peroxides further.

Our laboratory oxidations were carried out by bubbling dry air at 80 to 125 cc. per minute STP, at 110°C. through 500 ml. of olefin in a round-bottomed, 1-liter, standard-taper, three-necked flask equipped with magnetic stirrer, Therm-O-watch controller, electric heating mantle, and condenser. Alkenyl hydroperoxide numbers [Method I of Mair and Graupner (*11*)] and polymeric dialkyl peroxide numbers (Method III minus Method I of Ref. *11*) were determined on small aliquots of about 5 ml. withdrawn at various times.

Kinetic studies were made in a 10-gallon Pfaudler, Glasteel-stirred autoclave, equipped with a line for admitting air or oxygen, and an exhaust-gas bleed line. Oxygen partial pressure was monitored and controlled with an F & M gas chromatography apparatus. A measured quantity (about 6 gallons) of C_{15} to C_{18} alpha-olefins was charged to the nitrogen-filled reactor. A small quantity of "seed" containing previously formed alkenyl hydroperoxides and polymeric dialkyl peroxides was added to eliminate the induction period. The reactor contents were then brought to the reaction temperature and compressed air was admitted to give the desired exhaust gas rate. The oxygen partial pressure was controlled by chromatographic analysis. Periodically, samples were withdrawn and analyzed for alkenyl hydroperoxides, polymeric dialkyl peroxides, and olefin content. The ranges of the autoclave variables studied were 0 to 12 hours' reaction time; 3, 9, and 15 p.s.i.g. oxygen partial pressure; 107.2, 118.3, and 129.4°C.; and 0.2 and 1 standard cubic feet per minute exhaust air rate.

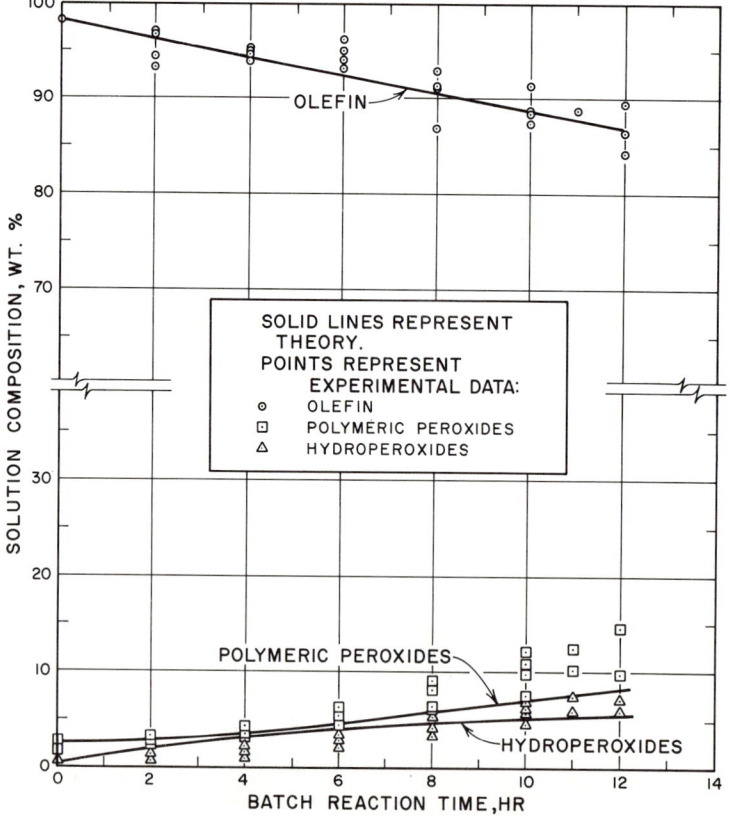

Figure 2. Comparison of theoretical and experimental data at 107.2°C.

Theory

The simplified kinetic mechanism postulated is one involving a sequence of reactions:

$$A + S \xrightarrow{k_1} B \tag{1}$$

$$B + S \xrightarrow{k_2} C \tag{2}$$

$$C \xrightarrow{k_3} D \tag{3}$$

where A = olefin, S = oxygen, B = hydroperoxides, C = polymeric peroxides, D = other products, and k_i = reaction velocity constant. Since oxygen is present at constant concentration and in excess, the following pseudo-first-order rate equations may be used:

$$\frac{-d[A]}{dt} = k_1' [A] \tag{4}$$

$$\frac{d[B]}{dt} = k_1' [A] - k_2' [B] \tag{5}$$

$$\frac{d[C]}{dt} = k_2' [B] - k_3' [C] \tag{6}$$

$$\frac{d[D]}{dt} = k_3' [C] \tag{7}$$

where k_i' is the observed pseudo-first-order rate constant and equals $k_i S$. The mathematical solutions of Equations 4, 5, and 6 are:

$$[A] = [A]_o\, e^{-k_1't} \tag{8}$$

$$[B] = [B]_o\, e^{-k_2't} + \frac{k_1'[A]_o}{k_2' k_1'} [e^{-k_1't} - e^{-k_2't}] \tag{9}$$

$$[C] = [C]_o\, e^{-k_3't} + \frac{k_1' k_2' [A]_o}{[k_2' - k_1'][k_3' - k_1']} [e^{-k_1't} - e^{-k_3't}]$$

$$+ \frac{k_2'[B]_o - \dfrac{k_1' k_2' [A]_o}{k_2' - k_1'}}{k_3' - k_2'} [e^{-k_2't} - e^{-k_3't}] \tag{10}$$

where $[A]_o$, $[B]_o$, and $[C]_o$ are the initial concentrations at $t = 0$ of species A, B, and C, respectively, and t is the batch reaction or contact time.

k_1' is easily evaluated using Equation 8. Evaluations of k_1' at several temperatures allow one subsequently to compute the values of A and E

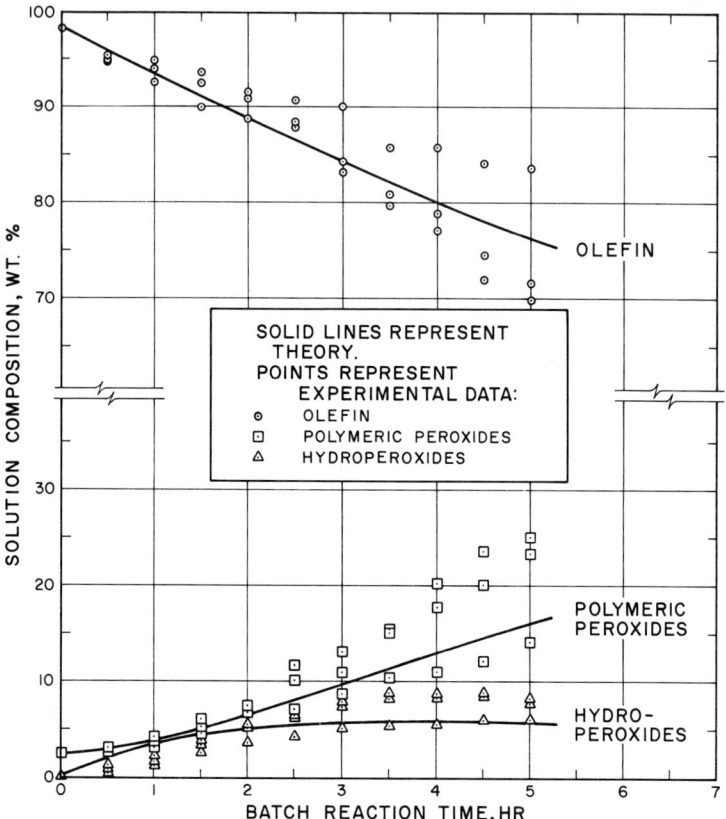

Figure 3. Comparison of theoretical and experimental data at 118.3°C.

in the Arrhenius equation, $k = Ae^{-E/RT}$; k_2' is easily calculated using Equation 5 for the case of $[B] = [B]_{max}$ where $\dfrac{d[B]}{dt} = 0$. A final adjustment of k_2' was made by making slight adjustments to the initial value and noting the standard deviation between experimental and calculated concentration values. The value of k_2' selected was based on the minimum standard deviation. Subsequently, values of A and E were determined. The value of k_3' could be evaluated in a manner similar to k_2'. However, our autoclave experiments were not run over a long enough time to show a maximum concentration of C—i.e., $\dfrac{d[C]}{dt}$ did not become zero. In this case, Equation 6 was used with values of $\dfrac{d[C]}{dt}$ determined by graphic techniques.

To assess the empirical validity of the derived rate constants, standard deviations between experimental and calculated concentration data were calculated by:

$$s = \left[\frac{(\text{Exptl.} - \text{Calcd.})^2}{n-1} \right]^{1/2} \quad (11)$$

where s = standard deviations in concentration units, Exptl. = experimental value of the concentration, Calcd. = calculated value of the concentration, and n = number of comparisons.

Results and Discussion

Substantial alkenyl hydroperoxide concentrations build up in 1-hexadecene, 2-octene, mixed internal olefins, and mixed C_{14}-C_{16} alpha-olefins

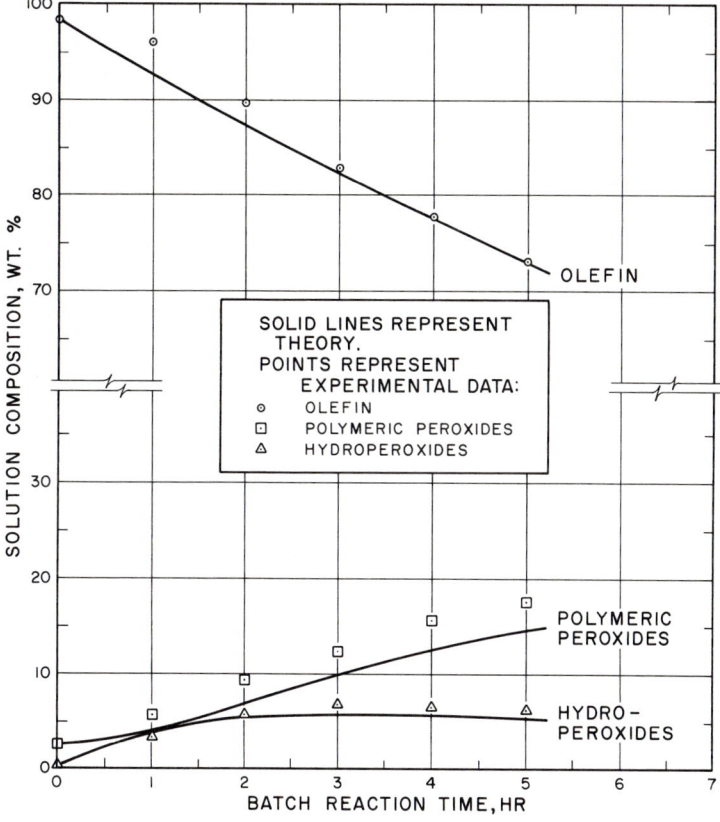

Figure 4. Comparison of theoretical and experimental data at 129.4°C.

within several hours at 110 to 117°C. These alkenyl hydroperoxides are fairly stable at room temperature after an initial drop of about 10 to 15% in hydroperoxide number in the first 2 to 6 days.

Results obtained in glass apparatus are summarized in Figure 1. The unsaturation falls off nearly linearly after a short induction period. After the hydroperoxide functional groups attain their maximum, the olefin disappearance decreases and becomes nonlinear as it is consumed by reaction to form polymeric dialkyl peroxide functions. The maximum concentration of polymeric dialkyl peroxide occurs well after the maximum alkenyl hydroperoxide concentration, giving the appearance of a sequential oxidation mechanism. Infrared and gas-liquid chromatographic analyses showed that hydroxylic derivatives, carbonyl derivatives, and lower molecular weight olefins continued to build up as by-products as the oxidation proceeded, as does the acidity titer.

The experimental autoclave results are summarized in Table I and Figures 2 through 4. Autoclave product compositions are expressed in weight percent, assuming that one peroxide function per olefin unit is incorporated in the polymeric dialkyl peroxides (*17, 19*).

Table II. Summary of Kinetic Results

Constant, $Hr.^{-1}$	Reaction Temperature, °C.			Frequency Factor A, $Sec.^{-1}$	Activation Energy, E, Kcal./Mole
	107.2	118.3	129.4		
k_1'	0.0103	0.0255	0.0596	1.81×10^8	24.0
k_2'	0.1342	0.3458	0.8455	1.14×10^{10}	25.2
k_3'	0.0058	0.0353	0.1940	6.72×10^{21}	48.1

The kinetic results are summarized in Table II. The autoxidation products in general are similar to those observed by Van Sickle at lower temperatures and conversions. Table III summarizes analyses made by Van Sickle at conditions approximating our levels of conversion and temperature. The polymeric dialkyl peroxides are included in the residue.

The alkenyl hydroperoxides and polymeric dialkyl peroxides are fairly stable at ambient temperature but decompose appreciably at the reaction temperatures studied. Thermal stabilities of the alkenyl hydroperoxides and dialkyl peroxides in the olefin solution were determined by heating the solution at 110°C. under nitrogen. The peroxide numbers were plotted *vs.* time to estimate the half-lives in solution. The thermal decomposition half-lives of these alkenyl hydroperoxides are compared with values from the literature for acyclic and cyclic hydroperoxides in Table IV. Secondary acyclic alkenyl hydroperoxides appear to be less

Table III. Products of 1-Hexadecene Oxidation (16)

Products	Millimoles Product			Millimoles O_2			Yield Based on Alkene, %		
Temperature, °C.	90	110	110	90	110	110	90	110	110
Conversion, %	6	18	28	6	18	28	6	18	28
$C_{13}H_{27}CH=CH—CH_2O_2H$ + $C_{13}H_{27}CHO_2H—CH=CH_2$	1.63	0.55	0.92	1.63	0.55	0.92	29.4	6.6	21.4
$C_{13}H_{27}COCH=CH_2$	0.08	0.18	0.11	0.08	0.18	0.11	1.4	2.2	2.6
$C_{13}H_{27}CH=CH—CHO$	0.31	0.21	0.21	0.31	0.21	0.21	5.6	2.5	4.9
$C_{14}H_{29}$ (epoxide, CH—CH$_2$ with O)	0.36	1.13	0.56	0.18	0.57	0.28	6.5	13.6	13.1
$C_{14}H_{29}CHO$	0.17	0.24	0.02	0.17	0.24	0.02	3.1	2.9	0.5
RCOOH (titrated)	0.61	1.69	2.21	0.91	2.54	3.31	11.0	20.3	27.6
Unknown volatile	0.36	0.40	0.30	0.36	0.40	0.30	6.5	4.8	7.0
Residue	1.02	1.96	0.49	1.90	3.92	0.98	36.6	47.1	23.4

thermally stable than saturated secondary cyclic and saturated tertiary hydroperoxides.

The derived kinetic expressions appear to represent experimental data adequately for synthetic control purposes, although the kinetic treatment may be oversimplified for theoretical understanding. Figures 2, 3, and 4 compare the calculated curves and experimental data points. The best fit of data with the curves occurs at the higher temperature of 129.4°C. Absolute standard deviations between calculated and experimental values over the entire temperature range studied are ±2.60 weight % for the olefin, ±1.47 weight % for the alkenyl hydroperoxides, and ±3.01 weight % for the polymeric dialkyl peroxides. Correlation errors involved in the olefin expression (Equation 8), are injected in the expressions for alkenyl hydroperoxide and polymeric dialkyl peroxide composition calculations. The expression for the polymeric dialkyl peroxide concentration, Equation 10, similarly includes the correlation errors for both the olefin and alkenyl hydroperoxide expressions.

The experimental activation energies given in the last column of Table II are in the anticipated order of magnitudes. The activation energy of 24.0 kcal. per mole for the oxidation of 1-hexadecene to hydroperoxide is close to the value of 25.3 kcal. per mole recently reported for "the constant velocity of peroxide accumulation . . . for butene-1" (9). The activation energy for the alkenyl hydroperoxide decomposition is reasonable. The activation energy of 48.1 kcal. per mole for the decomposition polymeric dialkyl peroxide is considerably higher than the value of about 37 kcal. per mole for *tert*-butyl peroxide decomposition. The frequency factor associated with the first two reactions is reasonable, but the value of 6.72 × 10^{21} per second for the third reaction is unusually large even for a very highly ordered transition state for the decomposition of the polymeric dialkyl peroxides.

It is generally agreed that alkenyl hydroperoxides are primary products in the liquid-phase oxidation of olefins. Kamneva and Panfilova (8) believe the dimeric and trimeric dialkyl peroxides they obtained from the oxidation of cyclohexene at 35° to 40° to be secondary products resulting from cyclohexene hydroperoxide. But Van Sickle and co-workers (20) report that, "The abstraction/addition ratio is nearly independent of temperature in oxidation of isobutylene and cycloheptene and of solvent changes in oxidations of cyclopentene, tetramethylethylene, and cyclooctene." They interpret these results to support a branching mechanism which gives rise to alkenyl hydroperoxide and polymeric dialkyl peroxide, both as primary oxidation products. This interpretation has been well accepted (7, 13). Brill's (4) and our results show that acyclic alkenyl hydroperoxides decompose extensively at temperatures above 100°C. to complicate the reaction kinetics and mechanistic interpretations. A simplified reaction scheme is outlined below.

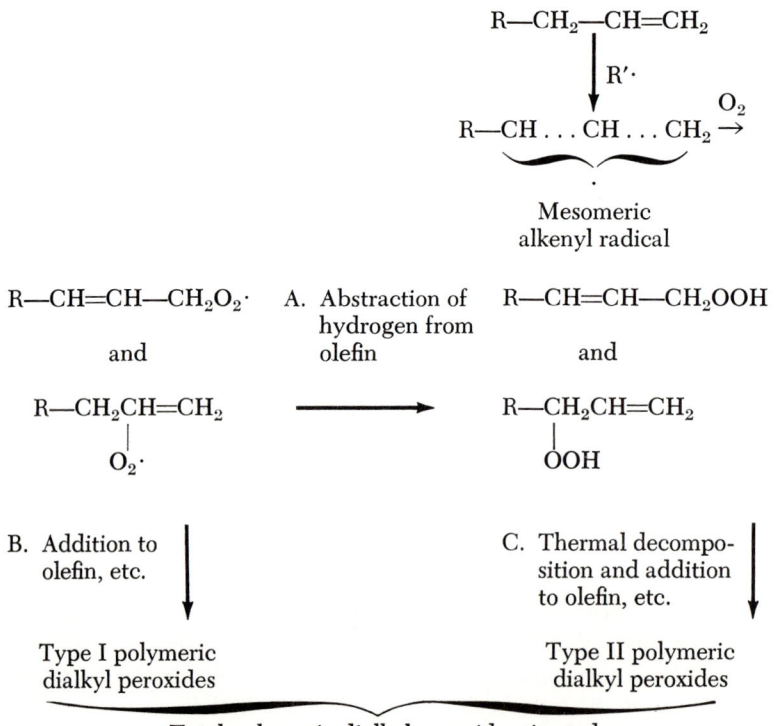

Table IV. Comparison of Peroxide Thermal Stabilities

Peroxide Type	Conditions	Half-Life at 110°C., Hrs.	Ref.
Hexadecenyl hydroperoxides	In 1-hexadecene under nitrogen	3.9 ± 1.5	Our work
3-Butene-2-hydroperoxide	0.07M benzene solution	11.5[a]	(4)
Cyclopentenyl hydroperoxide	In benzene	34.8	(18)
Cyclohexenyl hydroperoxide	In benzene	137[a]	(4)
Hexadecenyl polymeric peroxides	In 1-hexadecene	15 ± 4	Our work
tert-Butyl hydroperoxide	70% solution	40	(10)

[a] Extrapolated as half of value at 100°C.

At Van Sickle's conditions of low temperatures and low conversions, branching routes A and B appear to be dominant since there is little alkenyl hydroperoxide decomposition. In our work above 100°C., the branching routes are supported by the nearly linear initial portions at low conversions for alkenyl hydroperoxide and polymeric dialkyl peroxide curves (see Figures 2, 3, and 4). The polymeric dialkyl peroxides formed under our reaction conditions include those formed by the branching mechanism postulated by Van Sickle (routes A and B) and those formed by the reaction of the alkenoxy and hydroxy radicals from alkenyl hydroperoxide thermal decomposition reacting further and alternately with olefin and oxygen (step C). The importance and kinetic fit of the sequential route A to C appears to increase with temperature and extent of olefin conversion owing to the extensive thermal decomposition of the alkenyl hydroperoxides above 100°C.

Acknowledgment

The authors thank F. R. Mayo and D. E. Van Sickle for their inspiration and consultation. They also acknowledge the assistance of N. F. Seppi, J. P. Harris, M. J. Reuter, and K. W. Robinson in obtaining the data.

Literature Cited

(1) Bateman, L., "Chemistry and Physics of Rubber-Like Substances," pp. 595-609, Wiley, New York, 1963.
(2) Bateman, L., Quart. Rev. 8, 147 (1954).
(3) Brill, W. F., ADVAN. CHEM. SER. 51, 70 (1965).
(4) Brill, W. F., ADVAN. CHEM. SER. 75, 93 (1968).
(5) Brill, W. F., J. Am. Chem. Soc. 85, 141 (1963).
(6) Ibid., 87, 3286-7 (1965).
(7) Emanuel, N. M., private communication.

(8) Kamneva, A. I., Panfilova, Ya. S., in "Oxidation of Hydrocarbons in the Liquid Phase," by N. M. Emanuel, pp. 211-218, Macmillan, New York, 1965.
(9) Koucher, R. V., Tsherniak, B. I., Nikolayevsky, A. N., Abadsheva, R. N., Preprint, International Oxidation Symposium, Vol. I, p. I-84, Aug. 28–Sept. 1, 1967, San Francisco, Calif.
(10) Lucidol Division, Wallace & Tiernan, Inc., 1741 Military Road, Buffalo, N. Y., "Lucidol Peroxide Data Sheet."
(11) Mair, R. D., Graupner, A. J., *Anal. Chem.* **36**, 194-204 (1964).
(12) Mayo, F. R., *J. Am. Chem. Soc.* **80**, 2497 (1958).
(13) Mayo, F. R., private communication.
(14) Sergeyev, P. G., Ivanova, L. A., "Oxidation of Hydrocarbons in the Liquid Phase," by N. M. Emanuel, pp. 211-218, Macmillan, New York, 1965.
(15) Tobolsky, A. V., Metz, R. J., Mesrobin, R. G., *J. Am. Chem. Soc.* **72**, 1942 (1950).
(16) Van Sickle, D. E., private communication.
(17) Van Sickel, D. E., Mayo, F. R., Arluck, R. M., *J. Am. Chem. Soc.* **87**, 4824 (1966).
(18) *Ibid.*, p. 4832.
(19) Van Sickel, D. E., Mayo, F. R., Arluck, R. M., Syz, M. G., *J. Am. Chem. Soc.* **89**, 967 (1967).
(20) Van Sickel, D. E., Mayo, F. R., Gould, E. S., Arluck, R. M., *J. Am. Chem. Soc.* **89**, 977 (1967).
(21) Willson, C. L., Willson, D. W., "Comprehensive Analytical Chemistry," Vol. 1B, p. 762, Elsevier, New York, 1960.

RECEIVED October 10, 1967.

7

Butene Hydroperoxide

WILLIAM F. BRILL

Princeton Chemical Research, Inc., Princeton, N. J.

> *Butenes were subjected to photosensitized reaction with molecular oxygen in methanol. 1-Butene proved unreactive. A single hydroperoxide, 1-butene-3-hydroperoxide, was produced from 2-butene and isolated by preparative gas chromatography. Thermal and catalyzed decomposition of pure hydroperoxide in benzene and other solvents did not result in formation of any acetaldehyde or propionaldehyde. The absence of these aldehydes suggests that they arise by an addition mechanism in the autoxidation of butenes where they are important products. 1-Butene-3-hydroperoxide in the absence of catalyst is converted predominantly to methyl vinyl ketone and a smaller quantity of methyl vinyl carbinol —volatile products usually not detected in important quantities in the autoxidation of butene.*

Allylic hydroperoxides are primary products in the autoxidation of olefins, and lack of definite information on their reactivity and chemical behavior has hampered efforts to understand olefin oxidation mechanisms (2). This deficiency is most strongly felt in determining the relative rates of addition and abstraction mechanisms for acyclic olefins since assignment of secondary reaction products to the correct primary source is required. Whereas generalizations about the effect of structure on the course of hydroperoxide decompositions are helpful, most questions can be answered better by directly isolating the hydroperoxides involved and observing the products formed by decomposition of the pure compounds.

With the development of techniques for preparing and isolating acyclic allylic hydroperoxides, an attempt was made to examine the behavior of butene hydroperoxide and determine which secondary products arise from it during the oxidation of butenes. Butenes were of particular interest since indirect evidence had already indicated that none of the readily identified products can be ascribed unequivocally to a hydroperoxide source (4).

The preparation of acyclic allylic hydroperoxides has been described before (3, 7, 9), but it is not clear how the reactivities differ from the better known saturated hydroperoxides and cyclic allylic hydroperoxides. Dykstra and Mosher prepared allyl hydroperoxide by the reaction of allyl methanesulfonate with hydrogen peroxide and alcoholic potassium hydroxide and purified the hydroperoxide by gas chromatography. It detonated on heating and decomposed on exposure to light but was relatively stable in the cold and dark. The isomeric allylic hydroperoxides formed from the autoxidation of the branched olefin, 4-methyl-2-pentene, have also been isolated and were not abnormally reactive (3). In the present study, cis- and trans-2-butene were photooxidized in the presence of methylene blue as a sensitizer (14), and the product, 1-butene-3-hydroperoxide, was isolated by preparative chromatography. 1-Butene proved unreactive and 2-butene-1-hydroperoxide could be formed only by isomerization of the secondary hydroperoxide.

Experimental

Materials. Olefins were Phillips pure grade, appearing to be better than the reported 99% purity by gas chromatographic analysis on an $AgNO_3$ column. Methylene blue was U.S.P. basic blue No. 9, and methanol was Baker spectrophotometric reagent. All other solvents were either chromatographic or spectroscopic grade.

Photosensitized Oxidation. Rapidly stirred solutions of butenes were irradiated in methanol containing 0.5 gram per liter of methylene blue at 20°C. The reactor was fitted with a Hanovia 200-watt mercury lamp (654A036) placed in a water-cooled borosilicate glass well, and contained separate inlets for saturating the solution with butene and dispersing oxygen. Oxygen uptake was followed with wet-test meters, employing approximately a 100% excess of oxygen over that reacted and passing the effluent through an overhead reflux condenser followed by a trap, both kept at dry ice temperatures.

Oxidations were most conveniently run by first dissolving approximately 1 mole of butene in 1 liter of methanol and then irradiating while bubbling oxygen through the solution without adding more butene. Variations with 1-butene involved continuously or periodically saturating the solution with olefin while attempting oxidation in the absence or presence of the reflux condenser. In all cases, only traces of product evolved.

Isolation and Identification of Hydroperoxide. A solution produced by oxidizing trans-2-butene was concentrated at reduced pressure without heating from $0.085M$ to $1.5M$. The hydroperoxide was isolated from the concentrate by preparative chromatography under the following conditions: a 5-foot 3/8-inch column of aluminum containing 10% diisodecyl phthalate on Fluoropak 80; Autoprep 705 with flame ionization detector; carrier, 200 ml. per minute helium split 8 to 1 between trap and detector;

injection on column; temperature of injector, column, detector, and collector, 80°. Retention times were 14 minutes for 1-butene-3-hydroperoxide, 1.5 minutes for methanol, and 3 to 4 minutes for other products, which under the best conditions produce an area amounting to less than 10% of the hydroperoxide peak.

The trapped product gave an immediate test with KI in acetic acid. Its infrared spectrum was similar to that of 3-butene-2-ol with major absorption peaks at 3, 8.7, 9.5, 10.3, and 10.8 microns and minor peaks at 6.3, 7.2, 7.7, 11.6, 12.4, and 12.6 microns. There was no absorption arising from carbonyl. In a 25% solution of hydroperoxide in carbon tetrachloride, the hydroperoxide proton gave rise to a broad band at 8.7 p.p.m. (referred to TMS) in the NMR spectra.

The hydroperoxide could also be isolated in purities above 90% by careful distillation at low temperatures at 8 mm. Methanol was first removed through an 8-inch Vigreux at a 4 to 1 reflux ratio and a pot temperature of 20°C.

A sample of the methanolic oxidation product was treated directly with powdered sodium borohydride. Water was added, the solvent allowed to evaporate, and the aqueous solution extracted with carbon tetrachloride. Methyl vinyl carbinol was shown to be the major product, but about 10% impurities were present. Reduction by stirring overnight with 1.0M sodium sulfite followed by extraction with carbon tetrachloride produced a solution which analyzed as methyl vinyl carbinol with about 1% impurities.

Experiments with 1-Butene-3-hydroperoxide. From 0.5 to 2 ml. of solvent was added directly to the traps in which about 0.1 ml. of hydroperoxide had been collected by preparative gas chromatography, and the solutions were transferred to heavy-walled 8-ml. tubes and sealed with Teflon-lined caps. In several instances, it was necessary first to centrifuge the solutions to remove moisture apparently accumulated during the collection period.

The rate of thermal decomposition of a 0.070M solution in benzene at 100° was determined by following the decrease in the hydroperoxide area by gas chromatography, using the diisodecyl phthalate on a Fluoropak column.

The isomerization in hexane was allowed to proceed at room temperature until the ratio of the area of 1-butene-3-hydroperoxide (13.5 minutes) to the new peak (at 25.5 minutes) became constant in 14 days. The combined peak areas for a constant sample size (30 μliters) showed little decrease on longer standing. The isomerized mixture (0.7 ml.) was stirred with 0.3 ml. of 1.0M sodium sulfite for 16 hours. The hexane layer was separated, and the aqueous layer was saturated with sodium sulfate and extracted twice with 0.5-ml. portions of hexane. The combined extracts were found to contain 81% 3-butene-2-ol and 19% crotyl alcohol.

Analysis. Hydroperoxide was analyzed by gas chromatography under the sample conditions described for the isolation procedure; only

10- to 30-μliter samples were injected. Decomposition products were analyzed using a 12-foot, 1/8-inch column of 10% Carbowax 20M–terephthalic acid on Gas Chrom Q. The temperature was programmed from 60° to 190°C. at 8° per minute. Orders of the retention times for the solvents and more important possible products were: acetaldehyde, propionaldehyde, acetone, acrolein, methanol–methyl ethyl ketone, carbon tetrachloride, benzene–methyl vinyl ketone, water–2,3-butandione, crotonaldehyde, methyl vinyl carbinol, allyl alcohol, crotyl alcohol, and the "dimer" of methyl vinyl ketone. Since the quantities available did not allow isolation of the products, each peak was confirmed by adding the known compound to the solution being examined and observing the increase in peak height or the appearance of a new peak. Di-2-ethyl sebacate absorbant was used to separate benzene and methyl vinyl ketone and to provide additional confidence in product identity. Hallcomid separated benzene, methyl vinyl ketone, and methyl vinyl carbinol but gave poor resolution of the lower boiling carbonyl compounds.

Reactions in Methanol. By dropping 434 grams of the deep blue solution from the photosensitized oxidation of *trans*-2-butene onto glass beads heated to 200°C. in a Claisen flask, 414 grams of clear distillate were collected over 7 hours. Iodometric titration (15) showed the oxidate to be $0.089M$ and the distillate to be $0.073M$ in hydroperoxide, indicating that 21% of the hydroperoxide was lost in the flash distillation. Decomposition of the distillate in the vaporizer of the gas chromatograph produced the normal amounts of methyl vinyl ketone and methyl vinyl carbinol and a trace of crotonaldehyde. Considerable butene also remained in the distillate.

A $0.09M$ oxidation solution was refluxed in the presence of 1.0 mole % of cobaltous chloride hexahydrate for 20 hours. The concentration of hydroperoxide was reduced to $0.007M$. The products consisted of 3% acetone, 13% methyl vinyl ketone, 10% crotonaldehyde, 26% methyl vinyl carbinol, and 36% of a less volatile product. The major product appears to arise by the dimerization of methyl vinyl ketone since it also formed readily on refluxing methyl vinyl ketone in methanol containing cobaltous chloride and was present in commercial ketone. The refluxed solution was concentrated by removing the solvent through a semimicrocolumn, and the major products were isolated by preparative gas chromatography. Infrared spectra identical to those of authentic samples were obtained for crotonaldehyde and methyl vinyl carbinol. Trapped methyl vinyl ketone yielded a cloudy carbon tetrachloride solution which produced a poor but recognizable spectrum. The "methyl vinyl ketone dimer" absorbed strongly at 5.85, 6.1, and 8.95 microns.

Results

The photosensitized oxidation of either *trans*-2-butene or a mixture of *cis*- and *trans*-2-butene in methanol using methylene blue as a sensitizer produced a single hydroperoxide—3-butene-2-hydroperoxide—cleanly at atmospheric pressure. The hydroperoxide was reduced to 3-butene-2-ol by treating the methanolic reaction solution with sodium borohydride or sodium sulfite. Isolation of the hydroperoxide by preparative gas chromatography yielded a pure compound whose infrared spectra resembled that for 3-butene-2-ol, and the alcohol was produced by reducing the pure hydroperoxide. Whereas a 7% solution of hydroperoxide in methanol could be obtained from 2-butene in a 30-hour oxidation, 1-butene failed to yield detectable hydroperoxide on oxidation for 3 days. It was therefore necessary to isomerize 1-butene-3-hydroperoxide by allowing a dilute hexane solution to stand at room temperature until an equilibrium mixture with 2-butene-1-hydroperoxide was produced. Direct analysis by gas chromatography showed that the solution contained 86% of the secondary and 14% of the primary isomer. Reduction with sodium sulfite produced 81% of the secondary alcohol and 19% of the primary isomer, crotyl alcohol. No attempt was made to isolate the primary hydroperoxide, and the behavior of this isomer was not studied directly.

Neither the thermal nor the cobalt-catalyzed decomposition of 3-butene-2-hydroperoxide in benzene at 100°C. produced any acetaldehyde or propionaldehyde. In the presence of a trace of sulfuric acid, a small amount of acetaldehyde along with a large number of other products were produced on mixing. Furthermore, on heating at 100°C., polymerization is apparently the major reaction; no volatile products were detected, and only a slight increase in acetaldehyde was observed. Pyrolysis of a benzene or carbon tetrachloride solution at 200°C. in the injection block of the gas chromatograph gave no acetaldehyde or propionaldehyde, and none was detected in any experiments conducted in methanol.

Decomposition of a 0.07M solution of 2-butene-3-hydroperoxide in benzene at 100°C. had a half-life of 23 hours. The final products were 37% methyl vinyl ketone, 37% methyl vinyl carbinol, and 26% acetone. Attempts to substantiate the identity of the acetone were unsuccessful, but it was the only anticipated product with the correct gas chromatographic retention time on four different absorbents.

In the presence of approximately 1% cobalt naphthenate in benzene, only 4 hours at 100°C. were required to decompose the hydroperoxide almost completely. The yields of products from decompositions catalyzed by some commonly used cobalt and vanadium (1) compounds are given in Table I. Polymerization appears to be the major reaction.

The cleanest product composition may be effected by decomposition of the pure hydroperoxide or solutions in the injection block of the gas chromatograph. In carbon tetrachloride solution only methyl vinyl ketone and methyl vinyl carbinol were produced, the ratio of ketone to alcohol being 2.9. No definite traces of products from isomerized hydroperoxide were observed.

The hydroperoxide appears to be exceptionally stable in methanol, no isomer appearing after several months' standing. Flash distillation

Table I. Catalyzed Decomposition of 1-Butene-3-hydroperoxide in Benzene[a]

Catalyst-Conditions	Theoretical Yields, %		
	MVK	MVC	Other
1.84M hydroperoxide[b]			
0.5% Co naphthenate,[c] 85°C.	4	20	[d]
Vapor phase,[e] 200°C.	53	18	[f]
1.45M hydroperoxide			
1% Co $(C_5H_7O_2)_3$,[g] 55°C.	22	16	
$VO(C_5H_7O_2)$,[h] 55°C.	7	20	[i]

[a] Reaction time in liquid phase, 24 hours.
[b] Titrometric (15); calculated from wt. of hydroperoxide, 1.89M.
[c] Nuodex 6% cobalt.
[d] Approximately 4% acetone.
[e] On injection to gas chromatograph, est. 2-sec. reaction time.
[f] Approx. 7% acetone.
[g] MacKenzie Chemical Works cobaltic acetylacetonate; stirred flask, absorbed 0.3 mole O_2/mole peroxide.
[h] Saturated sol., <1%.
[i] Peak areas for products produced only with vanadium catalyst (epoxy alcohols) are 5.1% (double peak) and 3.5% of total products.

may be used to remove methylene blue from the oxidate and produce a clear methanolic hydroperoxide solution. Refluxing the oxidation solution in the presence of cobaltous chloride produced a large amount of methyl vinyl ketone "dimer" (13) in addition to the usual unsaturated ketone and alcohol products.

The neat hydroperoxide is also very stable and may be stored in dry ice for long periods. Surprisingly, after standing for 6 months at room temperature exposed to normal laboratory light, methyl vinyl ketone monomer may be readily detected in the hydroperoxide by infrared analysis.

Discussion

In oxidation studies it has usually been assumed that thermal decomposition of alkyl hydroperoxides leads to the formation of alcohols. However, carbonyl-forming eliminations of hydroperoxides, usually under the influence of base, are well known. Of more interest, nucleophlic rearrangements, generally acid-catalyzed, have been shown to produce a mixture of carbonyl and alcohol products by fission of the molecule (6). For 1-butene-3-hydroperoxide it might have been expected that a rearrangement (Reaction 1) similar to that which occurs with cumene hydroperoxide could produce two molecules of acetaldehyde.

$$CH_2=CHC(CH_3)-OOH \rightarrow CH_2=CHOH + CH_3CHO \quad (1)$$

Such a rearrangement was detected only in the presence of sulfuric acid, and furthermore at 100°C. it was supplanted by a homolytic breakdown. The products found in the purely thermal decomposition—methyl vinyl ketone and methyl vinyl carbinol—are in fact consistent with the behavior of alkyl hydroperoxides and are analogous to the products produced from the cyclic allylic hydroperoxide from cyclohexene (2).

It is clear that the aldehydes produced in the oxidation of butenes arise from addition reactions as already assumed by several workers (4, 16, 17). Nothing in the present work indicates whether the aldehyde-forming addition product is a peroxy alkyl (17), (Reaction 2) or an alkoxy alkyl (16) radical (Reaction 3) as has been postulated, or both.

$$ROO-CHCH-OOCHCH \rightarrow RO\cdot + 2CHO + \overset{O}{\overset{\frown}{CHCH}} \quad (2)$$

$$RO-CH_2-CH-R' \rightarrow R'CHO + ROCH_2\cdot \quad (3)$$
$$\underset{O\cdot}{|}$$

Of the many studies of the autoxidation of butenes, few (5, 11) have emphasized methyl vinyl ketone and methyl vinyl carbinol as major products. In the cumene hydroperoxide–initiated oxidation of 1-butene at 105°C. with 60 atm. of air, Chernyak (5) reported an average hourly rate of production of these two products approximately equal to the combined rates of formation of hydroperoxide and epoxide. The reported rates for hydroperoxide plus vinyl ketone and alcohol indicate that 60% of the products occur by abstraction, in agreement with Van Sickle (17).

The observed half life at 100°C. of 23 hours for a dilute solution of hydroperoxide in benzene indicates that significant decomposition may occur in the autoxidation of butene, depending on reaction conditions. No reliable evaluation can be made because of the known complications introduced on hydroperoxide decomposition by the effect of the solvent, the hydroperoxide concentration (2), the presence of oxygen (12), and the possibility of a strong acceleration in rate in the presence of oxidizing olefin, observed in at least one system (8). However, using the data reported by Bateman for a benzene solvent at 100°C. in the presence of air (2), 1-butene-3-hydroperoxide decomposes 13 times faster than cyclohexene hydroperoxide, a product which may be formed in extremely high yield by the oxidation of cyclohexene.

The free radical isomerization of allylic hydroperoxides (3) in dilute nonpolar solvents appears to give a reliable measure of their relative stabilities since almost no by-products form in this process. It was previously concluded from the isomerization of the hydroperoxides from 4-methyl-2-pentene that tertiary and secondary allylic hydroperoxide are equally stable. It now appears that a primary hydroperoxide represents the least stable allylic structure.

The unfortunately large difference in the reactivity of 1-butene and 2-butene to photosensitized oxidation is not surprising in view of the previous observed structural effects in such oxidations (*10*). Competitive oxidation experiments are underway to determine the exact magnitude of the rate difference involved and to evaluate the effect of impurities which might quench the reaction.

Literature Cited

(1) Allison, K., Johnson, P., Foster, G., Sparke, M., *Ind. Eng. Chem. Prod. Res. Develop.* **5,** 166 (1966).
(2) Bateman, L., Hughes, H., *J. Chem. Soc.* **1952,** 4594.
(3) Brill, W. F., *J. Am. Chem. Soc.* **87,** 3286 (1965).
(4) Brill, W. F., Barone, B. J., *J. Org. Chem.* **29,** 140 (1964).
(5) Chernyak, B. I., Koucher, R. V., Nikolayevsky, A. N., *Neftekhemiya* **4** (3), 452 (1964).
(6) Davies, A. G., "Organic Peroxides," Butterworths, London, 1961.
(7) Dykstra, S., Mosher, H. S., *J. Am. Chem. Soc.* **79,** 3474 (1957).
(8) Hiatt, R., Gould, G. W., Mayo, F., *J. Org. Chem.* **29,** 3461 (1964).
(9) Hoffman, J., *J. Org. Chem.* **22,** 1747-1749 (1957).
(10) Kopecky, K. R., Reich, H. J., *Can. J. Chem.* **43** (1965).
(11) Koucher, R. V., Tsherniak, B. I., Nikolayevsky, A. N., Abadsheva, R. N., International Oxidation Symposium, I-75, 1967.
(12) Morse, B. K., *J. Am. Chem. Soc.* **79,** 3376 (1957).
(13) Muller, A., *Ber.* **5,** 1142 (1921).
(14) Schenck, G. O., *Angew. Chem.* **69,** 579 (1957).
(15) Siggia, S., "Quantitative Organic Analysis," 2nd ed., Wiley, New York, 1954.
(16) de Roch, I. Seree, *Bull. Soc. Chim. France* **1962,** 1774.
(17) Van Sickle, D. E., Mayo, F. R., Arluck, R., Syz, M. G., *J. Am. Chem. Soc.* **89,** 967 (1967).

RECEIVED October 2, 1967.

Discussion

J. H. Knox: In liquid-phase olefin oxidation, epoxides increase with temperature relative to cleavage products and soon exceed them in yield. In gas phase oxidation at about 300°C. epoxides are minor products

(~10 to 20%) relative to cleavage products. These observations are not consistent with the hypothesis that epoxides arise from the end groups of polyperoxide chains. Any explanation?

W. F. Brill: For olefins producing high yields of epoxide in the liquid phase, much of the epoxide may be formed directly from the radical formed by the addition of peroxy radical to olefin without involving polyperoxide chains. While there are no indications that the addition is reversible in the liqud phase, it may be expected to become so at sufficiently high temperatures. The behavior of cis-trans isomers in the gas phase may clarify this matter.

H. Berger: Can one demonstrate an oxygen-pressure dependence of O_2 addition to give polyperoxide?

D. E. Van Sickle: Yes, such a dependency has been demonstrated.

C. J. Norton: Have you any indication of geometric isomers in the hydroperoxides from 2-butene? Would you expect differences in stabilities?

Dr. Brill: One would expect *trans*-2-butene-1-hydroperoxide to be the more stable isomer and the one formed in largest quantity by rearrangement of 1-butene-3-hydroperoxide. In a previous study, only *trans*-2-methyl-3-pentene-2-hydroperoxide was obtained by equilibration with its allylic isomer.

T. Traylor: In benzene solution, were products of butene hydroperoxide decomposition the same at very high and very low concentrations?

Dr. Brill: The major products were the same. Additional unidentified minor products were formed at high concentrations in the presence of catalyst.

8

Gamma-Radiation–Induced Oxidation of 2-Propanol

GORDON HUGHES and H. A. MAKADA
Donnan Laboratories, The University, Liverpool 3, England

> *The gamma-radiation–induced oxidation of 2-propanol has been investigated. Acetone and hydrogen peroxide are the principal products and arise via a chain reaction in aqueous acid solutions at high concentrations of 2-propanol. In neutral solutions of 2-propanol and in solutions of methanol and ethanol, no such chain reactions are observed. The reasons for this are discussed along with the implications of the results for the hydroxyl radical yield in water radiolysis.*

Although the radiation-induced oxidation of ethanol has been fully investigated (2, 22, 23), little work has been published on the oxidation of other alcohols. In connection with a project concerned with the relative rates of hydroxyl radical reactions using 2-propanol as reference solute, it was thought desirable first to investigate the radiation chemistry of 2-propanol–oxygen solutions both in aqueous solution and pure 2-propanol. The results of this investigation are presented here.

Experimental

Samples were irradiated with γ-rays from a 90-curie ^{137}Cs source at 20° ± 2°C. as previously described (11). Triply distilled water and analytical reagent grade chemicals were used in making up all solutions. In some cases Aristar 2-propanol was used (British Drug Houses). Solutions were aerated prior to irradiation by shaking in air. Oxygen-saturated solutions were prepared by bubbling through the solution oxygen, which had previously been washed with unirradiated solution, for at least 30 minutes before radiolysis. Hydrogen peroxide was determined by both the iodide ion (20) and titanium sulfate (14) methods. Acetone was determined by the salicylaldehyde method (5). Control experiments showed that hydrogen peroxide did not interfere with this determination. Formaldehyde was determined by the chromotropic acid method (6). Hydrogen peroxide, however, interfered slightly with this determination.

Results and Discussion

Acid Solution. Acetone and hydrogen peroxide are the principal oxidation products observed in the radiolysis of 2-propanol in sulfuric acid solution. Identical results were obtained for hydrogen peroxide estimation by both the titanium sulfate and iodide ion estimations, indicating that organic hydroperoxides were not formed or that, if formed, they rapidly decompose to acetone and hydrogen peroxide. The yields obtained are summarized in Table I. In all cases, it was clearly established that product yields were independent of dose initially.

$G(H_2O_2)$ is independent of oxygen concentration. Although $G(H_2O_2)$ does not depend greatly on 2-propanol concentration at low concentrations, there is a marked increase in $G(H_2O_2)$ at higher concentrations of 2-propanol. In oxygen-saturated solution $G(CH_3COCH_3)$ is lower than in air-saturated solution except at the lowest 2-propanol concentrations. There is again a marked increase in $G(CH_3COCH_3)$ at high concentrations of 2-propanol, the effect being slightly greater at higher acidity.

By analogy with the ethanol system, the following reactions might be expected to occur in the radiolysis of aqueous 2-propanol–oxygen solutions:

$$H_2O \longrightarrow H, e_{aq}^-, OH, H_2, H_2O_2 \quad (1)$$

$$e_{aq}^- + H^+ \rightarrow H \quad (2)$$

$$e_{aq}^- + O_2 \rightarrow O_2^- \quad (3)$$

$$O_2^- + H^+ \rightleftharpoons HO_2 \quad (4)$$

$$H + O_2 \rightarrow HO_2 \quad (5)$$

$$H + (CH_3)_2CHOH \rightarrow H_2 + (CH_3)_2\dot{C}OH \quad (6)$$

$$OH + (CH_3)_2CHOH \rightarrow H_2O + (CH_3)_2\dot{C}OH \quad (7)$$

$$(CH_3)_2\dot{C}OH + O_2 \rightarrow (CH_3)_2\underset{\underset{O-\dot{O}}{|}}{C}OH \quad (8)$$

$$(CH_3)_2\underset{\underset{O-\dot{O}}{|}}{C}OH \rightarrow (CH_3)_2CO + HO_2 \quad (9)$$

$$2HO_2 \rightarrow H_2O_2 + O_2 \quad (10)$$

Reaction of the hydrated electron *via* Reaction 2 or 3 and of the hydrogen atom *via* Reaction 5 or 6 will ultimately yield an HO_2 radical. From known values of the rate constants (*13*)—*viz.*, $k_2 = k_3 = 2 \times 10^{10}$ liters mole^{-1} sec.$^{-1}$—it can be calculated that under the experimental conditions of Table I, virtually all hydrated electrons react *via* Reaction 2.

Table I. Product Yields in Sulfuric Acid Solution
(Dose rate $\simeq 8 \times 10^{15}$ e.v. ml.$^{-1}$ min.$^{-1}$)

[(CH$_3$)$_2$CHOH] × 100, M	[H$_2$SO$_4$], M	Condition[a]	G(CH$_3$COCH$_3$)	G(H$_2$O$_2$)	G(CH$_3$COCH$_3$), Calcd.	G(H$_2$O$_2$), Calcd.
0.45	0.05	A	3.00	—	3.08	
	0.05	B	2.98	—		
1.0	0.05	A	3.67	4.39	3.28	3.97
	0.05	B	3.48	4.37		
10.0	0.05	A	6.30	4.84	4.87	3.97
	0.05	B	6.09	4.93		
100.0	0.05	A	11.1	10.0	6.18	3.97
	0.05	B	10.2	9.9		
1.0	0.20	A	3.60	4.44	3.37	4.08
10.0	0.20	A	6.67	5.48	5.02	4.08
100.0	0.20	A	14.5	13.4	6.37	4.08

[a] A. Air-saturated solution; B. Oxygen-saturated solution.

Under these conditions it follows that for the above reaction scheme

$$G(H_2O_2) = \tfrac{1}{2}(G_w(H) + G_w(e_{aq}^-) + G_w(OH)) + G_w(H_2O_2) \quad (11)$$

and

$$G(CH_3COCH_3) = G_w(OH) + \alpha(G_w(H) + G_w(e_{aq}^-)) \quad (12)$$

where

$$\alpha = \frac{k_6[(CH_3)_2CHOH]}{k_6[(CH_3)_2CHOH] + k_5[O_2]} \quad (13)$$

Using known values of the water radiolysis yields (18)—for example, $[G_w(H) + G_w(e_{aq}^-)] = 3.54$, $G_w(OH) = 2.89$, and $G(H_2O_2) = 0.75$ in 0.05M H$_2$SO$_4$ and 3.66, 2.97, and 0.76, respectively, in 0.2M H$_2$SO$_4$ and of the rate constants (17) $k_5 = 2 \times 10^{10}$ liter mole^{-1} sec.$^{-1}$ and $k_6 = 5 \times 10^7$ liter mole^{-1} sec.$^{-1}$—it is possible to calculate yields. The calculated values for air-saturated solution are compared with the observed values in Table I.

At the lowest concentrations of 2-propanol, $\leqslant 10^{-2}$ M, there is reasonable agreement between calculated and experimental results. These results then are consistent with the assumption that acetone is the only oxidation product from the radical produced from 2-propanol in aqueous solution containing oxygen. It has similarly been shown (23) that acetaldehyde is the only organic oxidation product in the radiation-induced oxidation of aqueous ethanol. However, our experiments indicate that in aqueous methanol, the observed yields of formaldehyde are always much lower than would be expected if

$$CH_2OH + O_2 \rightarrow CH_2O + HO_2 \quad (14)$$

were the only reaction of hydroxymethyl radicals. Thus, even in 1M CH_3OH in aerated 0.4M H_2SO_4, $G(CH_2O) < G_w(OH)$. The alternative reactions

$$CH_2OH + O_2 \rightarrow \underset{O\!-\!\dot{O}}{\underset{|}{CH_2OH}} \quad (15)$$

$$\underset{O\!-\!\dot{O}}{\underset{|}{CH_2OH}} + O_2^- \rightarrow \underset{O\!-\!O^-}{\underset{|}{CH_2OH}} + O_2 \quad (16)$$

$$\underset{O\!-\!O^-}{\underset{|}{CH_2OH}} + H^+ \rightarrow HCOOH + H_2O \quad (17)$$

yielding formic acid have been proposed in the radiolysis of pure methanol (19), and formic acid (10) has been identified as one of the products. We have not, however, been able to identify formic acid in the radiolysis of aqueous solutions.

However, at higher concentrations of 2-propanol, the yields of both acetone and hydrogen peroxide are higher than expected. Moreover, the high yields observed in 1M 2-propanol cannot be accounted for reasonably without invoking the occurrence of a chain reaction. The fact that yields in 1M 2-propanol depend on dose rate (Table II), whereas yields in 0.01M 2-propanol do not depend on dose rate, is consistent with this.

Table II. Dependence of $G(CH_3COCH_3)$ on Dose Rate in Air-Saturated 0.05M H_2SO_4

$(CH_3)_2CHOH$ $\times$ 100, M	Dose Rate, E.V. $Ml.^{-1}$ $Min.^{-1}$ $\times$ 10^{-16}	$G(CH_3COCH_3)$
1.0	2.30	3.63
	0.757	3.67
	0.225	3.67
100	2.30	9.41
	0.757	11.1
	0.225	13.4
"Pure"	1.84	32
	0.605	54
	0.180	93

The chain propagation step is probably

$$HO_2 + (CH_3)_2CHOH \rightarrow H_2O_2 + (CH_3)_2\dot{C}OH \quad (18)$$

Under these conditions the increase in $G(CH_3COCH_3)$ should be paralleled by an increase in $G(H_2O_2)$, as observed. If Reactions 8 and 9 are rapid relative to Reaction 18, then it follows from kinetic analysis of the above reaction scheme, assuming a stationary concentration of HO_2 radicals, that the yield of acetone arising from the chain reaction is given by

$$G(CH_3COCH_3) = k_{18}[(CH_3)_2CHOH]\left(\frac{6.02 \times 10^{25}}{Dk_{10}}\right)^{1/2} \times$$
$$(G_w(H) + G_w(e_{aq}^-) + G_w(OH))^{1/2} \quad (19)$$

where D is the dose rate in e.v. liter^{-1} sec.$^{-1}$ Some acetone is also produced as a result of Reactions 6 and 7, so that the total acetone yield is given by

$$G(CH_3COCH_3) = G_w(OH) + \alpha(G_w(H) + G_w(e_{aq}^-))$$
$$+ k_{18}[(CH_3)_2CHOH]\left\{\frac{6.02 \times 10^{25}(G_w(H) + G_w(e_{aq}^-) + G_w(OH))}{Dk_{10}}\right\}^{1/2} \quad (20)$$

For a constant 2-propanol concentration α is constant, and hence a plot of $G(CH_3COCH_3)$ against $D^{-1/2}$ should be linear. The results of Table III plotted according to Equation 20 are shown in Figure 1. Satisfactory agreement is observed and the intercept of Figure 1—$G(CH_3COCH_3) \simeq 6.5$—agrees well with the calculated value in Table I. From Figure 1 the ratio $k_{18}/k_{10}^{1/2} = 7.9 \times 10^{-5}$ may be calculated. Since k_{10} has previously been determined (9) as 2.7×10^6 liter mole^{-1} sec.$^{-1}$, $k_{18} = 0.13$ liter mole^{-1} sec.$^{-1}$

Table III. Bond Energies in Alcohols

Alcohol	$D(HRO{-}H)$, Kcal.	$\Delta H(HR\dot{O} \to \dot{R}OH)$, Kcal.	$D(H{-}ROH)$, Kcal.	
			This work	(28)
CH_3OH	102	−7.5	94.5	98
CH_3CH_2OH	102	−9.5	92.5	94.5
$(CH_3)_2CHOH$	102	−12	90	90.3

A similar chain process is observed in acidified "pure" 2-propanol—i.e., 24.5 ml. of 2-propanol + 0.5 ml. of 2.5M H_2SO_4—as may be seen from Table II. Figure 2 shows that $G(CH_3COCH_3)$ is again proportional to $D^{-1/2}$. No chain reactions have been observed in the radiolysis of oxygen-saturated aqueous solutions of ethanol, and we have been unable to find evidence of chain reactions in aqueous methanol at 20°C. Thus, in 0.4M H_2SO_4, $G(H_2O_2)$ is 3.5 and is independent of methanol concentration in the range 10^{-2} to 1M. It has been claimed (24) that in thoroughly dried oxygen-saturated methanol $G(CH_2O) \simeq 10$, indicating a short-chain reaction, but this result is at variance with other work (10,

19). Part of the high yield of formaldehyde may arise from the thermal-induced decomposition of hydrogen peroxide which is known (25) to occur in anhydrous methanol.

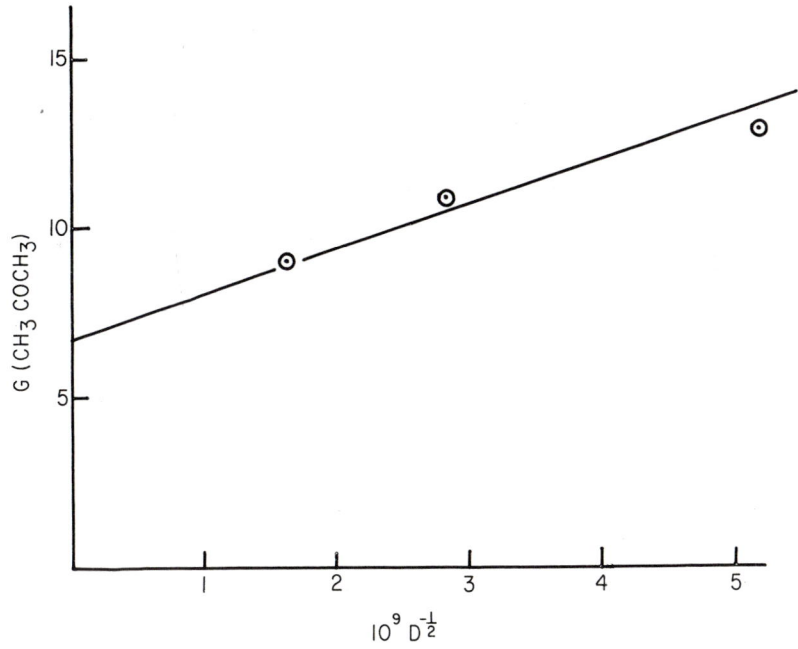

Figure 1. Effect of dose rate in 1M 2-propanol

The different behavior of the alcohols probably arises from differences in bond dissociation energies. Experiments show that radical attack on methanol (4) and ethanol (27) leads to rupture of the C—H rather than the O—H bond. There appear to be no direct measurements of C—H bond energies in alcohols. However, $D(R—OH)$ has been determined as 102 kcal. and does not appear to vary greatly with changes in R, provided R is a simple alkyl radical (16). Moreover, the heat of rearrangement of alkoxy radicals to hydroxyalkyl radicals has been determined from electron impact data (12). Considering, for example, 2-propanol and the following reactions

$$(CH_3)_2CHOH \rightarrow (CH_3)_2\dot{C}OH + H \qquad \Delta H_{21}$$

$$(CH_3)_2CHOH \rightarrow (CH_3)_2CH\dot{O} + H \qquad \Delta H_{22}$$

$$(CH_3)_2CH\dot{O} \rightarrow (CH_3)_2\dot{C}OH \qquad \Delta H_{23}$$

it follows that

$$\Delta H_{21} = D((CH_3)_2C(OH) - H)$$
$$= \Delta H_{22} + \Delta H_{23}$$
$$= D((CH_3)_2CHO - H) + \Delta H_{23}$$

Thus, the appropriate C—H bond strengths can be determined. These are given with the relevant thermochemical data in Table III. Also shown in Table III for comparison are alternative estimates of these bond energies (28) which appeared during the course of this work. The two sets of data agree fairly well.

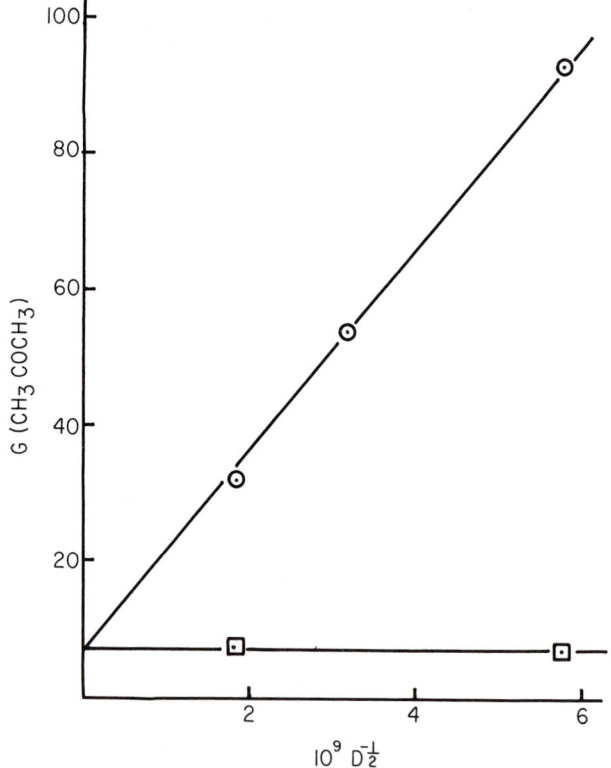

Figure 2. *Effect of dose rate in 13M 2-propanol.* ⊙—*acid solution;* □—*neutral solution*

$D(HO_2-H)$ is reported (7) to be 90 kcal., and hence abstraction by HO_2 of H from 2-propanol would be thermoneutral. It might, therefore, be expected to compete reasonably with dimerization of HO_2. The abstraction from methanol and ethanol would, however, have small but significant activation energies. The energies discussed above are for gas-phase reactions but are likely to be similar in the liquid phase.

Neutral and Alkaline Solutions. Product yields are shown in Tables IV and V.

Table IV. Product Yields in Neutral Solution
(Dose rate $\simeq 8 \times 10^{15}$ e.v. ml.$^{-1}$ min.$^{-1}$)

[$(CH_3)_2CHOH$] × 100, M	Condition[a]	$G(CH_3COCH_3)$	$G(H_2O_2)$
1.0	A	3.11	3.10
	B	2.98	3.08
10.0	A	3.90	3.20
	B	3.73	3.21
100	A	4.80	3.24
	B	4.75	3.04
Pure	C	7.00	
	D	6.94	

[a] A. Air-saturated solution; B. Oxygen-saturated solution; C. Air-saturated solution: dose rate 1.84×10^{16} e.v. ml.$^{-1}$ min.$^{-1}$; D. Air-saturated solution: dose rate 1.80×10^{15} e.v. ml.$^{-1}$ min.$^{-1}$.

Table V. Product Yields in Alkaline Solution
(Dose rate $\simeq 8 \times 10^{15}$ e.v. ml.$^{-1}$ min.$^{-1}$)

[$(CH_3)_2CHOH$] × 100, M	[NaOH], M	Condition	$G(CH_3COCH_3)$
1.0	0.10	A	2.83
	0.10	B	2.80
10.0	0.10	A	3.47
	0.10	B	3.38
100	0.10	A	3.87
	0.10	B	3.80
1.0	1.0	A	3.03
10.0	1.0	A	3.68
100	1.0	A	4.07

Pure 2-propanol was at its natural pH. Similar results were obtained for aqueous solutions at natural pH and using phosphate buffer to give pH = 6.98. In alkaline solution, hydrogen peroxide solutions are unstable and could not be measured accurately, but $G(H_2O_2) \simeq 3.5$ at all concentrations.

Product yields in both neutral and alkaline solution indicate that no chain reaction is occurring under these conditions. That the yields are independent of dose rate in pure 2-propanol is consistent with this and is in marked contrast to the results in acid solution.

The chain carrier in acid solution has been postulated to be the HO_2 radical. For the ionization (8)

$$HO_2 \rightleftharpoons H^+ + O_2^- \tag{24}$$

pKa is 4.4. Thus, in neutral and alkaline solutions this species will be present predominantly as the O_2^- species. Its failure to act as chain initiator may be caused by a lower energy of its bond with hydrogen. This can be calculated from the following reactions:

$$H_2O_2 \rightleftharpoons H^+ + HO_2^- \qquad \Delta H_{25}$$
$$HO_2 \rightleftharpoons H^+ + O_2^- \qquad \Delta H_{24}$$
$$HO_2 + H \rightarrow H_2O_2 \qquad \Delta H_{26}$$
$$H + O_2^- \rightarrow HO_2^- \qquad \Delta H_{27}$$

It follows that

$$\Delta H_{25} + \Delta H_{26} - \Delta H_{24} = \Delta H_{27}$$

and that

$$\Delta H_{27} = -D(H\text{---}O_2^-) = -90 + (\Delta H_{25} - \Delta H_{24})$$

In aqueous solution, it is probable that $\Delta S_{25} \simeq \Delta S_{24}$ and hence

$$\Delta H_{25} - \Delta H_{24} = \Delta G_{25} - \Delta G_{24}$$

$$= -RT \ln \frac{K_{25}}{K_{24}}$$

It is known (15) that pK_{25} is 11.75 and hence $\Delta H_{25} - \Delta H_{24}$ is $\simeq$ 10 kcal. Thus, $D(H\text{---}O_2^-)$ is $\simeq$ 80 kcal. The abstraction of H will be a much less favorable reaction for O_2^- than for HO_2 and is endothermic for O_2^- even with 2-propanol.

The occurrence of a chain reaction in acid but not in neutral solution has previously been observed, although no mechanism to account for this has been proposed (26).

There is much less agreement concerning product yields for water radiolysis in neutral and alkaline solutions (3). There is general agreement that $G_w(H) < 0.7$ at these pH. Thus, in 10^{-2} M oxygen-saturated aqueous 2-propanol, the maximum yield of acetone arising from Reaction 6 is 0.03. Under these conditions then $G(CH_3COCH_3) = G_w(OH) = 2.98$. However, from the radiolysis of oxygen-saturated neutral aqueous solutions of ethanol it was concluded (22) that $G_w(OH) = 2.3$. This has been confirmed by work on other systems (18). There is general agreement (18, 21) that $G_w(OH)$ is $\simeq$ 3.1 in 0.1M NaOH, and it has recently been suggested (1) that in both neutral and alkaline solutions $G_w(OH) = 3.2$. The yields obtained in this work agree more closely with the higher value of $G_w(OH)$ for neutral solution.

Literature Cited

(1) Adams, G. E., Boag, J. W., Michael, B. D., *Trans. Faraday Soc.* **61**, 492 (1965).
(2) Allan, J. T., Hayon, E. M., Weiss, J., *J. Chem. Soc.* **1959**, 3913.
(3) Allen, A. O., Proceedings of 5th Informal Conference on the Radiation Chemistry of Water, 1966, AEC Document No. **COO-38-519**, p. 6.
(4) Baxendale, J. H., Hughes, G., *Z. Physik. Chem. (Frankfurt)* **14**, 323 (1958).
(5) Berntsson, S., *Anal. Chem.* **28**, 1337 (1956).
(6) Bricker, C. E., Johnson, H. R., *Ind. Eng. Chem., Anal. Ed.* **17**, 400 (1945).
(7) Cottrell, T. L., "Strengths of Chemical Bonds," 2nd ed., p. 187, Butterworths, London, 1958.
(8) Czapski, G., Bielski, B. H. J., *J. Phys. Chem.* **67**, 2180 (1963).
(9) Czapski, G., Dorfman, L. M., *J. Phys. Chem.* **68**, 1169 (1964).
(10) Dobson, G., Hughes, G., *J. Phys. Chem.* **69**, 1814 (1965).
(11) Dobson, G., Hughes, G., *Trans. Faraday Soc.* **57**, 1117 (1961).
(12) D'Or, L., Collin, J., *Bull. Soc. Roy. Sci. Liége* **22**, 285 (1953).
(13) Dorfman, L. M., Matheson, M. S., in "Progress in Reaction Kinetics," G. Porter, ed., Vol. 3, p. 237, Pergamon, London, 1965.
(14) Eisenberg, G. M., *Ind. Eng. Chem., Anal. Ed.* **15**, 327 (1943).
(15) Evans, M. G., Uri, N., *Trans. Faraday Soc.* **45**, 224 (1949).
(16) Gray, P., Williams, A., *Chem. Revs.* **59**, 239 (1959).
(17) Hart, E. J., *Ann. Rev. Nuclear Sci.* **15**, 125 (1965).
(18) Hayon, E., *Trans. Faraday Soc.* **61**, 734 (1965).
(19) Hayon, E., Weiss, J. J., *J. Chem. Soc.* **1961**, 3970.
(20) Hochanadel, C. J., *J. Phys. Chem.* **56**, 587 (1952).
(21) Hughes, G., Willis, C., *Discussions Faraday Soc.* **36**, 223 (1963).
(22) Hummel, A., Allen, A. O., *Radiation Res.* **17**, 302 (1962).
(23) Jayson, G. G., Scholes, G., Weiss, J., *J. Chem. Soc.* **1957**, 1358.
(24) Lichtin, N. N., Rosenberg, L. A., Immamura, M., *J. Am. Chem. Soc.* **84**, 3587 (1962).
(25) Lichtin, N. N., Wilson, J. W., *J. Phys. Chem.* **69**, 3673 (1965).
(26) Saraeva, V. V., Natkh, B. S., *Kinetics Catalysis* **6**, 995 (1965).
(27) Taub, I. A., Dorfman, L. M., *J. Am. Chem. Soc.* **84**, 4053 (1962).
(28) Walsh, R., Benson, S. W., *J. Am. Chem. Soc.* **88**, 3480 (1966).

RECEIVED October 10, 1967.

9

Ionic Catalysis in Chain Oxidation of Alcohols

E. T. DENISOV, V. M. SOLYANIKOV, and A. L. ALEXANDROV

Institute of Chemical Physics, Academy of Sciences, Vorobyevskoye Chaussee 2-b, Moscow, USSR

> H_2O_2 decomposes to free radicals in 2-propanol by the action of H^+. Free radicals are also produced by the reaction between tert-BuOOH and Br^- in 1-propanol. The HCO_3^- ions inhibit the oxidation of cyclohexanol initiated by AIBN, destroying many oxyperoxide radicals—i.e., HCO_3^- is a negative catalyst. Appropriate reaction schemes and rate equations are proposed.

Catalysis by transition metals in liquid-phase oxidation has been thoroughly investigated. The roles of other ions have not been sufficiently studied. This paper is concerned with catalysis by hydrogen ions and some anions, in the chain oxidation of secondary alcohols such as cyclohexanol and 2-propanol. Secondary alcohols, because of their polarity, are convenient for studying ionic homolytic reactions and their role in chain oxidation.

The oxidation of secondary alcohols by oxygen may be represented by the following chain reaction mechanism.

$$\text{Initiator} \longrightarrow r\cdot \xrightarrow{\text{>C}\begin{smallmatrix}\text{OH}\\\text{H}\end{smallmatrix}} \text{>C}\begin{smallmatrix}\text{OH}\\\cdot\end{smallmatrix}$$

$$\text{>C}\begin{smallmatrix}\text{OH}\\\cdot\end{smallmatrix} + O_2 \longrightarrow \text{>C}\begin{smallmatrix}\text{OH}\\\text{OO}\cdot\end{smallmatrix}$$

$$\text{>C}\begin{smallmatrix}\text{OH}\\\text{OO}\cdot\end{smallmatrix} + \text{>C}\begin{smallmatrix}\text{OH}\\\text{H}\end{smallmatrix} \xrightarrow{k_p} \text{>C}\begin{smallmatrix}\text{OH}\\\text{OOH}\end{smallmatrix} + \text{>C}\begin{smallmatrix}\text{OH}\\\cdot\end{smallmatrix}$$

$$\text{>C}\begin{smallmatrix}\text{OH}\\\text{OO}\cdot\end{smallmatrix} + \text{>C}\begin{smallmatrix}\text{OH}\\\text{OO}\end{smallmatrix} \xrightarrow{k_t} \text{>C}\begin{smallmatrix}\text{OH}\\\text{OOH}\end{smallmatrix} + O_2 + \text{>C=O}$$

The rate of the reaction is

$$R = k_p k_t^{-1/2} \left[>C{<}_H^{OH} \right] \sqrt{R_i} \qquad (1)$$

For cyclohexanol, $k_p k_t^{-1/2} = 5.0 \times 10^3 \exp(-10,800/RT)$ l.$^{1/2}$ mole$^{-1/2}$ sec.$^{-1/2}$, $k_p = 1.1 \times 10^7 \exp(-11,900/RT)$ l. mole^{-1} sec.$^{-1}$, $k_t = 5.0 \times 10^6 \exp \times (-2200/RT)$ (1). For 2-propanol, $k_p k_t^{-1/2} = 3.0 \times 10^4 \exp(-12,000/RT)$ l.$^{1/2}$ mole$^{-1/2}$ sec.$^{-1/2}$ (2).

Homolytic Decomposition of Hydrogen Peroxide by Action of Hydrogen Ions

Strong mineral acids such as H_2SO_4, $HClO_4$, HCl, and HNO_3 were found to accelerate the chain oxidation of 2-propanol. Experiments on 2-propanol oxidation with an AIBN initiator showed that acids do not affect the reactions of peroxide radicals since they do not change the value of $k_p k_t^{-1/2}$. Oxidation is accelerated because of the reaction between H_2O_2 and H^+ involving the generation of free radicals.

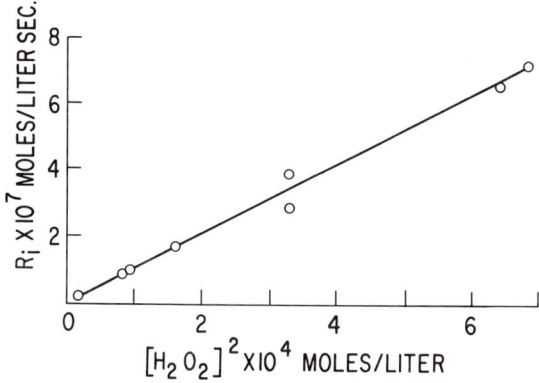

Figure 1. Relation of R_i to square of concentration of H_2O_2 in 2-propanol at 70°C. with 0.0925M H_2SO_4

The 2-propanol was oxidized at 70°C. The rate of oxygen consumption was measured manometrically. The rate of free radical generation, R_i, was calculated from Expression 1 (at 70°C. $k_p k_t^{-1/2} = 1.5 \ 10^{-3}$ l.$^{1/2}$ mole$^{-1/2}$ sec.$^{-1/2}$). In the absence of acid, free radicals were formed in the reaction between H_2O_2 and 2-propanol (2), $k_i = R_i/[H_2O_2] = 5 \times 10^{-8}$ sec.$^{-1}$ (at 70°C.); in the presence of 1.8×10^{-2} mole per liter $HClO_4$, $k_i = 6.3 \times 10^{-6}$ sec.$^{-1}$ (at 70°C.). It was found that $R_i \sim [H_2O_2]^2$ (Figure 1).

Ion concentration in the 2-propanol–H_2SO_4 system was measured by the electroconductivity method and calculated from the expression [ion] = κ/Λ_∞, where κ is the specific electroconductivity, Λ_∞ is the equivalent electroconductivity, and $\Lambda_\infty = 61$ sq. cm. ohm^{-1} mole^{-1} (70°C.).

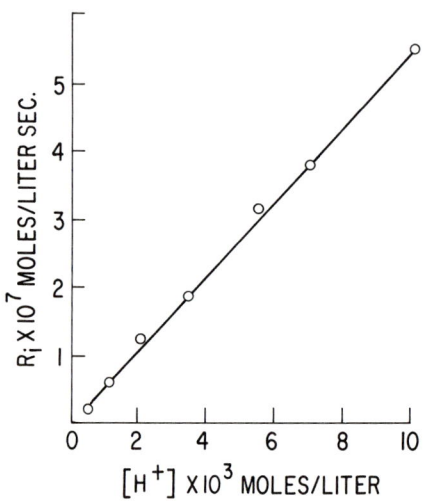

Figure 2. *Relation of* R_i *to concentration of* H^+ *in 2-propanol at 70°C. with 0.0181 mole/liter* H_2O_2. *Estimated by electroconductivity method*

The rate of initiation increases linearly with the concentration of hydrogen ions (Figure 2). It decreases when a base (pyridine) is introduced into the system 2-propanol + hydrogen peroxide + perchloric acid + oxygen. All the above is evidence that the reaction takes place between hydrogen peroxide and hydrogen ions. The rate of initiation is $R_i = k$ $[H_2O_2]^2[H^+]$. For sulfuric acid, $k = 0.14$ sq. liters mole^{-2} sec.$^{-1}$ (70°C.). The energy of activation, E, is 27 ± 2 kcal. per mole.

The total decomposition of hydrogen peroxide by the action of hydrogen ions was measured iodometrically in the absence of oxygen. The rate of hydrogen peroxide decomposition was found to be

$$R = k[H_2O_2][H^+]$$

In the case of perchloric acid $E = 24.4 \pm 2$ kcal. per mole; $k = 1.03 \times 10^{14}$ exp ($-24\ 400/RT$) liters mole^{-1} sec.$^{-1}$

Comparing R and R_i shows that decomposition of hydrogen peroxide into free radicals represents only a small part (from 1 to 5%) of the over-all decomposition. It follows that the greater part of hydrogen peroxide decomposes heterolytically under the action of acids. Two mechanisms may be proposed to explain the experimental results:

$$HA + ROH \rightleftharpoons A^- + ROH_2^+$$
$$K$$
$$ROH_2^+ + H_2O_2 \rightleftharpoons ROH + H_3O_2^+$$

$$I \begin{cases} H_3O_2^+ \xrightarrow{k_1} H_2O + OH^+ \\ OH^+ + {>}C{<}^H_{OH} \xrightarrow{k_2} H_2O + H^+ + {>}C{=}O \\ OH^+ + H_2O_2 \rightarrow HO\cdot + H^+ + HO_2\cdot \end{cases} \quad k_i = \frac{k_1 k_2 K}{k_3 + k_2[ROH]}$$

$$II \begin{cases} H_3O_2^+ + H_2O_2 \xrightarrow{k_4} H_3O^+ + HO\cdot + HO_2\cdot \\ H_3O_2^+ + ROH \xrightarrow{k_5} H_3O^+ + H_2O + {>}C{=}O \end{cases}$$

$$k_i = \frac{k_4 K}{k_4 + k_5[ROH]}$$

If OH^+ is formed in the system, ROOH would be expected to form by the reaction:

$$HO^+ + ROH \rightleftharpoons ROOH + H^+$$

It was found by iodometric analysis that hydroperoxide was not produced in the system $ROH + H_2O_2 + H^+$. Consequently, Mechanism II seems to be more probable.

Similar results were obtained for *tert*-butyl hydroperoxide and perchloric acid in 2-propanol. Thus, it is evident from the decomposition of hydrogen peroxide into free radicals that both heterolytic and homolytic reactions may be catalyzed by hydrogen ions. Further research is needed to investigate proton catalysis in certain homolytic reactions.

Initiation by Reaction between ROOH and Br⁻

Initiation by hydroperoxides may be intensified by anions capable of reducing other compounds. The formation of radicals in the reaction between *tert*-butyl hydroperoxide and tetraethylammonium bromide was studied at 55° to 77°C. in the presence of oxygen with 1-propanol as solvent. The rate of initiation was measured by consumption of α-naphthol that acted as acceptor of free radicals. Naphthol was determined colorimetrically. Two radicals were suggested to react with one molecule of α-naphthol. The rate of initiation was found to be (Figure 3).

$$R_i = k_o[ROOH] + k[ROOH][Et_4NBr]$$

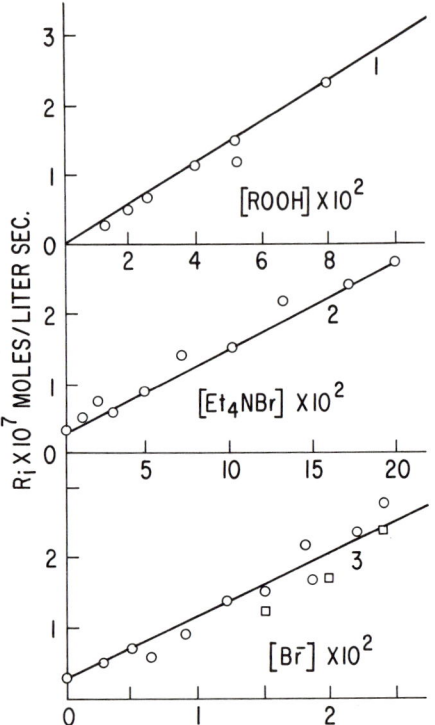

Figure 3. R_i dependence on concentrations at 70°C.
1. tert-BuOOH at $[Et_4NBr] = 0.1M$
2. Et_4NBr at $[ROOH] = 0.056M$
3. Br^- at $[ROOH] = 0.056M$

The degree of Et_4NBr dissociation into ions in 1-propanol was measured by the electroconductivity method. The rate of initiation was found to depend linearly on both $[Et_4NBr]$ and $[Br^-]$.

Experiments with a 1-propanol–water–hydroperoxide system at constant concentration of Et_4NBr were carried out to determine whether Et_4NBr or Br^- reacted with the hydroperoxide. Water increased the degree of Et_4NBr dissociation. The rate of initiation was found to increase linearly with Br^- at a constant concentration of Et_4NBr.

Thus, the reaction seems to take place between hydroperoxide and Br^-. The following mechanism may be proposed:

$$ROOH + Br^- \rightarrow RO\cdot + HO^- + Br\cdot$$

The rate constant is

$$k = 4.0 \times 10^8 \exp(-19500/RT) \text{ liters mole}^{-1} \text{ sec.}^{-1}$$

Negative Catalysis by HCO_3^- Ions in Chain Oxidation of Cyclohexanol

The inhibiting effect of $NaHCO_3$ on chain oxidation was established by studying the effect of ions on the oxidation of cyclohexanol. The latter was oxidized at 75°C. with AIBN as initiator ($R_i = 6.9 \times 10^7$ mole liter^{-1} sec.$^{-1}$). To dissolve $NaHCO_3$, 9% of water was added to cyclohexanol. The rate of oxidation was measured volumetrically.

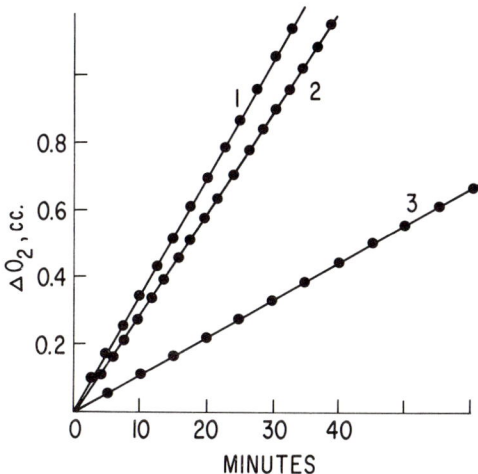

Figure 4. Kinetic curves of oxygen consumption in cyclohexanol oxidation at 75°C. with 0.01M AIBN
$R_i = 6.9 \times 10^{-7}$ mole/liter/second
1. Without inhibitor
2. With 1×10^{-4}M 1-naphthol
3. With 4×10^{-5}M $NaHCO_3$

The constancy of the low oxidation rate in the presence of HCO_3^- is surprising. As seen from Figure 4, the ordinary inhibitor of reaction, α-naphthol in a concentration of 10^{-4} mole per liter, disappears in a few minutes, and the reaction is rapidly accelerated. The HCO_3^- ion (in 4×10^{-5} mole per liter concentration) inhibits the oxidation of cyclohexanol for half an hour, and the rate of reaction does not increase during this period. Thus, the HCO_3^- ion inhibits the chain oxidation not as an ordinary inhibitor but as a negative homogeneous catalyst, and each ion of such a catalyst may terminate many chains.

The peroxy radicals of secondary alcohols are oxyperoxide radicals. Experiments using various systems show the relationship between the structure of peroxide radicals and negative catalysis by HCO_3^- (Table 1).

Table I. Effect of $NaHCO_3$ on Oxidation of Various Systems

75°C.; 0.01 mole per liter AIBN; $[NaHCO_3] = 1.4 \times 10^{-4}$ mole per liter
R, Rate in presence of $NaHCO_3$
R_o, Rate in absence of $NaHCO_3$
I. Cyclohexanol
II. Cyclohexanone

System	Structure of Peroxide Radicals		R_o, 10^{-6} Mole Liter^{-1} Sec.$^{-1}$	R/R_o
91% I 9% H_2O	>C<(OH)(OO·)		8.52	0.21
94.3% II 5.7% H_2O	>C<(H)(OO·)		3.30	0.95
89.3% II 5.7% H_2O 5.0% I	>C<(H)(OO·)	>C<(OH)(OO·)	4.74	0.64
94.3% II 5.7% H_2O 0.15 mole liter^{-1} H_2O_2	>C<(H)(OO·)	>C<(OH)(OO·)	2.92	0.61
74.3% II 5.7% H_2O 20.0% tert-BuOH	>C<(H)(OO·)	>C—O·	3.00	1.00

HCO_3^- ions inhibit chain oxidation only when oxyperoxide radicals appear in the system. The experimental dependence of oxidation rate on HCO_3^- (Figure 5) at 75°C. is

$$R_o^2/R^2 = 1 + a[HCO_3], \quad a = 1.8 \times 10^5 \text{ liters per mole}$$

The results obtained may be interpreted by using the following scheme:

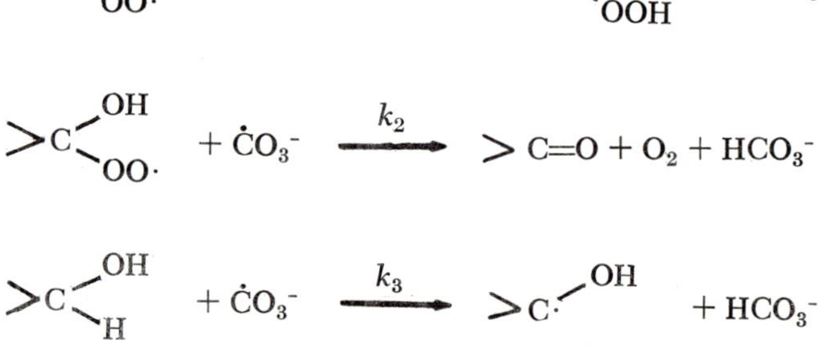

$$>\!C\!<^{OH}_{OO·} + HCO_3^- \xrightarrow{k_1} >\!C\!<^{OH}_{OOH} + \dot{C}O_3^-$$

$$>\!C\!<^{OH}_{OO·} + \dot{C}O_3^- \xrightarrow{k_2} >\!C\!=\!O + O_2 + HCO_3^-$$

$$>\!C\!<^{OH}_{H} + \dot{C}O_3^- \xrightarrow{k_3} >\!C\!\cdot^{OH} + HCO_3^-$$

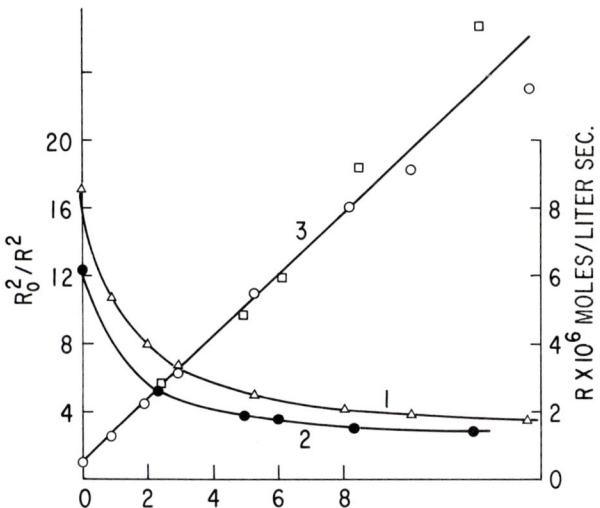

Figure 5. Dependence of rate of cyclohexanol oxidation with AIBN on concentration of HCO_3^-
$R_i = 6.9 \times 10^{-7}$ mole/liter/second
1. At 75°C.
2. At 65°C.
3. Plot of R_o^2/R^2 against HCO_3^-
   □ 65°C.
   ○ 75°C.

By assuming an equality $k_1[HCO_3^-][RO_2] = k_3[>CHOH][\dot{C}O_3^-]$ it is possible to obtain for small HCO_3^- ($k_p[>CHOH] > k_1[HCO_3^-]$)

$$\frac{R_o^2}{R^2} = 1 + \frac{k_1 k_2 [HCO_3^-]}{k_3 k_t [>CHOH]}$$

By comparison with the experimental function, we obtained

$$\frac{k_1 k_2}{k_3 k_t [>CHOH]} = 1.8 \times 10^5 \text{ liters per mole}$$

(at 75°C., 9% H_2O), $[>CHOH] = 8.75$ moles per liter, $k_t = 2.1 \times 10^5$ liters mole^{-1} sec.$^{-1}$. Therefore, $k_1 k_2 / k_3 = 3.3 \times 10^{11}$ liters mole^{-1} sec.$^{-1}$.

Thus, certain anions may be seen to serve as negative catalysts in the chain oxidation of alcohols.

Literature Cited

(1) Alexandrov, A. L., Denisov, E. T., *Izv. Akad. Nauk SSSR, Otd. Khim. Nauk* **1966**, 1737.
(2) Solyanikov, V. M., Denisov, E. T., *Neftekhimia* **3**, 360 (1963); **4**, 458 (1964).

RECEIVED November 16, 1967.

10

Liquid–Phase Oxidation of Acrolein
Action of Catalysts

AKIRA MISONO, TETSUO OSA,[1] and YASUKAZU OHKATSU

Department of Industrial Chemistry, Faculty of Engineering, University of Tokyo, Tokyo, Japan

> *The effects of a metal catalyst, particularly a cobalt acetylacetonate, in oxidizing acrolein were studied. Visible light spectra of the reaction solution showed that part of the ligands of the catalyst were exchanged by a reactant such as acrolein. Furthermore, judging from the spectra and the molecular weight of the recovered catalyst, the catalyst during oxidation existed as a mononuclear complex, which probably coordinated with acrolein through its aldehyde carbonyl. The rates of formation of carbon monoxide and carbon dioxide indicate that the acyl radical derived from the hydrogen abstraction of acrolein retains its coordination with the catalyst, resulting in the desired reaction.*

In oxidizing α,β-unsaturated aldehydes, the selective formation of the corresponding carboxylic acids is difficult because of the high reactivity of vinyl and carbonyl groups conjugating with each other. The liquid-phase oxidation of acrolein to acrylic acid has been reported only in a few patents (5, 11), while that of methacrolein has been described in other literature (5, 9, 29). Brill and Lister (7) investigated the kinetics and products of metal salt–catalyzed oxidation of methacrolein and reported that methacrolein was converted into peroxide, methacrylic acid, and soluble polymers. Brill and Barone (6) claimed that methacrolein or acrolein was oxidized with molecular oxygen in high yield to the corresponding unsaturated acid in the presence of a small quantity of iodine at a low temperature. Recently, Farberov and Kosheli (12, 21) investigated the kinetics and mechanism of the liquid-phase oxidation of methacrolein, using a silver catalyst prepared from silver nitrate. They

---

[1] Present address: Case Institute of Technology, Cleveland, Ohio.

stated that neither the yield nor the reaction rate was affected by adding inhibitors such as salts of phenols and polyphenols. However, the above literature refers mainly to the sclective formation of α,β-unsaturated carboxylic acids and does not emphasize the kinetics and mechanisms of catalyst action in the liquid-phase oxidation of the corresponding aldehydes.

The liquid-phase oxidation of acrolein (AL), the reaction products, their routes of formation, reaction in the absence or presence of catalysts such as acetylacetonates (acac) and naphthenates (nap) of transition metals and the influence of reaction factors were discussed in an earlier paper (22). The coordinating state of cobalt acetylacetonate in the earlier stage of the reaction depends on the method of addition to the reaction system (25, 26). The catalyst, $Co(acac)_2$-H_2O-acrolein, which was synthesized by mixing a solution of $Co(acac)_2$ in benzene with a saturated aqueous solution, decreases the induction period of oxygen uptake and increases the rate of oxygen absorption. The acrolein of the catalyst coordinated with its cobalt through the lone pair of electrons of the aldehyde oxygen. Therefore, it is believed that the coordination of acrolein with a catalyst is necessary to initiate the oxidation reaction (10).

The interaction of acrolein with a catalyst was investigated using a DPPH method (27), in which an acyl radical formed by the initiation was trapped by DPPH, and the rate of consumption of DPPH, corresponding to the rate of formation of the radical, was determined by visible light spectra. The rate equation for acyl radical formation is

$$\frac{d(CH_2=CH\dot{C}O)}{dt} = k(CO^{3+})(CH_2=CH\dot{C}OH)$$

The effects of solvents were also reported (24). Lower aliphatic carboxylic acids, such as butyric and valeric, gave the best results with respect to the selective oxidation of acrolein. For butyric acid, the conversion of acrolein and the selectivity of acrylic acid were 45.1 and 86.0%, respectively, with 5×10^{-4} mole of $Co(acac)_3$ per liter of solution for 4 hours at 35°C.

Experimental

Commercial acrolein (Shell Chemical Corp.) was distilled and shaken with an equal volume of anhydrous calcium sulfate for 30 minutes. The acrolein (containing an impurity of 3.5% of propionaldehyde) was distilled again just before use (b.p., 52.5–52.9°C.). Oxidation products identified using acrolein (99.9% purity) without propionaldehyde, which was removed by the Tischenko reaction (31). Solvents were used after purification (especially dehydration) by conventional methods.

Peracrylic Acid. To a mixture of 1.0 mole of acrylic acid and 1.0 mole of benzene, 1.2 moles of 80% hydrogen peroxide were added in the presence of 1.0 weight % sulfuric acid. The solution was stirred for 30 hours at room temperature, and then the benzene phase was separated and dried over anhydrous sodium sulfate to give 0.45 mole peracrylic acid per liter of benzene solution. The resulting peracrylic acid was used in its decomposition.

Cobalt(III) acetylacetonate [Co(acac)$_3$] (*14*), manganese(III) acetylacetonate [Mn(acac)$_3$] (*15*), iron(III) acetylacetonate [Fe(acac)$_3$] (*30*), chromium(III) acetylacetonate [Cr(acac)$_3$] (*13*), nickel(II) acetylacetonate [Ni(acac)$_2$] (*8*), and copper(II) acetylacetonate [Cu(acac)$_2$] (*18*) were prepared and purified. Cobalt, manganese, iron, chromium, nickel, and copper naphthenates were all commercially available.

Oxidation of Acrolein. The system consisted of a 150-ml. gas-sealed reactor, with a gas inlet, a gas outlet, a magnetic stirrer, a water-cooled condenser, and a dry ice–methanol-cooled condenser, a gas recycling pump, a 10-liter gas holder, and a 100-ml. gas buret. After purging the system with oxygen, 10 ml. of freshly distilled acrolein and 40 ml. of solvent were charged into the reactor, and a known quantity of a catalyst was added. Next, oxygen was bubbled through the reaction solution at a fixed temperature. Unreacted oxygen and other volatile products passed through the water-cooled condenser, where the solvent was condensed and returned to the reactor. Then the gas mixture was passed through the second condenser, where acrolein was condensed, into the gas holder, and was again recycled with fresh oxygen by the pump into the reactor. The amount of oxygen consumed was occasionally measured, and the same amount of new oxygen was charged into the system to keep the inner pressure constant. The amount of oxygen absorbed was obtained by subtracting the amount of generated gases. Some samples of the reaction solution were taken and analyzed. Reaction conditions are described below.

Peracrylic acid was decomposed catalytically with a gas-tight apparatus under a nitrogen atmosphere (*28*). The apparatus was purged with nitrogen, and charged with peracrylic acid and a solvent and then with a catalyst. The reaction solution was mixed by a magnetic stirrer, and some samples of the reaction solution were taken and analyzed.

The oxidation products of acrolein were analyzed mainly by gas chromatography (*22*). Peracrylic acid and peroxides were determined by the method of Greenspan and McKeller (*16*). The polymeric products of the oxidation were measured as follows. At the end of the reaction, the reactor was cooled rapidly and hydroquinone was added. Then insoluble polymers containing oligomers were collected on a filter, dried, and weighed.

The catalyst (*22*) in the earlier stage of the oxidation of acrolein was recovered by cooling the reaction solution below 5°C. and removing the volatile components under a high vacuum. This residue was dissolved in benzene and filtered. The volatile components of the filtrate were again removed under a vacuum below 5°C. This run was repeated once more, and the substantially purified catalyst which contained no solvent, peracrylic acid, decomposition products, or polymers was recovered. Its molecular weight was determined using a vapor pressure osmometer Model 301 A (Merchrolab. Co., Ltd.). The catalyst after the decomposition of peracrylic acid was recovered by the method mentioned here.

A Hitachi Model EPI-S2 infrared spectrometer was used to obtain the spectra of the catalysts in the range 600–4000 cm.$^{-1}$

Visible light spectra were obtained using a Hitachi Model EPI-2 spectrometer. The absorption spectra of a reaction mixture formed, when cobaltous or cobaltic acetylacetonate was used as a catalyst, were measured to detect the state of the catalyst quantitatively. The spectrum of the reaction solution had a maximum absorption at 500 to 600mμ and a strong absorption in the ultraviolet region. The concentration of the catalyst in the oxidation solution was suitable for measuring the absorption spectra. The variation of the spectra with time was measured during the oxidation.

Nuclear magnetic resonance spectra of the recovered catalysts were observed at room temperature on a high resolution spectrometer (Nihondenshi JEOL 3H-60) at a frequency of 60 Mc. Chemical shifts were measured with reference to TMS.

Results and Discussion

The liquid-phase oxidation of acrolein is estimated to proceed through the free radical mechanism since adding hydroquinone stopped oxygen absorption. The primary molecular product is peracrylic acid. This reacts with acrolein to form the peroxide complex (not isolated),

$$CH_2=CH-\underset{\underset{H}{|}}{\overset{\overset{OOH}{|}}{C}}-O-\overset{\overset{O}{\|}}{C}-CH=CH_2$$

which then decomposes into two molecules of acrylic acid. The main path to acrylic acid is, *via* the complex, which is similar to the oxidation of acetaldehyde (*4, 19*). The analytical result of the oxidation in benzene with Co(acac)$_3$ as a catalyst at 40°C. is shown in Figure 1. In the earlier stage of the reaction, the rate of oxygen absorption was high, acrolein was consumed smoothly, acrylic acid and polymers were formed, and carbon monoxide evolved. As time passed, acetic acid, which was formed by the oxidation of acetaldehyde as a by-product, and carbon dioxide were produced. In the middle stage of the reaction, acrylic acid was produced in satisfactory yields, and side reactions were considerably suppressed. In the last period of the reaction, the side reactions were continuously suppressed by the amount of polymers composed of acrolein and acrylic acid increased abruptly. Therefore, the yield of acrylic acid increased only slightly.

The oxidation proceeded smoothly when catalyzed by a metallic ion, with the exception of chromium(III) acetylacetonate, which is entirely inactive in promoting the reaction. This may be attributed to

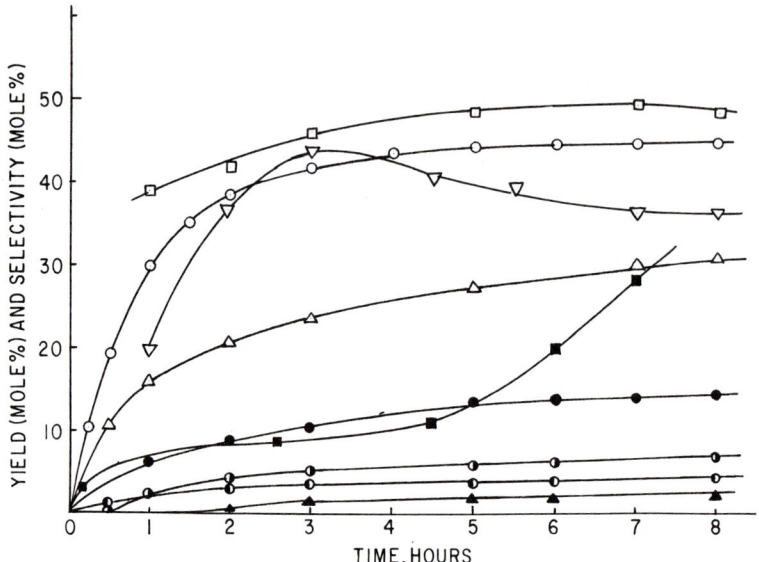

Figure 1. Oxidation products of acrolein in benzene solution at 35°C.

Acrolein, 3.0 moles per liter
$Co(acac)_3$, 5×10^{-1} mole per liter
Upper Curves:
○ Oxygen consumed × 2
▽ Peracrylic acid × 10
△ Acrolein consumed
■ Polymers × 2/5
● Acrylic acid
◐ Carbon dioxide
◑ Carbon monoxide
▲ Acetic acid

the stability of the complex. However, considerable differences, depending on the kind of catalyst, were observed, especially in the induction period of formation of acrylic acid and the maximum yield of acrylic acid. The induction periods, which were obtained with 3×10^{-3} concentration of a metal catalyst and $3M$ concentration of acrolein in benzene at 40°C. were as a whole longer for naphthenate catalyst than for the acetylacetonate of the same metal, except for the chromium catalyst. The orders of induction periods were:

Naphthenates. Nil(2.5 hr.) > Ni(1.2 hr.) > Cu(0.9 hr.) ≈ Fe (0.9 hr.) > Mn(0.7 hr.) ≈ Cr(0.6 hr.) > Co(0.2 hr.).

Acetylacetonates. Cr(∞) > Nil(2.5 hr.) > Ni(1.0 hr.) > Cu(0.9 hr.) ≈ Fe(0.8 hr.) >> Mn(0.3 hr.) > Co(0.1 hr.).

The ranking orders of the catalysts correlated well with the redox potential of a metallic ion in aqueous medium (22).

The liquid-phase oxidation of aldehydes with molecular oxygen may be generally explained by the following kinetic scheme (3).

$$M^{n+} + RCHO \xrightarrow{k_1} M^{(n-1)+} + R\dot{C}O + H^+ \qquad (1)$$

$$R\dot{C}O + O_2 \xrightarrow{k_2} RCOO_2\cdot \qquad (2)$$

$$RCOO_2\cdot + RCHO \xrightarrow{k_3} RCOO_2H + R\dot{C}O \qquad (3)$$

$$\left. \begin{array}{l} 2R\dot{C}O \xrightarrow{k_4} \\ R\dot{C}O + RCOO_2\cdot \xrightarrow{k_5} \\ 2RCOO_2\cdot \xrightarrow{k_6} \end{array} \right\} \text{inert compounds} \qquad \begin{array}{l} (4) \\ (5) \\ (6) \end{array}$$

Accepting the steady-state treatment of the above steps, for sufficiently high oxygen pressure, the rate of oxygen absorption may be represented as:

$$-dO_2/dt = (k_3/k_6^{1/2})Ri^{1/2}(RCHO)^{1.0}$$
$$= (k_3/k_6^{1/2})k_1^{1/2}(M^{n+})^{1/2}(RCHO)^{1.5} \qquad (I)$$

For low oxygen pressure, the rate of oxygen absorption is:

$$-dO_2/dt = (k_2/(2k_4)^{1/2})Ri^{1/2}(O_2)^{1.0}$$
$$= k_2(k_1/2k_4)^{1/2}(M^{n+})^{1/2}(RCHO)^{1/2}(O_2)^{1.0} \qquad (II)$$

The orders of each factor with respect to the liquid-phase oxidation of acrolein were determined under various experimental conditions

Table I. Orders of Rate Equations

Catalyst	$Cu(acac)_2$	$Co(nap)_n$	$Co(acac)_3$	$Co(acac)_3$
Solvent	Benzene	Benzene	Benzene	Butyric acid
Catalyst order	0.4	0.5	0.47	0.26
Concn. range	1.05×10^{-4} $\sim 1.0 \times 10^{-2}$ mole/mole AL[a]	4.5×10^{-5} $\sim 1.0 \times 10^{-3}$ mole/mole AL	3×10^{-5} $\sim 1 \times 10^{-2}$ mole/liter	4×10^{-5} $\sim 6 \times 10^{-3}$ mole/liter
Acrolein order	—	—	1.22	0.82
Concn. range, moles/liter			$3 \sim 10$	$0.5 \sim 10$
Oxygen order	—	1.04	1.13	0.95
Pressure range, mm. Hg	—	7.6×10^2 $\sim 3.32 \times 10^3$	2×10^2 $\sim 6 \times 10^2$	2×10^2 $\sim 7 \times 10^2$

[a] Acrolein.

(Table I), and the rate of oxygen absorption was approximately:

$$-dO_2/dt = k(CH_2\!\!=\!\!CHCHO)^{0.82\sim1.22}(M^{n+})^{0.26\sim0.5}(O_2)^{0.95\sim1.13}$$

Though it is usually reported (3) that the rate of oxidation of saturated aldehydes does not depend on the oxygen pressure if it is sufficiently high, Brill and Lister in kinetic study of the oxidation of methacrolein (7) reported that it depends on the first power of the oxygen pressure, and concluded that Steps 2 and 4 should be the rate-determining Steps.

Not only may the proposals of Brill and Lister be accepted, but part of the effect of oxygen may be ascribed to the abstraction of aldehyde hydrogen by the metal catalyst—i.e., the rate of consumption of acrolein in the initiation reaction can be represented (27) as:

$$-\frac{d(CH_2\!\!=\!\!CHCHO)}{dt} = K(CH_2\!\!=\!\!CHCHO)^{1.0}(M^{n+})^{1.0}(O_2)^{0.5}$$

Besides, taking into consideration what we have reported—i.e., that elementary Step 1 may follow a course of coordination such as (25):

$$\diagdown\!\!\!\!\diagup\!\!\!\mathrm{C}\!\!=\!\!\mathrm{O} \rightarrow \mathrm{Co}$$

a part of the abstraction of aldehyde hydrogen may be accelerated by the presence of oxygen.

$$\mathrm{Co} \leftarrow \mathrm{O}\!\!=\!\!\mathrm{C}\diagdown\!\!\!\!\diagup\!\!\!\!\times\!\!\!\diagdown\mathrm{H}\cdots\mathrm{O_2}$$

However, this could not fully explain the oxygen dependence of the rate equation.

To clarify the catalyst action, the change of catalyst species during the oxidation reaction was followed by the visible light spectrometer (Figure 2). $Co(acac)_3$ and $Co(acac)_2$ have absorptions at 600 and 495mμ, respectively. However, the characteristic bands of all the cobalt acetylacetonate catalysts, including $Co(acac)_2 \cdot 2H_2O$, in the reaction solution occurred at 600 mμ after oxygen absorption. Furthermore, the magnetic susceptibility showed that two divalent cobalt acetylacetonates ($\beta = 4.0$ to 5.0) changed to trivalent six-coordination forms ($\beta = 0.48$) like $Co(acac)_3$. This may indicate that all the catalysts are in substantially the same form during the oxidation reaction.

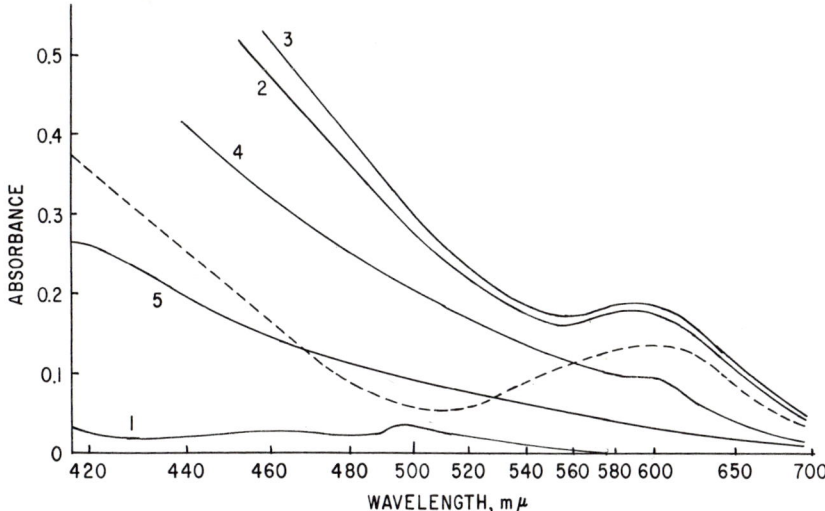

Figure 2. Visible light spectra of cobalt acetylacetonate catalysts in benzene solution

Co(acac)₂ after reaction times (minutes):
1. 0
2. 2
3. 20
4. 60
5. 240

- - - - Co(acac)₃

Catalyst concentration 1×10^{-3} mole per liter

The relations of molar extinction coefficients and oxygen absorption rates, plotted vs. time, are illustrated in Figures 3 and 4 (solvents, benzene and butyric acid, respectively). The variation of coefficients was parallel to the rate of oxygen absorption—i.e., the larger the coefficient, the higher the oxygen absorption rate. It is considered therefore that the catalyst at its maximum absorption coefficient is in a desirable form for oxidizing acrolein.

The mechanism by which the d-d transitions gain intensity still remains to be determined. For octahedral and square planar complexes which have a center of symmetry, the transitions are partly inhibited. Slightly distorted octahedral and square planar metal complexes may have a fractional part of a d-d transition allowed, and this static distortion mechanism may be responsible for some intensity in many cases (1, 2). The fact that when Co(acac)₃ catalyst was utilized in the oxidation, its absorption wavelength remained unchanged and its coefficient increased,

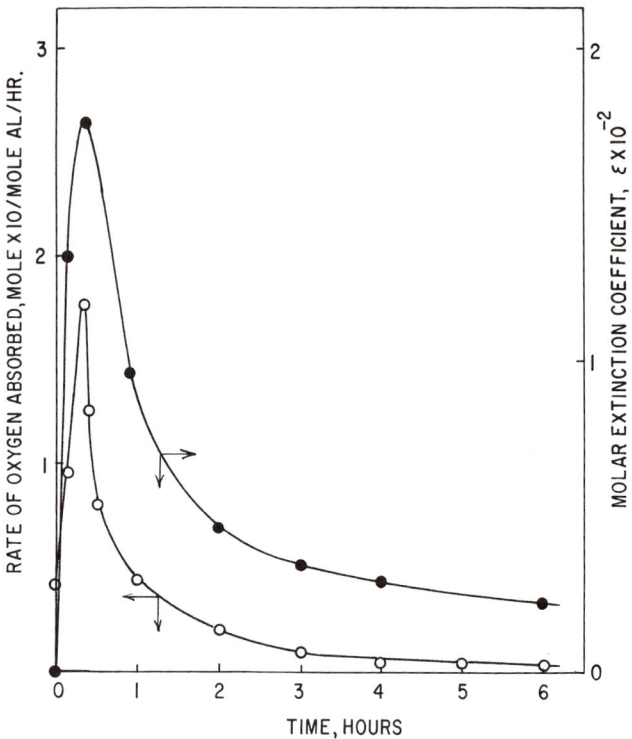

Figure 3. Rate of oxygen absorbed (mole × 10/mole AL/hr) and molar extinction coefficient (ϵ) at 600 mμ during oxidation of acrolein in benzene solution at 35°C.

Acrolein, 3.0 moles per liter
$Co(acac)_3$, 5.0×10^{-4} mole per liter

may be explained as follows. The former is ascribed to the unchangeability of the catalyst configuration or the constancy of the difference (Δ) between mutual energy levels of d-d transition even if a ligand exchange occurs. However, the unchangeability of the configuration may be denied if we take into account the fact that the other cobalt acetylacetonate catalysts, such as $Co(acac)_2 \cdot 2H_2O$ and $Co(acac)_2$, changed their absorption bands to 600 mμ during the oxidation. Thus, a substance —e.g., acrolein—similar to acetylacetone is estimated to be coordinated with the catalyst through its carbonyl group like acetylacetone, and Δ to become constant. This explains the increase of the molar extinction coefficient because the acrolein-coordinated catalyst will have a more distorted octahedral configuration. Therefore, the initiation reaction through Step 1 may also occur during the reaction.

The acrolein coordinated with catalyst then gives an acyl radical by abstracting its aldehyde hydrogen. In the general oxidation of aldehydes, the acyl radical is considered to be discontinuing coordination with catalyst, as was described by Bawn (3) and Hoare and Waters (17). However, in the acrolein oxidation, the acyl radical formed by hydrogen abstraction may not conform to this proposal, as described below.

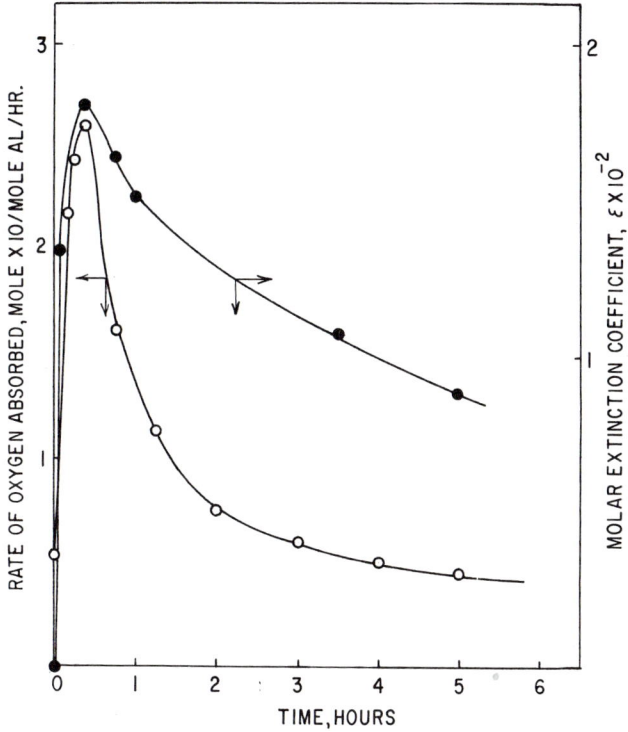

Figure 4. Rate of oxygen absorbed (mole $\times$ 10/mole AL/hr) and molar extinction coefficient (ϵ) at 600 mμ during oxidation of acrolein in benzene solution at 35°C.

Acrolein, 3.0 moles per liter
$Co(acac)_2$, 5.0 $\times$ 10^{-4} mole per liter

The formation of carbon monoxide and carbon dioxide during the oxidation of acrolein is shown in Figure 5. The facts that the decomposition of peracrylic acid produced not carbon monoxide but carbon dioxide (28) and that in the initial state of the oxidation of acrolein only carbon monoxide is formed, indicate that the evolution of carbon monoxide during oxidation may be ascribed to the decomposition of the acyl radical.

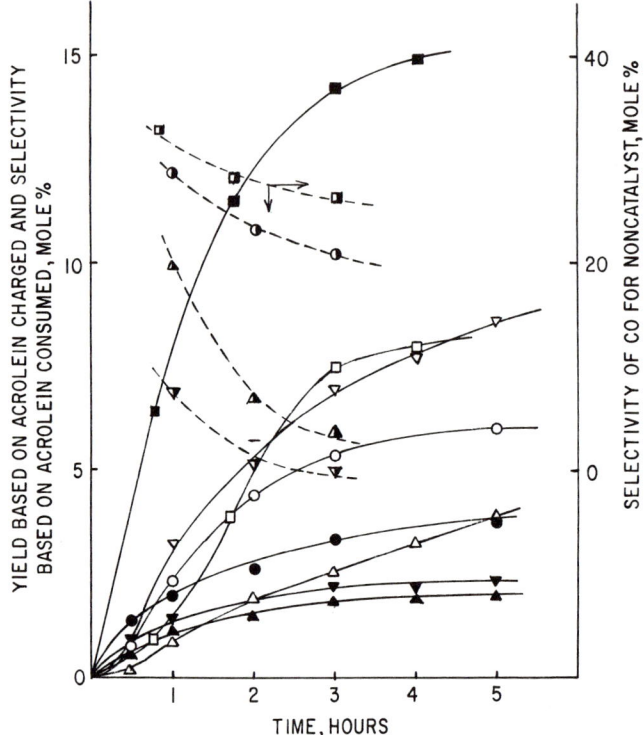

Figure 5. Formation of carbon monoxide and carbon dioxide and selectivity of carbon monoxide

Acrolein, 10 ml.
Benzene, 40 ml.
    □ *Noncatalyst, 35°C.*
    ○ *Co(acac)$_3$, 5 × 10^{-5} mole per mole AL, 35°C.*
    ▽ *Co(nap), 1 × 10^{-3} mole per mole AL, 40°C.*
    △ *Co(nap), 4.5 × 10^{-5} mole per mole AL, 40°C.*
Open symbols: CO_2
    filled: CO
  half-filled: selectivity of CO

On the other hand, carbon dioxide may be derived from the decomposition of the intermediates formed by the reactions of the acyl radical and oxygen, such as the acid radical, peroxy radical, or peracid-aldehyde complex. Therefore, they may be derived as follows:

$$\begin{array}{c}
CH_2{=}CHCHO \rightarrow CH_2{=}CH\dot{C}O \xrightarrow{O_2} CH_2{=}CHCOO_2\cdot \\
\quad\quad\quad\quad\quad\quad\quad\quad\quad\downarrow \quad\quad\quad\quad\quad\quad\quad\downarrow \\
\quad\quad\quad\quad\quad\quad\quad\quad CO \quad\quad CH_2{=}CHCOO_2H \rightarrow CH_2{=}CHCO_2\cdot \\
\quad\quad\quad\quad\quad\quad\quad\quad\quad\quad\quad\quad\quad\quad\downarrow \\
\quad\quad\quad\quad\quad\quad\quad\quad\quad\quad\quad\quad CH_2{=}CHCHO \\
\quad\quad\quad\quad\quad\quad\quad\quad\quad\quad\quad\quad\downarrow \\
\quad\quad\quad\quad\quad\quad\quad\quad\quad\quad\quad\quad Complex
\end{array} \Bigg\} \rightarrow CO_2$$

The rate of carbon monoxide formation to carbon dioxide in the case of the four reaction conditions is calculated from the slopes of the curves for the formation of carbon monoxide and carbon dioxide in Figure 5. The rate is approximately in inverse proportion to the reaction time.

$$dCO/dCO_2 = \alpha/t$$

The inverse constants, α, are shown in Table II.

The constant α is proportional to the selectivity of carbon monoxide. Therefore, it is regarded as having some relation to the reactivity of an acyl radical with oxygen since higher selectivity of carbon monoxide may signify that an acyl radical produces more carbon monoxide rather than reacting with oxygen. If the acyl radical exists alone without any bond with the catalyst, its reactivity with oxygen—namely, α—should remain constant, independent of the presence of the catalyst since the steps after the formation of the acyl radical must coincide in all cases. However, α decreases considerably in the presence of a catalyst. Therefore, the acyl radical during the oxidation is believed to combine chemically with the catalyst, resulting in its easier reaction with oxygen. This is supported by the fact that polymerization occurred in the oxidation of acrolein and that the polymers formed before the evolution of carbon dioxide precipitated the catalyst, which could not be separated from them by any treatment; the acyl radical's coordinating with the catalyst started the polymerization, resulting in co-precipitation of the catalyst.

It is concluded that the catalyst has in addition to acetylacetonates a reactant or its radical as one of its ligands, probably coordinated with the catalyst through its carbonyl group, enabling the oxidation reaction to proceed with the formation of acrylic acid by suppressing the side reactions as follows:

$$Co^{3+}L_6 + CH_2=CHCHO \rightarrow L_5Co^{3+}\!-\!O\!=\!\overset{H}{\underset{|}{C}}CH=CH_2$$

$$L_5Co^{3+}\!-\!O\!=\!\overset{H}{\underset{|}{C}}CH=CH_2 \rightarrow L_5Co^{2+}\!-\!O\!=\!\overset{\cdot}{C}CH=CH_2 + H^+$$

$$L_5Co^{2+}\!-\!O\!=\!\overset{\cdot}{C}CH=CH_2 + O_2 \rightarrow L_5Co^{2+}\!-\!O\!=\!\overset{\overset{O_2\cdot}{|}}{C}\!-\!CH=CH_2$$

In the metal salt–catalyzed oxidation of methacrolein, Brill and Lister proposed the following initiation steps (7):

$$RCOO_2H + M^{n+} \rightarrow M^{(n+1)} + RCO_2\cdot + OH^-, \text{ and} \tag{1'}$$

$$RCOO_2H + M^{(n+1)} \rightarrow M^{n+} + RCO_3\cdot + H^+ \tag{1''}$$

Table II. Inverse Constants

Catalyst	Catalyst Concn., Mole/Mole AL	Inverse Constant(α), Liters/Hr.	Selectivity of CO,[a] %
None	—	2.5	32.5
Co(acac)$_3$	5×10^{-4}	0.69	12.4
Co(nap)	4.5×10^{-5}	0.51	10.1
Co(nap)	1×10^{-3}	0.25	6.8

[a] Values 1 hour after start of reaction

$$\text{Selectivity of CO} = \frac{\text{CO formed (mole)}}{\text{AL consumed (mole)}} \times 100$$

Our investigation (27, 28) of the decomposition of peracrylic acid by various homogeneous metal catalysts helped to verify the above steps as the initiation reaction. But Step 1″ plays no fundamental role since it occurs very slowly. The metal catalysts also accelerate the decomposition of the peracid-aldehyde complex which forms the radicals of chain branching. Step 1, and the steps just mentioned, become comprehensible in view of the initiation of the catalyzed oxidation as described above.

The polymers, formed in the decomposition of peracrylic acid, precipitated and inactivated the catalyst, as in the polymerization by the acyl radical's coordinating with the catalyst. According to the literature (20), this polymerization can be considered a coordination radical polymerization; it was ascertained to be just that (23). In other words, acyl and peroxy radicals, coordinating with the catalyst, started polymerization and then precipitated and inactivated the catalyst.

However, this phenomenon varied when different solvents were used. For example, investigating the decomposition reaction of peracrylic acid with respect to aspects of catalysts and polymers indicates that the solvents can be classified into three types (Table III). Solvents, such as

Table III. Solution Color and Catalyst State during Decomposition of Peracrylic Acid Using Co(acac)$_3$

Type	Solvent	Earlier Stage	Middle Stage	Later Stage
1	Dimethyl formamide Dimethyl sulfoxide	Green, dissolved	Green, dissolved	Green, dissolved
2	Benzene Carbon tetrachloride	Green, dissolved	Pale green [a]	— [b]
3	Acetic acid Butyric acid Valeric acid	Green, dissolved	Greenish-yellow [c]	Green-yellow [b]

[a] Almost all the catalyst precipitated with polymers but remained green.
[b] Precipitated catalyst changed color to brown.
[c] Part of the catalyst precipitated with polymers.

dimethyl formamide and dimethyl sulfoxide, in which the catalyst remained dissolved throughout the reaction without the visible formation of polymers, belong to Type 1. In solvents of Type 2, such as benzene and carbon tetrachloride, the catalyst precipitated with polymers immediately after the earlier stage of the decomposition. Solvents of Type 3 are intermediate between solvents of Types 1 and 2.

The catalysts having the initial form of $Co(acac)_3$ were recovered after the decomposition of peracrylic acid in benzene and dimethyl sulfoxide solution; the catalyst recovered from benzene (catalyst B) could hardly be separated from the polymers, indicating a chemical bond between the catalyst and polymers. On the other hand, the catalyst from

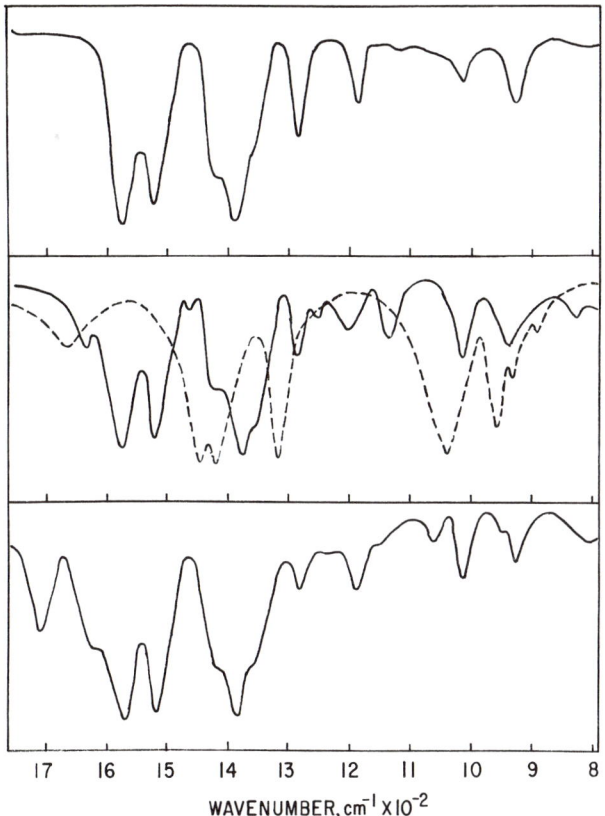

Figure 6. Infrared spectra of recovered catalysts

Upper curve, $Co(acac)_3$
Middle curves:
—— Catalyst D
--- DMSO
Lower curve, Catalyst B

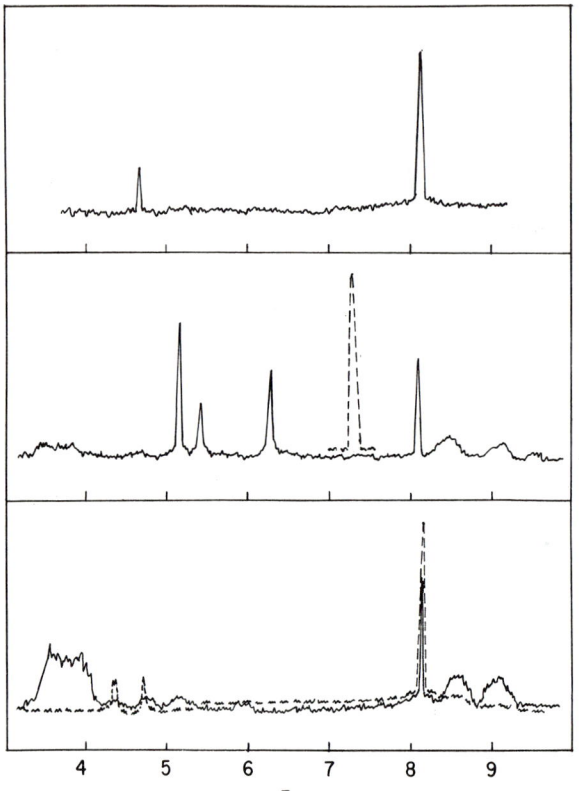

Figure 7. NMR spectra of recovered catalysts

Upper curve, Co(acac)₃
Middle curves:
——— Catalyst D
– – – DMSO
Lower curves:
——— Catalyst B
– – – Catalyst B*

dimethyl sulfoxide (catalyst D) was recovered easily by washing the polymers with benzene. The catalyst separated from the decomposition solution of peracetic acid in benzene is referred to as catalyst B*. Infrared and NMR spectra of the above catalysts are shown, with those of Co(acac)₃, in Figures 6 and 7.

As shown by NMR spectra (Figure 7), the signals of catalyst B differed from those of Co(acac)₃, and even those of catalyst B*. Though catalysts B and B* were recovered from decomposition solutions of similar peracids, the signals of catalyst B different from catalyst B*, which resembled Co(acac)₃, would probably be assigned to acrylic or peracrylic acid coordinated with the catalyst. On the other hand, catalyst D had no new absorption bands caused by acrylic acid or peracrylic acid in the infrared region but showed signals at $\tau = 6.29, 5.36$, and 5.17 different from those of $Co(acac)_3$ and catalyst B. These signals may be ascribed to the solvent—dimethyl sulfoxide—which is coordinated with the catalyst since catalyst D was recovered from the reaction solution when dimethyl sulfoxide was used as a solvent in place of benzene and may correspond to infrared absorptions at 1460, 1255, and 1135 cm.$^{-1}$. In these interpretations, more detailed discussion may be needed, but the shifts of infrared absorptions and NMR signals can be explained only after electron transfers and characters of bonds between the catalyst and ligands are fully clarified. Regrettably we have no such information. However, the above findings and interpretations can qualitatively explain the mechanism of the catalyst action in the decomposition of peracrylic acid (Table IV).

Table IV. Possible Actions of Catalyst in Decomposition of Peracrylic Acid without Acrolein

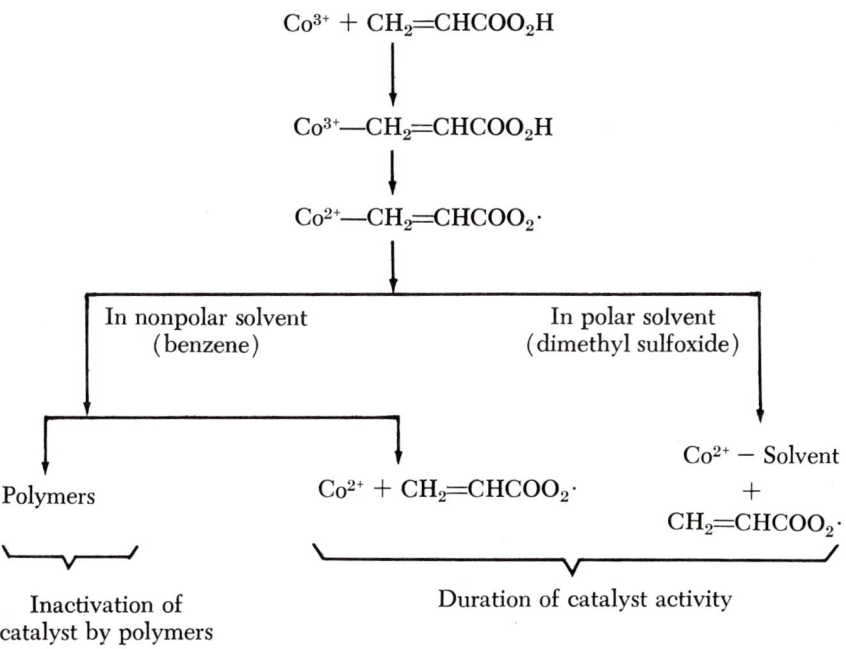

Conclusions

In the liquid-phase oxidation of acrolein, the metal ion with higher valence coordinates acrolein to produce an acyl radical by hydrogen abstraction.

The acyl radical formed from acrolein, maintaining its coordination with a catalyst, may react preferably with oxygen, rather than decompose to produce carbon monoxide, though it is generally believed that a free acyl radical is formed after the abstraction of aldehyde hydrogen by a metal. In such a case, the catalyst metal is considered as behaving as a mononuclear, not a binuclear complex. The molecular weight of the catalyst recovered from the oxidation solution was measured (Table V).

Table V. Molecular Weight of Soluble Catalysts

Formula of Original Catalyst	Molecular Weight	
	Original Catalyst	Soluble Catalyst
$Co(acac)_3$	356	402
$Co(acac)_2$	257	389
$Co(acac)_2 \cdot 2H_2O$	293	394

As Table V shows, all the reacted catalysts have molecular weights larger than those of the original catalysts. This may be ascribed to the fact that the catalyst recovered contains acrolein, acrylic acid, and/or their oligomer(s).

When the radicals, coordinated with a catalyst, produce polymers, the catalyst is precipitated and inactivated by forming an inseparable bond with polymers, especially in a nonpolar solvent such as benzene. However, a polar solvent such as butyric acid or dimethyl sulfoxide is believed to expel the coordinated peroxy radical from the metal, thereby keeping it active during the reaction.

Literature Cited

(1) Ballhausen, C. J., Liehr, A. D., *Mol. Phys.* **2**, 123 (1959).
(2) Ballhausen, C. J., Liehr, A. D., *J. Mol. Spectrosc.* **2**, 342 (1958).
(3) Bawn, C. E. H., Hobin, T., Raphael, L., *Proc. Roy. Soc. (London)* **237A**, 313 (1956).
(4) Bawn, C. E. H., Williamson, J. B., *Trans. Faraday Soc.* **47**, 721 (1951).
(5) Bauer, W., Weisert, P., U. S. Patent **1,911,219** (1933).
(6) Brill, W. F., Barone, F. J., U. S. Patent **3,253,025** (1966).
(7) Brill, W. F., Lister, F., *J. Org. Chem.* **26**, 565 (1961).
(8) Charles, R. G., *J. Phys. Chem.* **62**, 440 (1958).
(9) Church, J. M., Lynn, L., *Ind. Eng. Chem.* **42**, 768 (1950).
(10) Cooper, T. A., Waters, W. A., *J. Chem. Soc.* **1964**, 1538.
(11) Dietrich, W., Ritzenthaler, B., German Patent **869,950** (1953).
(12) Farberov, M. I., Kocheli, G. N., *Kinetika i Kataliz* **6**, 666 (1965).
(13) Fernelius, W. C., Blanch, J. E. *Inorg. Syn.* **5**, 130 (1957).
(14) Fernelius, W. C., Bryant, B. E., *Inorg. Syn.* **5**, 188 (1957).
(15) Fyfe, W. S., *Anal. Chem.* **23**, 174 (1951).
(16) Greenspan, E. P., McKeller, D. G., *Anal. Chem.* **20**, 1061 (1948).
(17) Hoare, D. G., Waters, W. A., *J. Chem. Soc.* **1964**, 2552, 2560.

(18) Jones, M. M., *J. Am. Chem. Soc.* **81**, 3188 (1958).
(19) Kagan, M. Ya., Lyubarskiĭ, G. D., *Zh. Fiz. Khim.* **6**, 536 (1935).
(20) Kenedy, J. P., Langer, A. W., *Fortschr. Hochpolymer. Forsch.* **3**, 305 (1964).
(21) Kosheli, G. N., Farberov, M. I., *Zh. Prikl. Khim.* **39**, 2101 (1966).
(22) Misono, A., Osa, T., Ohkatsu, Y., Takeda, M., *Kogyo Kagaku Zasshi* **69**, 2129 (1966).
(23) Ohkatsu, Y., Ph.D. thesis, University of Tokyo, p. 119 (1967).
(24) Ohkatsu, Y., Hara, T., Osa, T., Misono, A., *Bull. Chem. Soc. Japan* **40**, 1413 (1967).
(25) Ohkatsu, Y., Osa, T., Misono, A., *Bull. Chem. Soc. Japan* **40**, 2111 (1967).
(26) *Ibid.*, **40**, 2116 (1967).
(27) *Ibid.*, in press.
(28) Ohkatsu, Y., Takeda, M., Hara, T., Osa, T., Misono, A., *Bull. Chem. Soc. Japan* **40**, 1893 (1967).
(29) Shingu, H., *et al.*, "Preprints," p. 305, 13th Annual Meeting, Chemical Society of Japan, 1960.
(30) Steinbach, J. F., *J. Am. Chem. Soc.* **80**, 1839 (1958).
(31) Vinch, H. De V., U. S. Patent **2,991,233** (1961).

RECEIVED October 17, 1967.

11

Autoxidation of Chloroprene

H. C. BAILEY

Research and Development Division, BP Chemicals (U.K.), Ltd., Great Burgh, Epsom, Surrey, England

> *The rate of autoxidation of chloroprene has been measured over the range 0 to 40°C. From oxidations with an added initiator the energy of activation for propagation was found to be 9.6 kcal./mole. The oxidation product is an unstable copolymer with oxygen formed by both 1,2 and 1,4 addition, the latter predominating. The molecular weight of the polyperoxide fell throughout the oxidation, to about 1000 at 20% oxidation. The molecular weight distribution was measured by gel permeation chromatography. The rates of oxidation of some monosubstituted vinyl monomers have been correlated by the Q,e reactivity scheme.*

Chloroprene is known to autoxidize at a measurable rate at temperatures as low as 0°C., yielding a polymeric peroxide as the principal product. Kern, Jockusch, and Wolfram (10) found that the peroxide is unstable and an effective catalyst for polymerizing chloroprene (11). They considered the most likely structure for the peroxide to be a 1,4-copolymer with oxygen. At low extents of oxidation Klebanskiĭ and Sorokina (12) obtained an amount of saponifiable chlorine which equaled the amount of oxygen absorbed and concluded this was most likely an allylic chlorine resulting from the formation of a 1,2-copolymer with oxygen. The present work was intended to extend these investigations of the structure of chloroprene peroxide and provide additional rate measurements.

Experimental

Chloroprene was fractionally distilled under a reduced pressure of nitrogen. It was stored at −80°C. *in vacuo*, and when required small amounts were distilled *in vacuo* into a subsidiary reservoir and from thence directly into the oxidation reactor. In this way chloroprene could be obtained completely free of peroxide, dimers, and higher polymers.

The sole major impurity in the monomer was 0.5 mole % of the isomer 1-chlorobutadiene.

Oxidations were carried out using about 2 ml. of chloroprene in an apparatus similar to that described by Bolland (5), modified for automatic recording. In some instances a gas-circulating system was used, in which oxygen from the reaction vessel passed through water in a conductivity cell, which was used to record the formation of hydrogen chloride (2).

Solutions of peroxide were prepared by oxidizing to the required extent, quenching the oxidation by cooling, and adding an excess of an inert diluent such as toluene. More than half the toluene was then pumped off while the oxidate was kept at −20°C. After this procedure had been repeated twice, solutions of peroxide in toluene could be prepared in which the residual chloroprene concentration was about 0.5% (w./w.) of the peroxide. Complete removal of solvent gave faintly yellow viscous peroxidic material which was mildly explosive at room temperature.

Total peroxide was estimated by reaction in acetic acid with hydrogen iodide generated *in situ* from sodium iodide and hydrochloric acid, a method we have found to be quantitative for some other polyperoxides and for di-α-cumyl peroxide. The less vigorous reagent of Skellon and Wills (16) was also used, which in our experience estimates hydroperoxides quantitatively. This reagent reacts very slowly with dialkyl peroxides and therefore gives an upper limit to the hydroperoxide content of a peroxide sample. A much more sensitive reagent was ferrous thiocyanate in methanol (19), which again generally estimates hydroperoxides and not dialkyl peroxides.

Results

Autoxidation of Chloroprene. The oxidation was autocatalytic and up to about 5 mole % oxidation—i.e., 5 moles of oxygen absorbed per 100 moles of chloroprene initially present—the quantity (mole % oxidation)$^{1/2}$ was a linear function of time, as observed by Kern (10). Beyond this extent the oxidation continued at a rather greater rate than given by this relation and was still accelerating at 25 mole % oxidation. Values of K in the expression

$$\text{Mole \% oxidation} = (Kt)^2$$

where t = time in minutes, were computed from experiments carried out over the range 20° to 45°C., at a total pressure of 700 mm. of Hg, and are given in Table I, together with the value obtained at 0°C. by Kern *et al.* (10). An Arrhenius plot of the results gave a line lying close to Kern's value; the equation of the least squares line through all the results over the range 0° to 45°C., with 95% fiducial limits on the slope, is

$$K = 1.381 \times 10^{10} \exp\left[(-16{,}310 \pm 640)/RT\right] \text{ (mole \%)}^{1/2}/\text{min.}$$

Table I. Constant K for Autoxidation of Chloroprene

Temp., °C.	$10^2 K$ (Mole %)$^{1/2}$/Min.
45	9.07; 8.52
35	3.99; 3.74; 3.71; 3.31
20	1.01; 0.896
0	0.124 (10)

The partial pressure of oxygen was varied between 12 and 476 mm. of Hg at 25°C.; all values of K in this pressure range lay between 1.54 and 1.75×10^{-2} (mole %)$^{1/2}$/min., so that at the lower limit the rate of autoxidation was still substantially independent of the partial pressure of oxygen.

Using the circulating apparatus it was found that only traces of hydrogen chloride were evolved during the initial stages of autoxidation; thus, at 2 mole % oxidation the mole ratio of oxygen absorbed to hydrogen chloride evolved was 435.

Initiated Oxidation. The initial rates of oxidation of chloroprene, initiated with 2,2'-azobisisobutyronitrile, were measured in the range 20° to 40°C. at a total pressure of 700 mm. of Hg. The difficulty with these measurements was that chloroprene autoxidizes so readily that even when the initiator is used at the fairly massive concentration of 0.462M, the rate of oxidation is constant for only a few minutes before acceleration, resulting from a contribution to initiation from chloroprene peroxide.

Initial rates of oxidation were estimated from tangents at the origin of the oxygen absorption plots (Table II). The equation expressing the dependence of these rates on temperature is

$$\text{Rate} = 5.453 \times 10^{13} \exp\left[(-25{,}080 \pm 2{,}030)/RT\right] \text{ mole/liter/sec.}$$

Table II. Initial Rates of Oxidation of Chloroprene Initiated with 0.462M Azobisisobutyronitrile

Temp., °C.	10^2 Rate, Mole/Liter/Hr.
20	4.26; 4.31
25	6.92
28	10.0
35	31.0; 31.1
40	59.6; 64.9; 64.1

Inhibition of Oxidation. Several antioxidants were tested in chloroprene at 45°C. Those which can be classified as mainly suppressors of initiation (1), because of their ability to destroy hydroperoxides—namely, zinc dialkyldithiophosphates, zince dialkyldithiocarbamates, triphenylphosphine, and the like—had no inhibiting effect at the 100-p.p.m. level.

Various aromatic secondary amines, substituted phenols, and pyrazolidones (*3*) that function as traps for the propagating peroxy radicals gave dead-stop induction periods when used at a concentration of 50 p.p.m. An indication of the ease of oxidation of chloroprene is that 50 p.p.m. of 2,6-di-*tert*-butyl-4-methylphenol gave an induction period of only 15 minutes, while the same concentration of antioxidant prevented *n*-hexadecane from oxidizing for 2 hours at 160°C.

In chloroprene containing $0.05M$ azobisisobutyronitrile and $0.02M$ 2,2,6,6-tetramethyl-4-piperidone-1-oxyl an induction period of 22 minutes was observed, followed by retarded oxidation. In the absence of the initiator 110 p.p.m. of *N,N*-dimethyl-4-nitrosoaniline inhibited oxidation for 1 hour. Nitroxide radicals and their nitroso precursors (*17*) do not function as peroxy radical traps since they cause no inhibition and little retardation of the initiated oxidation of cumene at 60°C.

During the induction periods caused by adding antioxidants, a small contraction in volume occurred because of the formation of dimers of chloroprene (*14*). This reaction occurs during the oxidation but was most easily studied by dilatometry in the absence of oxygen. A few values of the initial rate of dimerization of chloroprene, inhibited against polymerization with 2,2,6,6-tetramethylpiperidine-1-oxyl, are given in Table III. Their dependence on temperature is given by

$$\text{Rate} = 3.69 \times 10^9 \exp(-21{,}650/RT) \text{ mole/liter/sec.}$$

Table III. Initial Rates of Dimerization

Temp., °C.	10^6 Rate, Mole/Liter/Sec.
30	1.03
40	2.49; 2.50
50	8.46; 9.48

Chloroprene Peroxide. The efficiency of conversion of oxygen to total peroxides and "hydroperoxide" at various extents of oxidation was determined by iodometric methods. At up to 12% oxidation the proportion of "hydroperoxide" was constant at 20% of the whole. Ferrous thiocyanate likewise estimated a constant proportion (40%) of the total peroxide. Direct analysis of oxidates was somewhat difficult since the chloroprene tended to continue oxidizing during manipulation. Total peroxide estimates on chloroprene-free solutions of peroxide in toluene showed that at 20% oxidation 84% of the oxygen absorbed was present as peroxide groups. This is a minimum value since a small amount of the peroxide may have decomposed while chloroprene was being removed at −20°C.

It was confirmed that no volatile peroxides were formed (*10*). Chloroprene was oxidized to 10% at 45°C. and then flash-distilled *in vacuo* onto a trace of 2,6-di-*tert*-butyl-4-methylphenol. The amount of peroxide found in the distillate by the ferrous thiocyanate method corresponded to only 0.01% of the oxygen absorbed.

The number average molecular weight of the peroxide formed by autoxidation at 35°C. was measured with a Mechrolab 301A vapor phase osmometer using solutions in toluene. The concentration of peroxide was calculated from the amount of oxygen absorbed during oxidation, on the assumption that the repeating unit in the peroxide had the formula $C_4H_5ClO_2$. The peroxide solutions were not entirely stable, so molecular weights were estimated at intervals over about 1 hour and extrapolated back to zero time. The molecular weights obtained at extents of oxidation from 4 to 26 mole %, together with three values obtained by depression of the freezing point of benzene, are shown in Figure 1. The molecular weight fell from over 4000 at 4% oxidation to less than 1000.

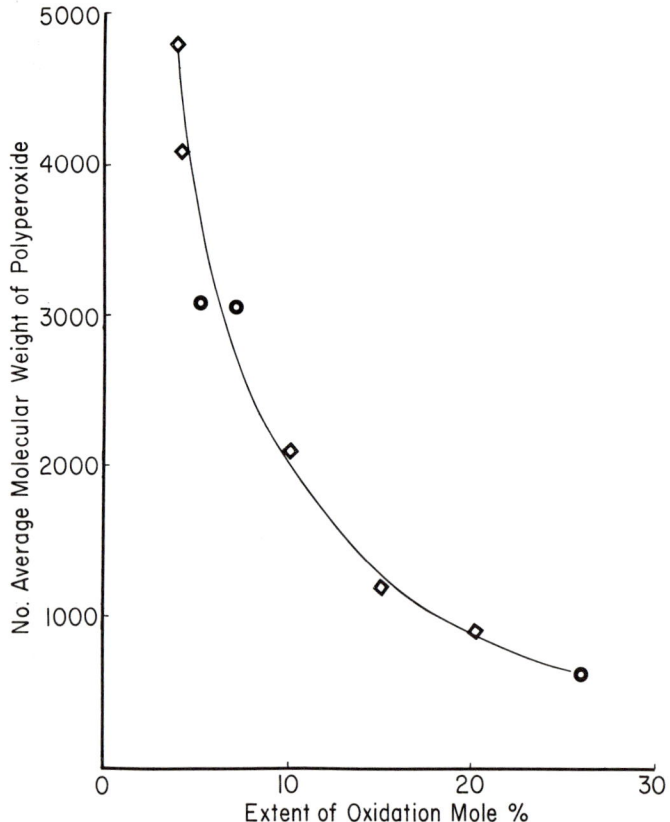

Figure 1. Dependence of molecular weight on extent of oxidation at 35°C.

The molecular weight distribution of peroxide formed at 4% oxidation was determined with a Waters gel permeation chromatograph. The peroxide was prepared as a 0.7% (w./v.) solution in tetrahydrofuran, and the molecular weight distribution then obtained is shown in Figure 2. By analogy with polychloroprene count 25 is equivalent to about 140 monomer units in the peroxide, and the peak maximum is at about 18 units—*i.e.*, a molecular weight of 2000. The incipient peaks at counts 34, 36, and between 32 and 33 result from products of peroxide decomposition.

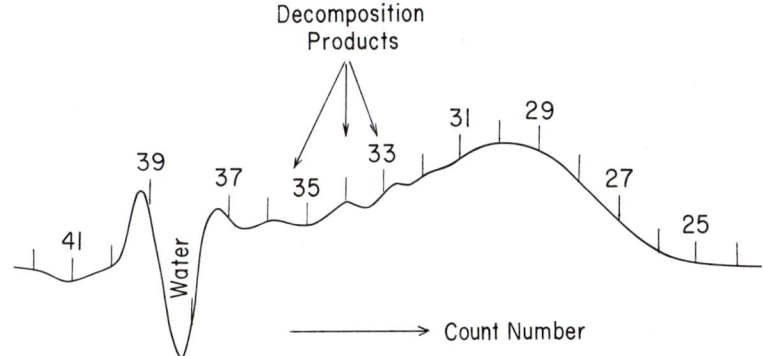

Figure 2. *Molecular weight distribution of polyperoxide at 4% oxidation*

On running the same solution of peroxide 20 hours later the peroxide of highest molecular weight, before count 26, had disappeared, while the three decomposition products were much more in evidence, particularly at count 36, which corresponds to about 3 monomer units. Chloroprene dimer, used as a reference compound, gave a peak at count 37. The molecular weight distribution curve of the peroxide from 18% oxidized chloroprene also showed three pronounced decomposition product peaks, but there was also a small amount of high molecular weight material containing 300 monomer units.

It was not possible to determine the composition of the peroxide by microanalysis since samples tended to explode at room temperature; however, it was possible to measure the atomic ratio of chlorine to oxygen in a concentrated solution in toluene of peroxide prepared by oxidation to 10 mole % at 35°C. This ratio was 0.51, close to that required by the repeating unit $(C_4H_5ClO_2)_n$.

An infrared spectrum of a film of peroxide between rock salt plates showed the presence of carbonyl and hydroxyl absorption from decomposition products. There was a wide region of C—O stretching absorption between 1000 and 1100 cm.$^{-1}$; no absorption by vinyl or substituted vinyl groups was apparent, but absorption at 830 cm.$^{-1}$ indicated the presence of a triply substituted double bond. The C—H stretching absorption was low in intensity, which could result from the presence of an adjacent oxygen atom in the group CH$_2$—O. The spectrum was thus consistent with that of a 1,4-copolymer with oxygen.

$$-O(-OCH_2\overset{\overset{\displaystyle Cl}{|}}{C}=CHCH_2-O-)_nO-$$

When the NMR spectrum of a 30% (w./v.) solution of peroxide in toluene was recorded at 34°C., absorption was observed between δ 2.74 and 5.46. There were seven main resonances, all multiplets, which were interpreted in terms of aliphatic hydrogen shifted by oxygen. Resonance from ethylenic hydrogen amounted to only a fraction of a proton. However, the sample darkened while in the instrument and probably decomposed extensively. When the spectrum of a solution of peroxide prepared by oxidation to 10.4 mole % was recorded using a cold probe at −35°C. a different picture was obtained. There was complex absorption from both ethylenic and saturated hydrogen which was interpreted as arising from a mixture of 1,2 and 1,4 oxygen copolymers in an approximate ratio of 1 to 2. In this sample the residual chloroprene amounted to 0.15% of the monomer units in the peroxide and dimers of chloroprene to 0.6% of the peroxide.

The decay of the peroxide was studied over a period of 6 hours in the infrared, using a 10% (w./v.) solution in toluene of peroxide obtained from 13% oxidized chloroprene. Carbonyl compounds were present initially, with more α,β-unsaturated C═O than saturated. This ratio altered until finally the saturated C═O predominated. In 3 hours the intensity of the combined bands increased by a factor of 3. There were initially three C═C stretching absorptions, comprising conjugated C═C at 1628 cm.$^{-1}$ and nonconjugated at 1667 an 1684 cm.$^{-1}$ Absorption decreased with time at 1667 cm.$^{-1}$ In the OH region there was initially no absorption, but this soon appeared and grew to a final value equivalent to a 20% yield of a C$_4$ alcohol. The broad C—O absorption between 1000 and 1100 cm.$^{-1}$ remained throughout the decomposition, although there were changes in the relative intensities at different frequencies in this range.

A shoulder at 950 cm.$^{-1}$, which might well have been vinyl absorption shifted by oxygen and chlorine, disappeared during the first 30 minutes.

The decomposition of a 0.45M solution of peroxide in toluene at 35°C. was also followed by total peroxide estimation. Nearly 10% of the peroxide decomposed in the first 2.5 hours; thereafter the decomposition was slower and first-order, with $k = 5.8 \times 10^{-7}$ sec.$^{-1}$.

Discussion

In the early stages of the autoxidation of chloroprene the amount of oxygen absorbed increased as the square of the time. This dependence on time is frequently observed in autoxidations and is an approximation to that expected for an oxidation of long chain length, initiated by the first-order decomposition of the peroxidic product and terminated by a bimolecular reaction of the propagating peroxy radicals.

If the oxidations initiated with azobisisobutyronitrile can be described by the usual rate equation

$$-d(O_2)/dt = (2ek_dI/k_t)^{1/2} k_p \text{ [chloroprene]}$$

where e is the fraction of radicals effective in initiation, I the concentration of initiator, k_d its first-order decomposition constant, k_t the second-order termination constant, and k_p the rate-determining propagation constant for the addition of peroxy radicals to chloroprene

$$RO_2{}^{\cdot} + CH_2\!=\!\underset{\underset{Cl}{|}}{C}\!-\!CH\!=\!CH_2 \xrightarrow{k_p} ROOCH_2\!-\!\underset{\underset{\cdot}{|}}{\underset{Cl}{C}}\!-\!CH\!=\!CH_2 \longleftrightarrow$$

$$ROOCH_2\!-\!\underset{\underset{Cl}{|}}{C}\!=\!CH\!-\!CH_2{}^{\cdot}$$

then the activation energy of the initiated oxidation is given by

$$E_{ox} = \tfrac{1}{2} E_d + E_p - \tfrac{1}{2} E_t$$

For the decomposition of azobisisobutyronitrile in styrene monomer Howard and Ingold (7) give the equation:

$$k_d = 1.99 \times 10^{15} \exp(-30{,}900/RT) \text{ sec.}^{-1}$$

This decomposition usually shows little dependence on solvent, so if E_d for decomposition in chloroprene is likewise 30.9 kcal. per mole, then since $E_{ox} = 25.1$ kcal. per mole $E_p = 9.6$ kcal. per mole, assuming termination to require no energy of activation. This is 1.2 kcal. per mole larger than k_p for styrene oxidation (8). Values of e for azobisisobutyronitrile in oxidation systems usually lie in the range 0.6 to 0.8; if $e = 0.7$, the above equation for the decomposition of the azonitrile and that given earlier for the initiated oxidation of chloroprene permit calculation of $k_p/k_t^{1/2}$ for chloroprene and also the kinetic chain lengths of the oxidations (Table IV).

Table IV. Values of $k_p/k_t^{1/2}$ and Kinetic Chain Length of Initiated Oxidations

Temp., °C.	$10^3\ k_p/k_t^{1/2}$ (Liters/Mole/Sec.)$^{1/2}$	Chain Length
0	2.77	1920
20	9.28	930
35	20.76	570

If the autoxidation of chloroprene is initiated by the first-order decomposition of the polyperoxide, P, the rate equation will be of the form

$$-d(O_2)/dt = (2k_i P/k_t)^{1/2} k_p\ [\text{chloroprene}]$$

From the equation for the rate of oxidation of chloroprene and the values of $k_p/k_t^{1/2}$ given in Table IV the decomposition constant, k_i, for the peroxide can be obtained. Some values are given in Table V, together with kinetic chain lengths calculated for 5 mole % oxidation. The latter are much larger than the number average molecular chain lengths, so there may well be a chain transfer reaction occurring during oxidation, as with styrene, α-methylstyrene, indene, and butadiene.

Table V. Calculated Values for k_i for Decomposition of Chloroprene Peroxide and Kinetic Chain Lengths at 5 Mole % Oxidation

Temp., °C.	$10^7\ k_i$, Sec.$^{-1}$	Chain Length
0	1.01	90
20	5.39	130
35	16.50	170

Our studies of the product of oxidation are consistent with its being a mixture of 1,2 and 1,4-copolymers with oxygen, the latter predominating. A minor proportion decomposes more rapidly than the remainder, and this may be the decomposition of the peroxy groups formed by 1,2 addition since vinyl absorption in the infrared decayed rapidly. In oxidates the level of hydroxyl absorption in the infrared was below the limit of detection; this excludes hydroperoxides as a major part of the product but not as chain-ending groups. The inability of peroxide-destroying antioxidants to inhibit the oxidation of chloroprene is consistent with initiation by dialkyl peroxide groups rather than hydroperoxides. The ability of nitroxide radicals and their precursors to inhibit oxidation is interesting since they do not react with peroxy radicals (13) and are competing successfully with oxygen for the propagating allyl-type radicals, which exist in equilibrium with the allyl peroxy radicals (4).

The effect of structure on the rates of oxidation of monomers has been investigated by Mayo, Miller, and Russell (15) by comparing their rates of oxidation at 50°C., at a fixed monomer concentration of $1M$, when initiated with 0.01M azobisisobutyronitrile. In Table VI the rates of oxidation of six monosubstituted vinyl monomers and chloroprene are given for these conditions. The rates for styrene, butyl acrylate, and methyl vinyl ketone are taken from the literature (6, 15); the others have been remeasured. The rate found for vinyl acetate is only half that given in the literature (15). The Q and e values have been taken from the

Table VI. Rates of Oxidation of 1M Monomer Initiated with 0.01M Azobisisobutyronitrile at 50°C.

Monomer	Q	e	10^4 Oxidn. Rate, Moles/Liter/Hr.	log (Rate Relative Styrene)	log Q −0.5 (e − 0.8)
Chloroprene	7.26	−0.02	290	0.611	0.471
Styrene	1.0	−0.8	71	0	0
Acrylonitrile	0.6	1.2	14	$\bar{1}$.149	$\bar{2}$.778
Methyl vinyl ketone	0.69	0.68	9.6	$\bar{1}$.130	$\bar{1}$.099
Vinyl acetate	0.026	−0.22	2.3	$\bar{2}$.511	$\bar{2}$.125
Butyl acrylate	0.5	1.06	1.9	$\bar{2}$.431	$\bar{2}$.769
Acrylic acid	0.3	0.77	1.8	$\bar{2}$.405	$\bar{2}$.692

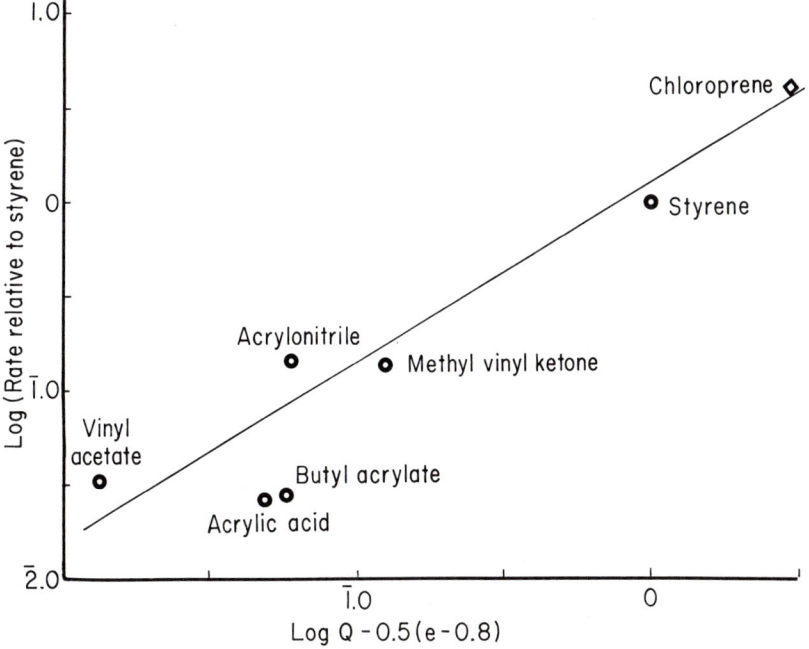

Figure 3. *Relative rates of initiated oxidation of monomers*

compilation by Young (20) except Q for acrylic acid, for which $Q = 0.3$ has been chosen from a correlation of Q values of monomers with the wavelength of their $\pi \rightarrow \pi^*$ ultraviolet absorptions and seems consistent with the high rate of polymerization of acrylic acid.

The Q,e system of describing the reactivity of monomers in copolymerization gives the rate constant for the addition of monomer 2 to the radical of monomer 1 as

$$k_{12} = P_1 Q_2 \exp(-e_1 e_2)$$

where P_1 is characteristic of radical 1, Q_2 represents the reactivity of the double bond of monomer 2, and e_1, e_2 represent the polarity of species 1 and 2. When monomers are oxidized, the rate-determining step is addition of a peroxy radical to the double bond. If to a first approximation all peroxy radicals are assumed to have the same reactivity and polarity, P_1 and e_1 are constants for oxidation. Making the further approximation that primary and secondary peroxy radicals all terminate at similar rates, so that k_{12}—i.e., k_p—is proportional to the rate of oxidation at a fixed rate of initiation, the rate of oxidation of any monomer related to that of styrene is given by

$$\text{(Rate relative to styrene)} = Q/Q_{\text{styrene}} \exp - e_1 (e - e_{\text{styrene}})$$
$$= Q \exp - e_1 (e + 0.8)$$

since $Q = 1$ and $e = -0.8$ for styrene

Therefore, $\log_{10}$ (relative rate) $= \log_{10} Q - \Delta(e + 0.8)$

where $\Delta = e_1/2.303$

Δ can be obtained from the slope and intercept of a plot of $[\log_{10}$ (relative rate) $-\log_{10} Q]$ as a function of e for the series of monomers and is found to be about 0.5; thus, e_1 for a peroxy radical is about 1.2. The electrophilic peroxy radicals thus have a similar polarity to monomers with electron-withdrawing groups, such as acrylonitrile and acrylates (15).

Figure 3 shows the correlation obtained by plotting $\log_{10}$ (relative rate) as a function of $\log Q - 0.5(e + 0.8)$ for the seven monomers. Over a range of oxidation rates varying by a factor of 100 the relation predicts the rate from Q,e values to less than a factor of 3. This is less precise than the correlation with excitation energies used for alkyl-substituted ethylenes (18), but is probably all that can be expected, since the Q,e system is an empirical relation and the assumption of equal reactivities and termination rate constants for primary and secondary peroxy radicals is imprecise (9).

Acknowledgment

I thank my colleagues at the Research Department for doing most of the analytical and experimental work.

Literature Cited

(1) Bailey, H. C., *Ind. Chem.* **38**, 315 (1962).
(2) Bailey, H. C., *Polymer Preprints* **5**, No. 2, 525 (1964).
(3) Bailey, H. C., Godin, G. W., Brit. Patent **930,565** (1963).
(4) Benson, S. W., *J. Am. Chem. Soc.* **87**, 972 (1965).
(5) Bolland, J. L., *Proc. Roy. Soc. (London)*, Ser. A, **186**, 218 (1946).
(6) Dyer, E., Brown, S. C., Medeiros, R. W., *J. Am. Chem. Soc.* **81**, 4243 (1959).
(7) Howard, J. A., Ingold, K. U., *Can. J. Chem.* **40**, 1851 (1962).
(8) *Ibid.*, **43**, 2729 (1965).
(9) *Ibid.*, **45**, 793 (1967).
(10) Kern, W., Jockusch, H., Wolfram, A., *Makromol. Chem.* **3**, 223 (1949).
(11) *Ibid.*, **4**, 213 (1950).
(12) Klebanskiĭ, A. L., Sorokina, R. M., *Zh. Prikl. Khim.* **35**, 2735 (1962).
(13) Khloplyankina, M. S., Buchachenko, A. L., Neiman, M. B., Vasil'eva, A. G., *Kinetika i Kataliz* **6**, 347 (1965).
(14) Leeming, P. A., Lehrle, R. S., Robb, J. C., *Soc. Chem. Ind. (London) Monograph* **20**, 203 (1966).
(15) Mayo, F. R., Miller, A. A., Russell, G. A., *J. Am. Chem. Soc.* **80**, 2500 (1958).
(16) Skellon, J. H., Wills, E. D., *Analyst* **73**, 78 (1948).
(17) Tudos, F., Kende, I., Berezhnykh, T., Solodovnikov, S. P., Voevodskiĭ, V. V., *Kinetika i Kataliz* **6**, 203 (1965).
(18) Van Sickle, D. E., Mayo, F. R., Arluck, R. M., Syz, M. G., *J. Am. Chem. Soc.* **89**, 967 (1967).
(19) Wagner, C. D., Clever, H. L., Peters, E. D., *Ind. Eng. Chem., Anal. Ed.* **19**, 980 (1947).
(20) Young, L. J., *J. Polymer Sci.* **54**, 411 (1961).

RECEIVED October 9, 1967.

12

Effect of Solvents on Rates and Routes of Oxidation Reactions

G. E. ZAIKOV and Z. K. MAIZUS

Institute of Chemical Physics, Academy of Sciences, Vorobyevskoye chaussee 2-b, Moscow V-334, USSR

The effect of the medium on the rates and routes of liquid-phase oxidation reactions was investigated. The rate constants for chain propagation and termination upon dilution of methyl ethyl ketone with a nonpolar solvent—benzene— were shown to be consistent with the Kirkwood equation relating the constants for bimolecular reactions with the dielectric constant of the medium. The effect of solvents capable of forming hydrogen bonds with peroxy radicals appears to be more complicated. The rate constants for chain propagation and termination in aqueous methyl ethyl ketone solutions appear to be lower because of the lower reactivity of solvated $RO_2 \cdot \ldots HOH$ radicals than of free $RO \cdot$ radicals. The routes of oxidation reactions are a function of the competition between two $RO_2 \cdot$ reaction routes. In the presence of water the reaction selectivity markedly increases, and acetic acid becomes the only oxidation product.

Recent research has established that the medium has an important effect on the rates and routes of oxidation reactions (4, 5, 10, 12). Chain oxidation reactions occur *via* peroxy radicals. The mechanism of liquid-phase oxidation at low extents of conversion is known to involve the following elementary reactions:

$$2RH + O_2 \xrightarrow{k_0} R\cdot + H_2O_2 + R\cdot \tag{0}$$

$$R\cdot + O_2 \xrightarrow{k'} RO_2\cdot \tag{1}$$

$$RO_2\cdot + RH \xrightarrow{k_2} ROOH + R\cdot \tag{2}$$

$$\text{ROOH} + \text{RH} \xrightarrow{k_3} \text{RO}^\cdot + \text{H}_2\text{O} + \text{RO}_2^\cdot \qquad (3)$$

$$\text{R}^\cdot + \text{R}^\cdot \xrightarrow{k_4} \text{products} \qquad (4)$$

$$\text{R}^\cdot + \text{RO}_2^\cdot \xrightarrow{k_5} \text{products} \qquad (5)$$

$$\text{RO}_2^\cdot + \text{RO}_2^\cdot \xrightarrow{k_6} \text{alcohol} + \text{ketone} + \text{O}_2 \qquad (6)$$

At sufficiently high oxygen concentrations only $\text{RO}_2^\cdot$ radicals are present in the system, and chain termination occurs solely by Reaction 6.

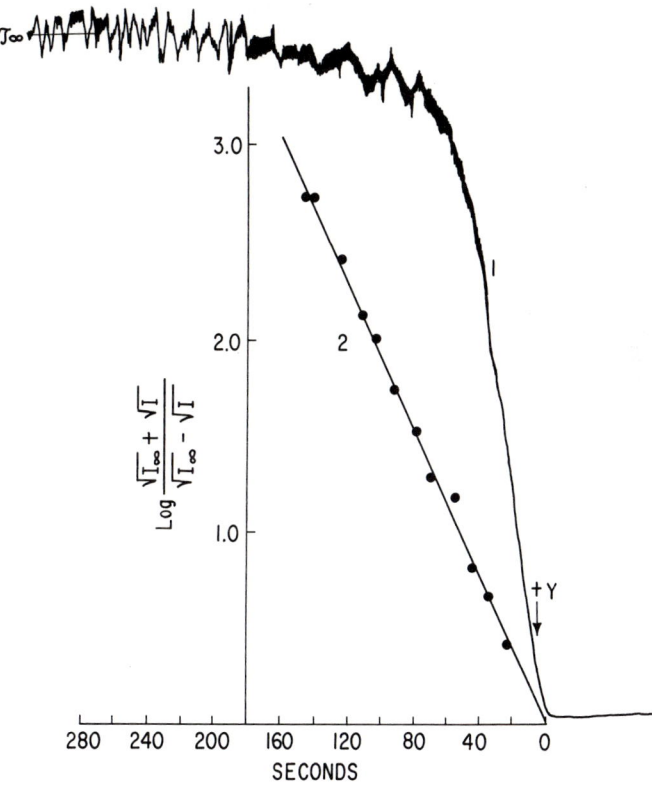

Figure 1. Chemiluminescence exhibited by methyl ethyl ketone oxidation in benzene solution at 60°C. Arrow shows time of adding initiator Y

1. At molar ratios of $[C_4H_8O] : [C_6H_6] = 1 : 4$
   Rate of chain initiation, 3.8×10^{-9} mole/liter sec.
2. Semilogarithmic anamorphose of 1

Table I. Rate Constants for Reactions of Chain Methyl Ethyl Ketone in Benzene

$k_2 \times 10^2$ Mole^{-1} Liter Sec.$^{-1}$

T, °C.	Molar Ratio $C_4H_8O:C_6H_6$			
	1:0	1:1	1:4	1:9
35	13.9 ± 0.6	6.6 ± 0.7	4.8 ± 0.9	4.9 ± 1
40	17 ± 1.5	8.5 ± 0.8	6.2 ± 1.1	5.7 ± 1.3
50	26 ± 1	14.3 ± 1.8	13 ± 0.2	12.8 ± 2.5
60	39 ± 2	26 ± 3	25 ± 3	25 ± 5
70	55.5 ± 1	39.5 ± 3	40.5 ± 7	43.5 ± 9
75	66 ± 3	51 ± 4	59 ± 9	58 ± 12

Table II. Pre-exponential Factors and Activation Oxidation of C_4H_8O in a

Parameters	Dilution in Molar		
	1:0	9:1	4:1
$A_6 \times 10^{-7}$ l. mole^{-1} sec.$^{-1}$	2.0 ± 0.2	1.8 ± 0.1	1.6 ± 0.1
E_6, kcal./mole	1.6 ± 0.8	1.8 ± 0.8	2.0 ± 0.8
$A_2 \times 10^{-5}$ l. mole^{-1} sec.$^{-1}$	1.26 ± 0.06	2.8 ± 0.1	6.0 ± 0.2
E_2, kcal./mole	8.4 ± 0.6	9.0 ± 0.7	9.6 ± 0.7

The rate of oxidation under these conditions is

$$W = \frac{k_2}{\sqrt{k_6}} [RH] \sqrt{w_i} \qquad (1a)$$

where w_i is the rate of radical formation (1).

The effect of solvents on the rate constants for chain propagation and termination was investigated for methyl ethyl ketone oxidation. The rate of the latter at temperatures of 35° to 75°C. was determined from that of oxygen consumption (11). Azoisobutyronitrile was used as initiator. The rates of chain initiation, w_i, for various solvents were measured using the inhibitor technique (1). Knowing W, the methyl ethyl ketone concentration, and w_i, it was possible to calculate the $k_2/\sqrt{k_6}$ ratio.

The rate constants for recombination of radicals, k_6, were determined using the chemiluminescence method, which consists in recording the intensity of luminescence induced by recombination of $RO_2\cdot$ radicals, when the system passes from one steady state to another (6). A change in steady state was induced by introducing an initiator (Figure 1). The k_6 value was obtained from

$$\ln \frac{\sqrt{I_\infty} + \sqrt{I}}{\sqrt{I_\infty} - \sqrt{I}} = 2t \sqrt{k_6 w_i}$$

Propagation and Termination in Oxidation of Solutions at Various Temperatures

$k_6 \times 10^{-5}$ Mole^{-1} Liter Sec.$^{-1}$

Molar Ratio $C_4H_8O:C_6H_6$			
1:0	4:1	1:1	1:9
14.4 ± 1.3	6.1 ± 0.6	2.6 ± 0.3	0.8 ± 0.2
15.8 ± 3	6.9 ± 0.6	2.8 ± 0.3	0.75 ± 0.05
16 ± 1.4	7.1 ± 0.6	3.1 ± 0.3	0.94 ± 0.09
18 ± 2	7.8 ± 1.3	3.2 ± 0.35	1.0 ± 0.2
18.7 ± 1	8.4 ± 0.4	3.8 ± 0.4	1.15 ± 0.07
20.4 ± 2.1	8.7 ± 0.8	4.0 ± 0.2	1.24 ± 0.1

Energies for Chain Propagation and Termination in Solution of C_6H_6 (50°C.)

Ratios $C_4H_8O:C_6H_6$

2:1	1:1	1:2	1:4	1:9
1.6 ± 0.1	1.30 ± 0.05	1.17 ± 0.03	0.71 ± 0.03	0.60 ± 0.02
2.2 ± 0.8	2.4 ± 0.8	2.5 ± 0.8	2.6 ± 0.8	2.7 ± 0.8
14.5 ± 0.5	43 ± 1	900 ± 20	1750 ± 90	2400 ± 100
10.2 ± 0.7	11.0 ± 0.7	13 ± 0.8	13.5 ± 0.8	13.7 ± 0.8

where t is the time that has passed from the moment the initiator was introduced, and I and I_∞ are chemiluminescence intensities at time t and at the moment of attaining the steady state.

The k_2 values were determined from those for k_6 and $k_2/\sqrt{k_6}$.

The rate of oxidation decreases with dilution of methyl ethyl ketone by a nonpolar solvent—namely, benzene. The rate constants for both chain propagation and chain termination also drop (Table I), and the activation energies increase (Table II) because the elementary reactions of chain propagation and termination represent interaction between two dipoles occurring at a rate which depends on the dielectric constant of the medium.

A good description of variations in k_2 and k_6 values is given by the Kirkwood equation (3):

$$\log K_i = \log K_i^\circ - \frac{1}{2.3\,kT} \frac{\epsilon - 1}{2\epsilon + 1} \sum \frac{\mu^2}{r^3} \qquad (2a)$$

where K_i° is the rate constant for a bimolecular reaction (at $\epsilon = 1$), and k is the Boltzmann constant. The $\sum \frac{\mu^2}{r^3}$ value is

$$\sum \frac{\mu^2}{r^3} = \frac{\mu_1^2}{r_1^3} + \frac{\mu_2^2}{r_2^3} - \frac{(\mu^*)^2}{(r^*)^3} \qquad (3a)$$

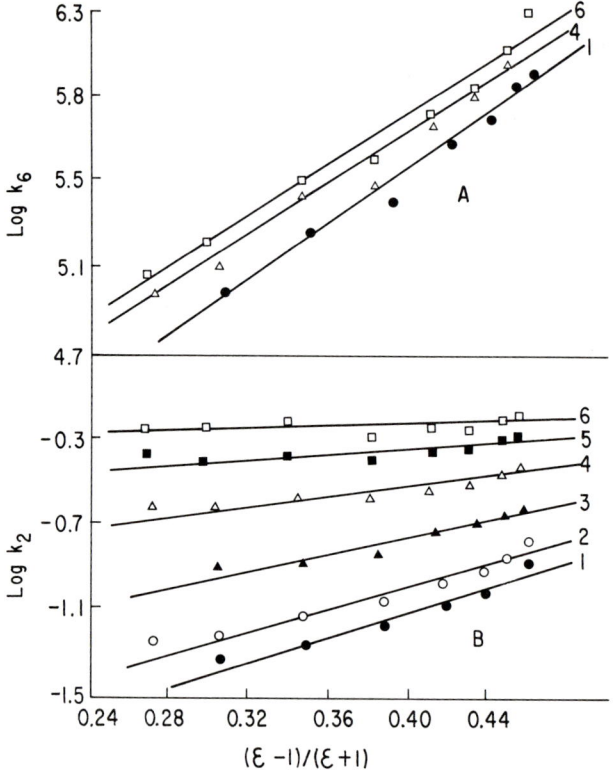

Figure 2. log k_6 and log k_2 as functions of $(\epsilon - 1)/(2\epsilon + 1)$ in oxidation of methyl ethyl ketone in benzene solutions

1. At 35°C.
2. At 40°C.
3. At 50°C.
4. At 60°C.
5. At 70°C.
6. At 75°C.

where μ_1, μ_2, r_1, and r_2 are dipole moments and effective radii of the reactants, and μ^* and r^* denote those for an activated complex.

The experimental values of k_2 and k_6 for various methyl ethyl ketone dilutions by benzene fall on straight lines plotted as log k_2 and log k_6 vs. $(\epsilon - 1)/(2\epsilon + 1)$ (Figure 2).

The dipole moments of activated complexes in reactions of chain propagation and termination calculated from the data of Figure 2, using Equations 2a and 3a, are $(\mu^*)_2 = (8.1 \pm 0.1)$ debyes and $(\mu^*)_6 = (11.1 \pm 0.1)$ debyes. This is consistent with the dipole moments of activated complexes we calculated for two possible structures of the activated complex (7, 10). Thus, the variations in k_2 and k_6 constants with methyl ethyl ketone dilution by benzene agree with the concept of strictly electrostatic interaction between the solvent and the reaction particles.

Similar dependences were obtained for the oxidation of methyl ethyl ketone in a solution of p-dichlorobenzene, n-decane, and CCl_4 (Figures 3 and 4) and in acetic acid solutions.

The $\mu_2{}^*$ and $\mu_6{}^*$ values for all solvents investigated are given in Table III.

The close values of $\mu_2{}^*$ and $\mu_6{}^*$ obtained for various solvents provide additional evidence that variations in the k_2 and k_6 values are related solely to the dielectric constant of the medium, while specific solvation—e.g., the formation of π-complexes of $RO_2{}^\cdot$ radicals and aromatics in the course of methyl ethyl ketone oxidation—has no essential importance.

Considerably more complicated is the effect of solvents capable of forming hydrogen bonds with the reacting particles. Water is one of these solvents. When methyl ethyl ketone is oxidized in aqueous solutions, the rate constants for chain propagation and termination do not increase, as would be consistent with the concept of the medium dielectric constant effect; instead, they decrease. At high dilution of methyl ethyl ketone by water the rate constants of these reactions no longer depend on water concentration (Figure 5). The activation energies for these reactions

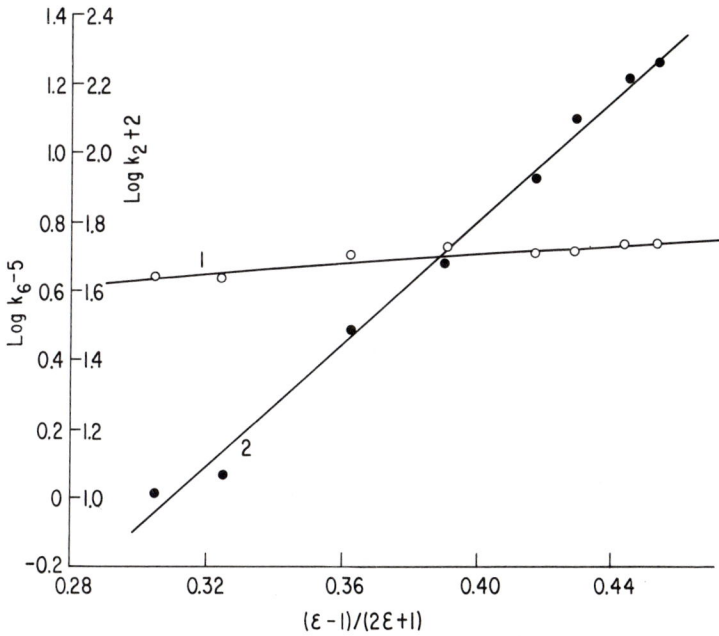

Figure 3. log k_2 and log k_6 as functions of $(\epsilon - 1)/(2\epsilon + 1)$ in oxidation of methyl ethyl ketone in p-dichlorobenzene solutions at 70°C.

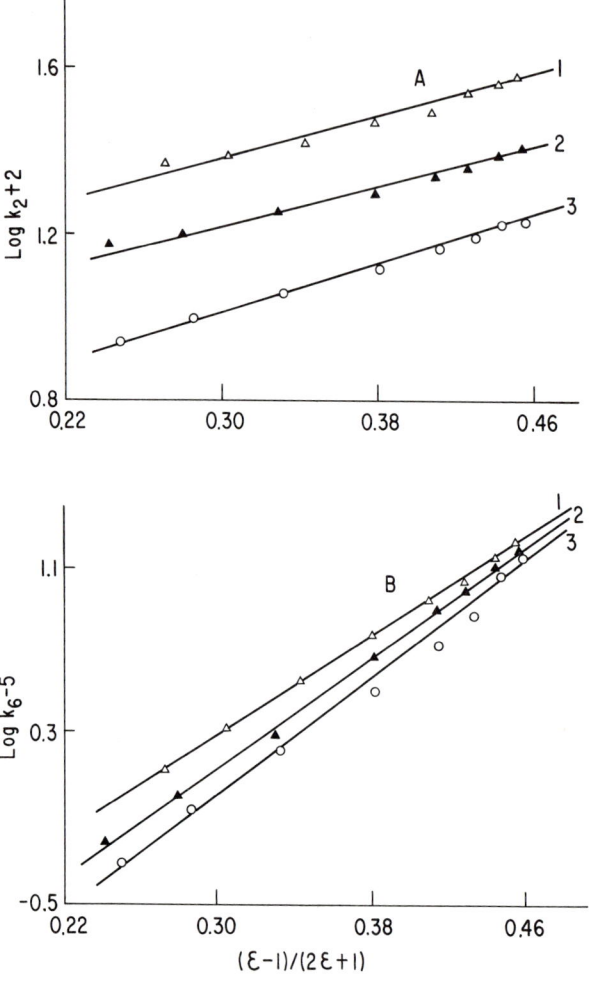

Figure 4. log k_2 and log k_6 as functions of $(\epsilon - 1)/(2\epsilon + 1)$ in oxidation of methyl ethyl ketone in solution

1. CCl_4, 60°C.
2. Decane, 50°C.
3. Decane, 40°C.

increase to a certain limit only. Variations in the pre-exponential factor values are evidence of a compensating effect of solvents in these reactions (Table IV). These experimental data clearly demonstrate the decrease in RO_2· reactivity as a result of solvation by water (2, 8).

Two types of peroxy radicals, nonhydrated and hydrated by formation of hydrogen bonds between H_2O molecules and unshared electrons

Table III. Dipole Moments of Activated Complexes for Chain Propagation (μ_2^*) and Termination (μ_6^*) in Oxidation of Methyl Ethyl Ketone in Various Solvents

Solvent	μ_2^*, Debyes	μ_6^*, Debyes
Benzene	8.1 ± 0.2	11.1 ± 0.1
Acetic acid	8.4 ± 0.3	11.4 ± 0.2
n-Decane	8.0 ± 0.1	11.5 ± 0.1
CCl_4	8.0 ± 0.1	11.3 ± 0.1
p-Dichlorobenzene	8.1 ± 0.2	12.5 ± 0.1

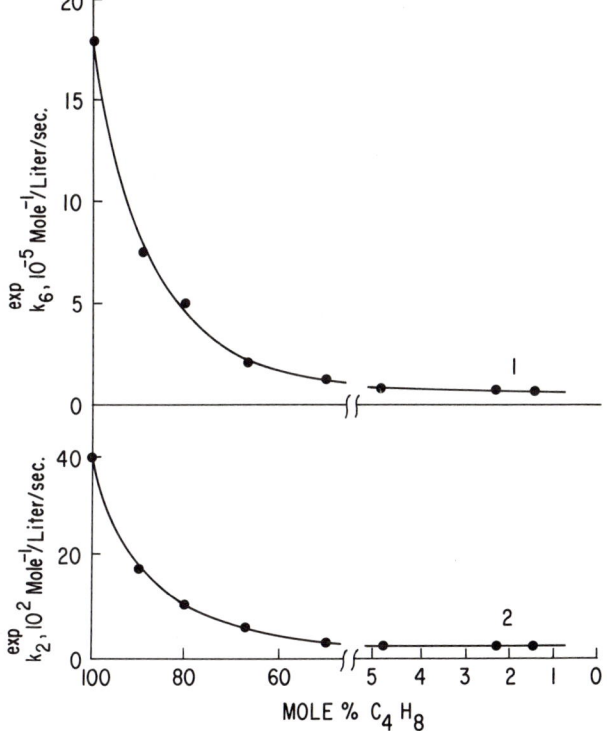

Figure 5. k_2^{exp} and k_6^{exp} as functions of C_4H_6O content in mixtures of C_4H_8O and H_2O at $60°C$.

of the peroxy radicals, appear in the oxidation of methyl ethyl ketone in aqueous solutions (2, 8).

It was shown by experiment that equilibrium

$$RO_2\cdot + mH_2O \overset{K_I}{\rightleftarrows} RO_2\cdot \ldots mH_2O \qquad (1)$$

Table IV. Pre-exponential Factors and Activation Energies of Rate Methyl Ethyl Ketone

	1:0	Molar Ratios 9:1	4:1
E_6, kcal./mole	1.6 ± 0.8	1.6 ± 0.6	2.5 ± 1.0
$A_6 \times 10^{-7}$, l. mole^{-1} sec.$^{-1}$	2.02 ± 0.21	0.83 ± 0.08	2.2 ± 0.2
E_2, kcal./mole	8.4 ± 0.5	10.8 ± 0.5	13.1 ± 0.7
$A_2 \times 10^{-6}$, l. mole^{-1} sec.$^{-1}$	0.126 ± 0.007	2.2 ± 0.12	41.0 ± 2.5

is rapidly attained in the system. In this case the mechanism of chain propagation and termination is

$$RO_2 \cdot + RH \xrightarrow{k_2} ROOH + R \cdot \quad (2)$$

$$RO_2 \cdot \ldots mH_2O + RH \xrightarrow{k_2'} ROOH + R \cdot + mH_2O \quad (2')$$

$$RO_2 \cdot + RO_2 \cdot \xrightarrow{k_6} \left. \begin{matrix} \\ \\ \end{matrix} \right\} \quad (6)$$

$$RO_2 \cdot + RO_2 \cdot \ldots mH_2O \xrightarrow{k_6'} \text{nonactive products} \quad (6')$$

$$RO_2 \cdot \ldots mH_2O + RO_2 \cdot \ldots mH_2O \xrightarrow{k_6''} \quad (6'')$$

Thus, the experimental rate constants for chain propagation and termination in the oxidation of methyl ethyl ketone in aqueous solutions appear to be effective and are

$$k_2^{exp} = \kappa k_2 + \kappa' k_2'$$

$$k_6^{exp} = \eta k_6 + \eta' k_6' + \eta'' k_6''$$

where κ, κ', η, η', and η'' are the fractions of hydrated and nonhydrated radicals.

At weak dilution of methyl ethyl ketone by water $k_2^{exp} = \kappa k_2$. Since

$$\kappa = \frac{[RO_2 \cdot]}{[RO_2 \cdot] + K_I [RO_2 \cdot][H_2O]^m}$$

$$k_2/k_2^{exp} = 1 + K_I[H_2O]^m$$

log k_2/k_2^{exp} as a function of log [H_2O] falls on a straight line (Figure 6). The $m = 1$ value was obtained from the slope of this line. This means that association of only one H_2O molecule is essential for the reactivity of peroxy radicals. Further bonding of H_2O molecules with $RO_2 \cdot$ radicals has no essential effect. The equilibrium constant is

$$K_I = (7.6 \pm 0.6) \times 10^{-4} \exp(4800/RT) \text{l./mole}.$$

Constants for Chain Propagation and Termination in Oxidation of in Presence of H_2O

$C_4H_8O:H_2O$					
2:1	1:1	1:20	1:40	1:60
3.4 ± 1.2	3.6 ± 1.4	4.6 ± 1.0	4.6 ± 1.0	4.6 ± 1.0
3.8 ± 0.5	3.4 ± 0.6	7.2 ± 0.3	7.2 ± 0.3	7.2 ± 0.3
14.5 ± 0.8	15.7 ± 0.9	16.5 ± 1.0	16.5 ± 1.0	16.5 ± 1.0
170 ± 13	620 ± 60	1300 ± 70	1300 ± 70	1300 ± 70

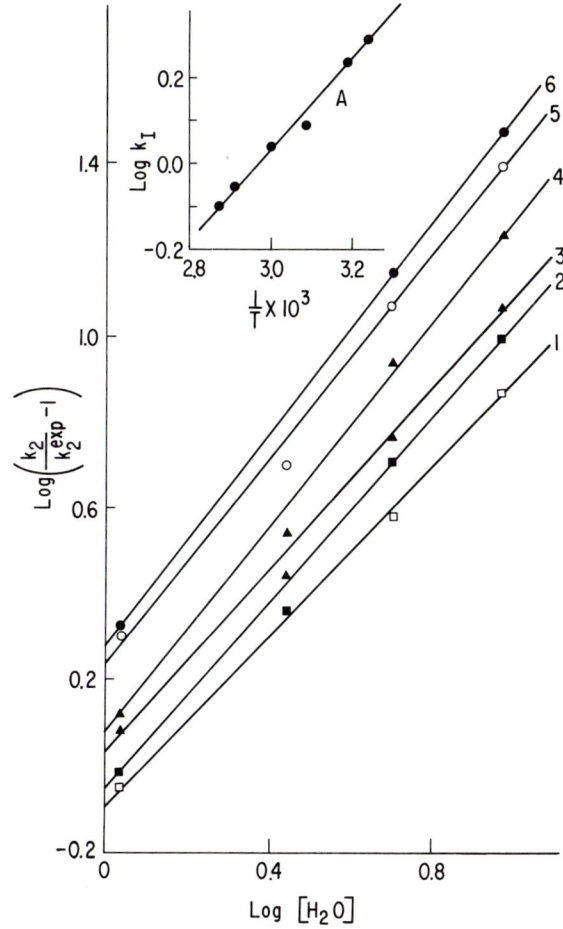

Figure 6. Log $(k_2/k_2^{exp} - 1)$ as function of log $[H_2O]$ in oxidation of methyl ethyl ketone at different temperatures. A: dependence of log K_I on $1/T$.

  1. 75°C.   4. 50°C.
  2. 70°C.   5. 40°C.
  3. 60°C.   6. 35°C.

Table V. Rate Constants (60°) for Chain Propagation and Termination and Pre-exponential Factors and Activation Energies for Methyl Ethyl Ketone Oxidation in Aqueous Solutions

Type of Reaction	Eq.	K, L. Mole^{-1} Sec.$^{-1}$	A, L. Mole^{-1} Sec.$^{-1}$	E, Kcal./Mole
$RO_2\cdot + RH$	2	$(39 \pm 4) \times 10^{-2}$	$(1.26 \pm 0.07) \times 10^5$	8.4 ± 0.5
$RO_2\cdot \ldots H_2O + RH$	2^I	$(2.0 \pm 0.2) \times 10^{-2}$	$(1.30 \pm 0.07) \times 10^9$	16.5 ± 1.0
$RO_2\cdot + RO_2\cdot$	6	$(180 \pm 20) \times 10^4$	$(2.02 \pm 0.21) \times 10^7$	1.6 ± 0.8
$RO_2\cdot + RO_2\cdot \ldots H_2O$	6^I	$(45 \pm 15) \times 10^4$	—	—
$RO_2\cdot \ldots H_2O + RO_2\cdot \ldots H_2O$	6^{II}	$(7.0 \pm 0.8) \times 10^4$	$(7.2 \pm 0.3) \times 10^7$	4.6 ± 1.0

Table VI. Rates of Methyl Ethyl Ketone Oxidation Products[a] Formed by Bimolecular (W_2) and Unimolecular (W_1) Routes in Reaction Involving $RO_2\cdot$ in Benzene Solution (145°C., 50 atm.)

Rates, Mole % Per Hour

Reaction Products	Molar Ratios of Methyl Ethyl Ketone to Benzene									
	1:0		2:1		1:1		1:2		1:3	
	W_2	W_1	W_2	W_1	W_2	W_1	W_2	W_1	W_2	W_1
Acetic acid	30.6	0.7	25.1	1.1	25.1	1.4	22.4	2.3	14.9	2.6
Formic acid	—	—	—	0.1	—	0.2	—	0.3	—	0.6
Acetone	—	0.1	—	0.1	—	0.1	—	0.2	—	0.3
Acetaldehyde	—	0.1	—	0.5	—	0.6	—	0.7	—	1.6
Formaldehyde	—	—	—	—	—	—	—	0.1	—	0.2
Diacetyl	10.2	—	10.0	—	9.8	—	8.2	—	9.5	—
Ethanol	0.1	—	0.1	—	0.1	—	0.1	—	0.5	—
Methanol	—	—	—	—	—	—	—	—	—	0.1
Ethyl acetate	7.6	—	8.6	—	8.4	—	6.2	—	6.1	—
Methyl acetate	—	—	—	—	—	—	—	—	0.2	0.1
CO_2	0.3	0.1	1.1	0.1	1.1	0.1	1.1	0.1	1.6	0.2
CO	—	0.2	—	0.1	—	0.2	—	0.2	—	0.3
Peroxide	1.0	—	0.8	—	0.8	—	1.0	—	1.2	—
Total	49.8	1.2	45.7	2.0	45.5	2.6	39.0	3.9	34.0	6.0

[a] For maximum consumption of C_4H_8O, which corresponds to a degree of conversion of 4 to 6%.

Since at a high dilution of methyl ethyl ketone with water the experimental rate constants obtained do not depend on the amount of water, it may be considered that only $RO_2\cdot \ldots H_2O$ radicals are present in the system under given conditions, and the results obtained represent the k_2^I and k_6^{II} constants.

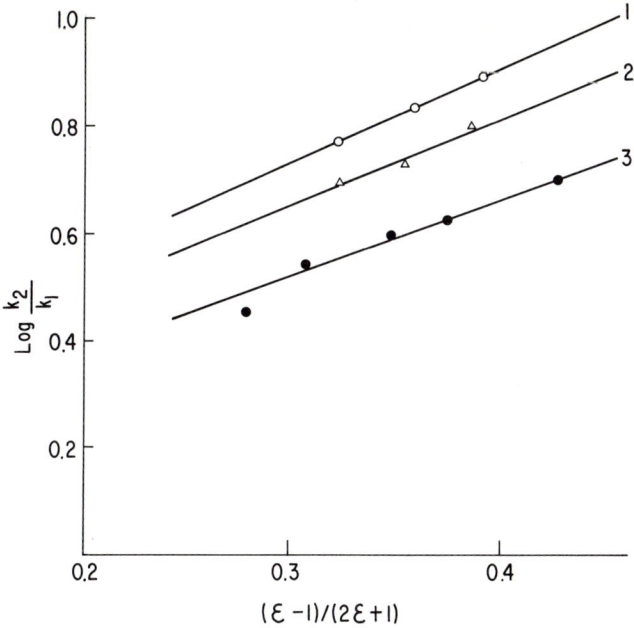

Figure 7. $\log k_2/k_1 = \log \dfrac{w_2 [RH]}{w_1}$ as function of $(\epsilon - 1)/(2\epsilon + 1)$ in oxidation of methyl ethyl ketone at 50 atm. and different temperature
1. 100°C. 2. 122°C. 3. 145°C.

Table VII. Rates of Products of Methyl Ethyl Ketone Accumulation in Water Over W_1 and W_2 Routes at a Molar Ratio of $C_4H_8O : H_2O = 5 : 1$
(Degree of Conversion 5%)

Reaction Products	$W_{1.1} + W_{1.2}$	$W_{1.3}$	$W_{1.4}$	$W_{1.5}$	W_2
Propionic acid	—	—	—	1.8	—
Acetic acid	5.22	1.1	—	—	20.4
Formic acid	—	0.55	0.55	0.3	—
Diacetyl	—	—	—	—	3.1
Acetone	—	—	1.1	—	—
Acetaldehyde	—	—	—	—	0.3
Formaldehyde	—	—	—	1.2	—
Methanol	5.22	—	—	—	—
Ethanol	—	—	—	—	1.2
Ethyl acetate	—	—	—	—	1.5
CO_2	5.22	1.1	—	—	0.9
CO	—	0.55	0.55	0.3	—
Peroxide	—	—	—	—	0.15
Total	15.66	3.3	2.2	3.6	27.55

For a general case (2, 8)

$$k_6^{exp} = \frac{k_6 + 2K_I[H_2O]k_6' + K_I^2[H_2O]^2 k_6''}{1 + 2K_I[H_2O] + K_I^2[H_2O]^2}$$

The k_6^I value may be calculated from this expression. At 60°C. $k_6^I = (0.45 \pm 0.15) \times 10^6$ liters/mole sec.

The rate constants for solvated and nonsolvated radicals are shown in Table V. The reactivity of solvated $RO_2 \cdot \ldots H_2O$ is lower than that of free RO_2.

The effect of solvents on the oxidation route as regards the composition of reaction products was investigated for methyl ethyl ketone at 145° to 160°C. at a pressure of 50 atm. (9, 10). The composition of products under these conditions is a function of the competition between two routes of the peroxy radical reactions:

$$RO_2 \cdot \xrightarrow{w_1} R'CHO + R''O \cdot \quad (1)$$

(where R' and R'' are radicals with a lesser number of carbon atoms than that in methyl ethyl ketone) and

$$RO_2 \cdot + RH \xrightarrow{w_2} ROOH + R \cdot \quad (2)$$

The reaction rates for forming the products of methyl ethyl ketone oxidation in benzene solution, by the unimolecular and bimolecular routes of $RO_2 \cdot$ conversion, are shown in Table VI. The yield of products formed by the bimolecular route decreases and that for the unimolecular route increases with dilution of methyl ethyl ketone by benzene.

Knowing which of the routes involves the formation of a given product, it is possible to estimate the ratio of rates w_1 and w_2 in the absence of a solvent and at various methyl ethyl ketone concentrations in the solution.

Table VIII. Rates of Reactions 1 and 2 in Oxidation of Methyl Ethyl Ketone in Presence of Water at 160°C., 50 atm.
(Degree of Conversion 5% [a])

Rates, Mole % of Overall Reaction Per Hour	Dilutions					
	1:0	5:1	2:1	1:5	1:10	1:20
$W_{1.1} + W_{1.2}$	25.1	15.7	1.11	1.89	0.00	0.00
$W_{1.3}$	2.1	3.3	1.26	1.23	0.48	0.00
$W_{1.4}$	4.2	2.2	0.54	0.14	0.00	0.00
$W_{1.5}$	4.8	3.6	0.94	0.00	0.00	0.00
W_1	36.2	24.8	3.85	3.26	0.48	0.00
W_2	38.7	27.6	22.4	15.7	11.1	9.9
W_1/W_2	0.94	0.90	0.17	0.20	0.04	0.00

[a] Similar results obtained for 10, 15, and 20% conversion.

Zaikov (9) showed that isomerization and decomposition of a methyl ethyl ketone peroxy radical may occur as shown below.

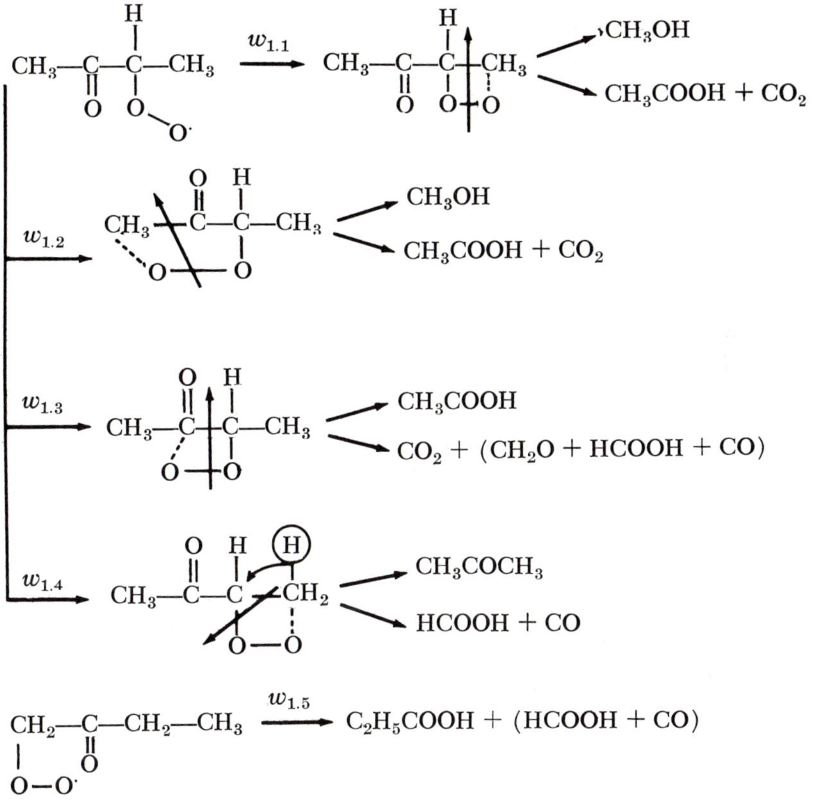

$$w_1 = w_{1.1} + w_{1.2} + w_{1.3} + w_{1.4} + w_{1.5}$$

The bimolecular reaction $RO_2{}^{\cdot} + RH$ yields the following products (9).

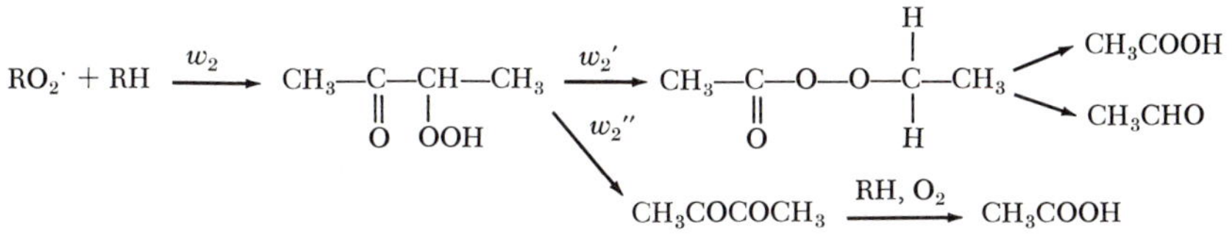

When methyl ethyl ketone is diluted by benzene, the ratio w_2/w_1 becomes lower. The amount of products formed by $RO_2{}^{\cdot}$ decomposition increases. The ratio of constants for elementary steps $k_2/k_1 = (w_2/w_1)\cdot[RH]$ changes according to the Kirkwood equation (Figure 7). The $\mu_2{}^*$

values obtained from the data in Figure 7 completely coincide with those obtained from rate constants for the elementary steps of chain propagation and termination (Figure 2).

When methyl ethyl ketone is oxidized in aqueous solutions, the over-all reaction rate drops because of solvation of the peroxy radicals, and w_1 decreases more than w_2. The reaction rates for formation of methyl ethyl ketone oxidation products in aqueous solutions are shown, as an example, in Tables VII and VIII.

When C_4H_8O is diluted by water to 80 volume %, the only product of C_4H_8O oxidation is acetic acid (99% per methyl ethyl ketone reacted) formed by ketone hydroperoxide conversion. The reason for this increase in the reaction selectivity is that the rate of decomposition of the radical complex $RO_2\cdot \ldots HOH$ is lower than that of free $RO_2\cdot$, while the decrease in the rate of reaction of $RO_2\cdot \ldots HOH$ with methyl ethyl ketone is somewhat offset by the higher dielectric constant of the medium.

Literature Cited

(1) Andronov, L. M., Zaikov, G. E., *Kinetika i Kataliz* **8**, 270 (1967).
(2) Andronov, L. M., Zaikov, G. E., Maizus, Z. K., *Zh. Fiz. Khim.* **41**, 1122 (1967).
(3) Glasstone, S., Laidler, K., Eyring, G., "Theory of Rate Processes," p. 400, Princeton University Press, New York-London, 1941.
(4) Hendry, D. G., Russell, G. A., *J. Am. Chem. Soc.* **86**, 2368 (1964).
(5) Howard, J. A., Ingold, K. U., *Can. J. Chem.* **42**, 1024, 1250 (1964); **43**, 2737 (1965).
(6) Vichutinskiĭ, A. A., *Dokl. Akad. Nauk SSSR* **157**, 150 (1964).
(7) Zaikov, G. E., *Izv. Akad. Nauk SSSR, Ser. Khim.*, No. 8, 1175 (1967).
(8) Zaikov, G. E., Andronov, L. M., Maizus, Z. K., Emanuel, N. M., *Dokl. Akad. Nauk SSSR* **174**, No. 1, p. 127 (1967).
(9) Zaikov, G. E., Kazancheva, S. D., Maizus, Z. K., *Zh. Teor. Exptl. Khim.* **2**, 60 (1966).
(10) Zaikov, G. E., Maizus, Z. K., *Izv. Akad. Nauk SSSR, Otd. Khim. Nauk* **1175**, No. 7 (1962); *Neftekhimiya* **3**, 381 (1963); *Dokl. Akad. Nauk SSSR* **150**, 116 (1963).
(11) Zaikov, G. E., Maizus, Z. K., Emanuel, N. M., *Kinetika i Kataliz* **7**, 401 (1966).
(12) Zaikov, G. E., Vichutinskiĭ, A. A., Maizus, Z. K., Emanuel, N. M., *Dokl. Akad. Nauk SSSR* **168**, 1096 (1966).

RECEIVED November 8, 1967.

Base-Catalyzed and Heteroatom Compound Oxidations

CHEVES WALLING
Session Chairman

13

The Role of Heteroatoms in Oxidation

CHEVES WALLING

Columbia University, New York, N. Y. 10027

> *The effects of heteroatoms on autoxidation reactions are reviewed and discussed in terms of six phenomena: (1) the effect on reactivity of α-hydrogens in the hydroperoxide chain mechanism in terms of electron supply and withdrawal; (2) the effect on α-hydrogen acidity in base-catalyzed oxidation; (3) the effect on radical ion stability in base-catalyzed redox chains; (4) the possibility of heteroatom hydrogen bond attack and subsequent reactions of the resulting heteroradical; (5) the possibility of radical attack on higher row elements via valence expansion; (6) the possibility of radical addition to electron-deficient II and III group atoms.*

Most of our knowledge of the autoxidation of organic molecules has developed from the study of hydrocarbons, and our reaction schemes are concerned primarily with processes involving oxidative attack either on carbon-bound hydrogen or on olefinic double bonds. Introducing heteroatoms into organic molecules perturbs these processes and introduces the possibility of others involving reaction at the heteroatom itself. This paper surveys these effects in general terms of electronic structures and reaction energetics. While I have presented no new experimental material and have not attempted a comprehensive review or exhaustive bibliography, I hope it brings various observations together into a relatively consistent pattern.

Effects of Heteroatoms on Neighboring C-H Bonds

In the familiar hydroperoxide chain mechanism for hydrocarbon autoxidation, with propagation steps,

$$ROO\cdot + RH \rightarrow ROOH + R\cdot \qquad (1)$$

$$R\cdot + O_2 \rightarrow ROO\cdot \qquad (2)$$

Reaction 1 determines the point of oxidative attack on a molecule, and since it is slow and rate determining, it strongly influences the over-all rate. Even when they play no other role, heteroatoms should be expected to affect the reactivity of neighboring C—H bonds, either by resonance stabilization of the resulting radical, or by altering the polar properties of the transition state. In practice, the latter is usually the dominant effect. Competitive autoxidations of substituted benzaldehydes (23), cumenes (19), and toluenes (19) all show negative Hammett ρ values and indicate that the peroxy radical is electrophilic as expected from its structure. The facile autoxidation of aldehydes to peracids and of ethers to a mixture of peroxide products occur similarly at points where unshared electron pairs on the heteroatom can contribute electrons, as does the less familiar oxidation of hydrazones (17) and some amine derivatives (14). Autoxidation of sulfides also occurs preferentially at α-hydrogens, although over-all reaction paths are relatively complex (1).

Conversely, groups of electronegative heteroatoms which should act to decrease electron availability, deactivate neighboring C—H bonds towards attack, even though the resulting radicals should be markedly stabilized by delocalization—e.g., R—ĊH—COOR, RĊH—NO$_2$. Thus, autoxidation of long chain esters occurs at points remote from the ester function.

Electron-withdrawing groups, however, facilitate another pathway for oxidation by increasing the ease of carbanion formation *via* base attack on neighboring hydrogen. The ready autoxidation of some carbanions has been known for many years. In some systems the reaction corresponds, at least kinetically, to simple fast addition of oxygen to the carbanion (20). In others, possessing groups such as —NO$_2$ which undergo easy one-electron reduction, autoxidation proceeds at high rates through complex oxidation reduction chains, sometimes involving radical anion intermediates (21). Our knowledge of this field has largely been opened up by the work of G. A. Russell. Although its ramifications are still being explored, some of the steps involved resemble those occurring in the autoxidation of hydroquinones studied some time ago by Weissburger and his colleagues (11).

Attack on Heteroatom-Hydrogen Bonds

The feasibility of reactions similar to Reaction 1, but involving attack on hydrogen bonded to an atom other than carbon, depends initially on the over-all energetics of the process. Since $D(\text{ROO—H})$ seems to be approximately 90 kcal. per mole, regardless of R, a bond strength for hydrogen of this order or lower becomes a requirement for fast reaction. Table I lists some bond dissociation values for isoelectronic molecules.

Table I. Bond Dissociation Energies, kcal. per mole

CH_3CH_2—H	CH_3NH—H	CH_3O—H
98	(102)	102
CH_3SiH_2—H	CH_3PH—H	CH_3S—H
83	(85)	88

In the first row, $D(N-H)$ and $(O-H)$ are large (actually greater than C—H), and since, as we have already noted, these atoms facilitate peroxy radical attack on neighboring carbon, it is not surprising that direct oxidation of amino and alcohol functions is not usually accomplished by molecular oxygen. In phenols and aromatic amines, where the resulting radicals are stabilized by delocalizing the odd electron over the aromatic system, bond strengths are lowered, and peroxy radical attack readily occurs—e.g.,

$$ArOH + ROO\cdot \rightarrow ROOH + ArO\cdot \qquad (3)$$

$$Ar_2NH + ROO\cdot \rightarrow ROOH + Ar_2N\cdot \qquad (4)$$

However, since there is no analog of Reaction 2 to propagate the reaction chain, the over-all autoxidation loses its chain character. In short, we see a simple explanation of the well known fact that phenols and aromatic amines with N—H bonds are powerful inhibitors of hydrocarbon autoxidation. It is noteworthy that Reactions 3 and 4 are much faster than 1, even where over-all energetics appear comparable. Apparently reactions forming or breaking C—H bonds tend to be slower than isoenergetic processes involving hydrogen bonds to most other elements. It is also interesting that at high temperatures where Reaction 2 becomes reversible for benzylic hydrogen (2),

$$ArCH_2\cdot + O_2 \rightleftharpoons ArCH_2O_2\cdot \qquad (5)$$

molecules such as toluene and cumene become effective inhibitors for alkane oxidation (8).

Since autoxidations of phenols and aromatic amines are non-chain radical processes, they require some rapid radical-generating step. In a few systems—e.g., hydroquinone autoxidation, this is supplied by a direct redox reaction with oxygen (11).

$$\text{(1,4-}O^-\text{-C}_6H_4\text{-}O^-\text{)} + O_2 \longrightarrow \text{(1,4-}O\cdot\text{-C}_6H_4\text{-}O^-\text{)} + O_2^{\bar{\cdot}} \qquad (6)$$

In others, a transition metal catalyst may play the role. The over-all reactions are often complex, but with suitable control they may yield interesting products, as in the oxidative polymerization of 2,6-dimethylphenol (9).

Returning to Table I, we see that bonds to hydrogen are significantly weaker among second-row elements. This suggests facile peroxy radical attack, but examples of simple hydroperoxide chain processes are relatively few.

Autoxidation of mercaptans gives rise to thiyl radicals, but these, like phenoxy radicals, are inert toward oxygen and normally dimerize to disulfides. Their participation in a chain reaction can be achieved in the co-oxidation of olefins and mercaptans, first demonstrated by Kharasch (12), which takes advantage of the rapid addition of thiyl radicals to double bonds.

$$RS\cdot + CH_2{=}CHR \rightarrow RSCH_2\dot{C}HR \qquad (7)$$

$$RSCH_2\dot{C}HR + O_2 \rightarrow RSCH_2\overset{\overset{\displaystyle OO\cdot}{|}}{C}HR \qquad (8)$$

$$RSCH_2\overset{\overset{\displaystyle OO\cdot}{|}}{C}HR + RSH \rightarrow RSCH_2\overset{\overset{\displaystyle OOH}{|}}{C}HR + RS\cdot \qquad (9)$$

The resulting β-alkythiohydroperoxides can be isolated, but they normally rearrange to β-hydroxy sufoxides, or in the presence of amines they are reduced to alcohols with concomitant oxidation of two further molecules of mercaptan to disulfide (16).

Alternatively, the fast autoxidation of mercaptans is achieved by working in basic solution and in the presence of transition metal catalysts, *via* reactions involving RS⁻ anions and redox steps, much as for phenols (22).

Primary and secondary alkyl phosphines autoxidize rapidly. In fact lower members may ignite on exposure to air. However, reaction paths appear to involve both P—H bond cleavage and participation of phosphoranyl radicals (discussed below).

Higher group IV hydrides also autoxidize readily. Since Curtice, Gilman, and Hammond have shown that the autoxidation of triphenylsilane is a chain process (6), an initial hydroperoxide chain seems plausible. Trialkyltin hydrides similarly take up oxygen on exposure to air (13), yielding species with Sn—O bonds, but the reactions have not been studied in much detail.

Valence-Shell Expansion

The presence of unoccupied d-orbitals makes valence shell expansion of second row elements possible, and some of their radical reactions with oxygen appear to involve such a step.

Trialkyl phosphites react rapidly with oxygen, particularly in the presence of radical sources, to yield trialkyl phosphates. Results of several studies are consistent with a sequence first proposed by Walling and Rabinowitz (24),

$$ROO\cdot + P(OR)_3 \rightarrow [ROO\dot{P}(OR)_3] \rightarrow RO\cdot + OP(OR)_3 \quad (10)$$

$$RO\cdot + P(OR)_3 \rightarrow [RO\dot{P}(OR)_3] \rightarrow R\cdot + OP(OR)_3 \quad (11)$$

$$R\cdot + O_2 \rightarrow RO_2\cdot \quad (12)$$

The structures in brackets are phosphoranyl radicals with nine electrons on phosphorus and are considered to be transient intermediates. The radical, $R\cdot$, presumably represents some initiator fragment, but since the phosphoranyl radical in Reaction 11 is symmetric, $R\cdot$ exchanges can (and do) take place. In this regard, triaryl phosphites autoxidize much more slowly, and it has been suggested (3) that here phenoxy radicals are generated via another exchange (Reaction 13) and then terminate chains.

$$RO\cdot + P(OAr)_3 \rightarrow [RO\dot{P}(OAr)_3] \rightarrow ROP(OAr)_2 + ArO\cdot \quad (13)$$

The same reaction would account for the known efficiency of triaryl phosphites as inhibitors of hydrocarbon oxidation.

The autoxidation of trialkyl phosphines occurs similarly, except that here the second intermediate phosphoranyl radical can cleave by two paths:

$$RO\dot{P}R_3 \begin{array}{l} \nearrow OPR_3 + R\cdot \\ \searrow ROPR_2 + R\cdot \end{array} \quad (14)$$

to yield a mixture of alkylated and partially dealkylated products (4). Secondary phosphines initially yield phosphine oxides $R_2P(O)H$, presumably by a similar course, and only later are they oxidized to phosphinic acids $R_2P(O)OH$ (18).

In contrast to phosphines, dialkyl sulfides are relatively stable towards autoxidation. As we have pointed out elsewhere (25), the simplest explanation is an energetic one: the over-all transfer of an oxygen atom from carbon to phosphorus (Reaction 11) is exothermic by some 33 kcal. per mole. Transfer to sulfur, in contrast, is approximately thermoneutral and apparently too slow to maintain a rapid chain. Consequently, autoxidation of sulfides becomes a complex process involving both attack on α-hydrogen and oxidation to sulfoxide. It has been studied by Bateman and his group (5), who note that intermediates formed in the autoxidation are strong inhibitors for further reactions *via* hydroperoxide chains. Thus, sulfides (and other sulfur compounds) may serve as inhibitors of hydrocarbon autoxidation—an important matter in stabilizing oils and plastics.

Little is known about valence-shell expansion reactions among still higher row elements, but the considerable oxidative stability of silicones and of tetraalkyl derivatives of group IV elements suggests that valence shell-expansion doesn't provide an easy path for oxygen attack in this group.

Electron-Deficient Molecules

Trialkyl derivatives of boron, and in fact many other molecules such as boroxines with carbon-boron bonds, react readily with oxygen. The initial products are peroxy derivatives with BOOR bonds, which tend to react further to form borate esters. The ease of the initial reaction is shown by the fact that reported examples of vinyl polymerization induced by trialkyl borons require oxygen and are actually radical processes induced by the boron oxygen reaction or intermediate peroxides (7).

Although a simple polar insertion reaction

$$R_3B + O_2 \rightarrow R_2BOOR \qquad (15)$$

has been suggested for the initial autoxidation, Davies has recently reported evidence (5) that a radical chain process is actually involved

$$R\cdot + O_2 \rightarrow RO_2\cdot \qquad (16)$$

$$RO_2\cdot + BR_3 \rightarrow [RO_2\dot{B}R_3] \rightarrow RO_2BR_2 + R\cdot \qquad (17)$$

Reaction 17 resembles the displacements on phosphorus discussed earlier, except that since trialkyl borons are electron deficient with only six electrons around boron, no valence-shell expansion is required. It appears to be a general reaction for boron and has been suggested by Matteson (15) to explain the conversion of trimethyl boroxine by *tert*-butyl hypo-

$$\text{Me-boroxine} + 3\ tert\text{-BuOCl} \rightarrow \text{BuO-boroxine} + 3\ \text{MeCl} \qquad (18)$$

chlorite to tri-*tert*-butyl metaborate. Here the *tert*-butoxy radical is the attacking species.

Other alkyl derivatives of group III elements also react readily with oxygen, presumably *via* the same sort of path. As we move to the left in the periodic table, high reactivity with oxygen continues, but carbon–metal bonds become increasingly ionic; hence, at least in ionizing solvents, we approach the carbanion reactions considered earlier. Grignard reagents may provide us with a final example in this discussion.

The autoxidation of Grignard reagents was shown to be a two-step process:

$$RMgX + O_2 \rightarrow ROOMgX \qquad (19)$$

$$ROOMgX + RMgX \rightarrow 2\ ROMgX \qquad (20)$$

by Walling and Buckler, who demonstrated that it could be stopped at the peroxide stage by inverse addition of the Grignard reagent to excess oxygen (26). The reactions parallel the boron case, and several other organometallic compounds behave similarly. Thus, with zinc and cadmium the analog of Reaction 2 is slower, and as shown by Hock and Ernst (10), yields of peroxides from Grignard reactions may be improved by first adding zinc or cadmium halides to the system.

At least three mechanisms are available for Reaction 19: simple insertion as written (presumably *via* an initial complex); a redox chain (considering the R—MgX bond as ionic):

$$R\cdot + O_2 \rightarrow RO_2\cdot \qquad (21)$$

$$RO_2\cdot + R^- \rightarrow RO_2^- + R\cdot \qquad (22)$$

and a displacement chain, where Reaction 22 is replaced by

$$RO_2 + RMgX \rightarrow [ROOMg\overset{R}{\underset{X}{\diagdown}}] \rightarrow ROOMgX + R\cdot \qquad (23)$$

The simple insertion was favored by Walling and Buckler (25) since they could find no evidence for a chain process. However, Lamb (27) has recently reported that during autoxidation 5-hexenylmagnesium bromide undergoes cyclization to give eventually cyclopentylmethanol—a cyclization which is typical of the 5-hexenyl free radical. If this is true, either Reactions 21 and 22 or 21, 22, and 23 are indicated. In view of the close parallel between the autoxidations of all these electron-deficient molecules, some of which are essentially covalent in structure, the displacement scheme (Reactions 21 to 23) may have wider application than has been realized.

Literature Cited

(1) Barnard, D., Bateman, L., Cunneen, J. E., "Organic Sulfur Compounds," Vol. I, Chap. 21, N. Kharasch, ed., Pergamon Press, New York, 1961.
(2) Benson, S. W., *J. Am. Chem. Soc.* **87,** 971 (1965).
(3) Bentrude, W. G., *Tetrahedron Letters* **1965,** 3543.
(4) Buckler, S. A., *J. Am. Chem. Soc.* **84,** 3093 (1962).
(5) Coffee, E. C. J., Davies, A. G., *J. Chem. Soc., C. Org.* **1966,** 1493.
(6) Curtice, J., Gilman, H., Hammond, G. S., *J. Am. Chem. Soc.* **79,** 4754, (1957).
(7) Fordham, J. W. L., Sturm, C. L., *J. Polymer Sci.* **33,** 503 (1958).
(8) Giammaria, J. J., Norris, H. D., *Ind. Eng. Chem. Prod. Res. Develop.* **1,** 16 (1962).
(9) Hay, A. S., Blanchard, H. S., Endres, G. F., Eustance, J. W., *J. Am. Chem. Soc.* **81,** 6335 (1959).
(10) Hock, H., Ernst, F., *Chem. Ber.* **92,** 2716 (1959).
(11) James, T. H., Weissberger, A., *J. Am. Chem. Soc.* **60,** 98 (1938).
(12) Kharasch, M. S., Nudenberg, W., Mantell, G. J., *J. Org. Chem.* **16,** 524 (1951).
(13) Kuivila, H. G., "Advances in Organometallic Chemistry," Vol. I, p. 47, F. G. H. Stone, R. West, eds., Academic Press, New York, 1964.
(14) Lock, M. V., Sager, B. F., *J. Chem. Soc. B, Phys. Org.* **1966,** 690.
(15) Matteson, D. S., *J. Org. Chem.* **29,** 3399 (1964).
(16) Oswald, A. A., Griesbaum, K., Hudson, B. E., Jr., *J. Org. Chem.* **28,** 2351 (1963).
(17) Pausacker, K. H., *J. Chem. Soc.* **1950,** 3478.
(18) Rauhut, M. M., Currier, H. A., *J. Org. Chem.* **26,** 4626 (1961).
(19) Russell, G. A., *J. Am. Chem. Soc.* **78,** 1047 (1956).
(20) Russell, G. A., Bemis, A. G., *J. Am. Chem. Soc.* **88,** 5491 (1966).
(21) Russell, G. A., *et al.,* ADVAN. CHEM. SER. **51,** 112, (1965).
(22) Tarbell, D. S., "Organic Sulfur Compounds," Vol. I, Chap. 10, N. Kharasch, ed., Pergamon Press, New York, 1961.
(23) Walling, C., McElhill, E. A., *J. Am. Chem. Soc.* **73,** 2927 (1951).
(24) Walling, C., Rabinowitz, R., *J. Am. Chem. Soc.* **81,** 1243 (1959).
(25) Walling, C., Mintz, M. J., *J. Org. Chem.* **32,** 1286 (1967).
(26) Walling, C., Buckler, S. A., *J. Am. Chem. Soc.* **77,** 6032 (1955).

RECEIVED January 23, 1968.

14

Oxidation of Carbanions

Oxidation of Diarylmethanes and Diarylcarbinols in Basic Solution

GLEN A. RUSSELL, ALAN G. BEMIS, EDWIN J. GEELS, EDWARD G. JANZEN, and ANTHONY J. MOYE

Department of Chemistry, Iowa State University, Ames, Iowa 50010

Oxidation of diphenylmethane in basic solutions involves a process where rate is limited by and equal to the rate of ionization of diphenylmethane. The diphenylmethide ion is trapped by oxygen more readily than it is protonated in dimethyl sulfoxide–tert-butyl alcohol (4 to 1) solutions. Fluorene oxidizes by a process involving rapid and reversible ionization in tert-butyl alcohol solutions. However, in the presence of m-trifluoromethylnitrobenzene, which readily accepts one electron from the carbanion, the rate of oxygen absorption can approach the rate of ionization. 9-Fluorenol oxidizes in basic solution by a process that appears to involve dianion or carbanion formation. Benzhydrol under similar conditions oxidizes to benzophenone by a process not involving carbanion or dianion formation.

We have reported that the triphenylmethide ion in dimethyl sulfoxide (DMSO) solution reacts with oxygen at a rate approaching the diffusion-controlled limit ($k > 10^9$ liters/mole sec.) (*16*). The triphenylmethide ion is actually more reactive toward molecular oxygen than the triphenylmethyl radical. Because of the reactivity of the triphenylmethyl anion toward molecular oxygen, it is possible to measure the rate of ionization of triphenylmethane in basic solution by the rate of oxygenation.

$$Ar_3CH + B^- \xrightarrow{slow} Ar_3C:^-$$

$$Ar_3C:^- + O_2 \xrightarrow{fast} Ar_3COO^-$$

The more acidic fluorene in *tert*-butyl alcohol solution, or in DMSO solution, reacts by a process that involves the carbanion in equilibrium with hydrocarbon. Thus, fluorene and 9,9-dideuteriofluorene oxidize at identical rates. We have established that the oxidation of the anion of fluorene can be catalyzed by a variety of electron acceptors (π), including various nitroaromatics (*18*). The catalyzed oxidation rates were found to follow the rates of electron transfer measured by ESR spectroscopy in the absence of oxygen. These results established the catalyzed reaction as a free radical chain process without shedding light upon the mechanism of the uncatalyzed reaction.

$$Ar_2CH_2 \overset{fast}{\rightleftarrows} Ar_2CH^-$$

$$Ar_2CH^- + \pi \rightarrow Ar_2CH\cdot + \pi\cdot^-$$

$$Ar_2CH\cdot + O_2 \rightarrow Ar_2CHOO\cdot$$

$$Ar_2CHOO\cdot + Ar_2CH^- \rightarrow Ar_2CHOO^- + Ar_2CH\cdot$$

$$\pi\cdot^- + O_2 \rightarrow \pi + O_2\cdot^-$$

$$Ar_2CHOO\cdot + O_2\cdot^- \rightarrow Ar_2CHOO^- + O_2$$

The regeneration of the nitroaromatic catalyst from the radical anion ($\pi\cdot^-$) has been studied recently in an isolated system, and the fate of the superoxide ion ($O_2\cdot^-$) was discussed (*15*).

The present work demonstrates that the oxidation of diphenylmethane in basic solution follows a pattern similar to triphenylmethane and not to fluorene. At high concentrations of good electron acceptors it is possible to realize a situation wherein the rate of oxidation of fluorene is limited by and equal to the rate of ionization. The oxidations of benzhydrol and 9-fluorenol in basic solution are considered; the difference in acidity of the methine hydrogens has a pronounced effect on the course of these oxidations.

Other studies of the oxidation of diarylmethanes or diarylcarbinols pertinent to this work involve the work of Sprinzak on the oxidation of fluorene and 2,3-diphenylindene (*27*), of Pratt and Trapasso using alumina impregnated with sodium methoxide (*11*), of Pauson and Williams on tetraphenylcyclopentadienyllithium (*10*), of Barton and Jones on 1,1,3-triphenylprop-1-ene and 9,10-dihydroanthracene (*1*), of Cairns, McKusick, and Wienmayr using dithienylmethane (*2*), of Wooster using diphenylmethylsodium in liquid ammonia (*30*), and of the dilithium adducts of tetraphenylethylene and anthracene (*7, 24*). The oxidation in basic solutions of benzhydrol and other diarylcarbinols has also been studied, primarily by Étienne and LeBerre (*3, 4, 8, 9*).

Results

Oxidation of Diphenylmethane in Basic Solution. Diphenylmethane reacts with an excess of oxygen in the presence of potassium *tert*-butoxide in various solvents to produce nearly quantitative yields of benzophenone. In DMSO (80%)–*tert*-butyl alcohol (20%) a 96% yield of the benzophenone–DMSO adduct [1,1-diphenyl-2-(methylsulfinyl)ethanol] was isolated at complete reaction (*17*).

If the oxidation of diphenylmethane in DMSO (80%)–*tert*-butyl alcohol (20%) is interrupted after the absorption of one mole of oxygen per mole of diphenylmethane, one obtains an 86% yield of benzhydrol, 10% yield of unreacted diphenylmethane, and a few percent of the benzophenone–DMSO adduct. The over-all course of the reaction follows Reactions 1 to 5.

$$(Ar)_2CH_2 + (CH_3)_3COK \xrightarrow{k_i} (Ar)_2CH:^- \quad (1)$$

$$(Ar)_2CH:^- + O_2 \xrightarrow{k_0} (Ar)_2CHOO^- \quad (2)$$

$$(C_6H_5)_2CHOO^- + DMSO \rightarrow (C_6H_5)_2CHO^- + CH_3SO_2CH_3 \quad (3)$$

$$(C_6H_5)_2CHO^- \xrightarrow{B^-, O_2} (C_6H_5)_2CO \quad (4)$$

$$(Ar)_2CO + CH_3SOCH_3 \xrightarrow{B^-} (Ar)_2C(OH)CH_2SOCH_3 \quad (5)$$

Independent evidence for Reactions 3 and 5 has been presented (*16, 17*).

The oxidation of diphenylmethane in DMSO solution does not involve Reaction 6. The stoichiometry and product isolation as defined in Figure 1 exclude Reaction 6.

$$Ar_2CHOO^- \rightarrow Ar_2CO + OH^- \quad (6)$$

In the absence of DMSO the conversion of diphenylmethyl hydroperoxide to benzophenone apparently does follow Reaction 6, at least in alcohol-containing solvents. The stoichiometry becomes nearly one mole of oxygen per mole of diphenylmethane, and the carbinol is eliminated as an intermediate. Table I lists the observed stoichiometries and initial rates of oxidation of diphenylmethane. In pyridine-, DMF-, or HMPA-containing solvents, a high yield of benzophenone was isolated upon hydrolysis after oxygen absorption had ceased. In pure HMPA there was considerable evolution of oxygen upon hydrolysis.

The high stoichiometry in pure HMPA, and to a lesser degree in pyridine solution, suggests that Reaction 6 may require the presence of a proton donor to yield the free hydroperoxide, which can then undergo

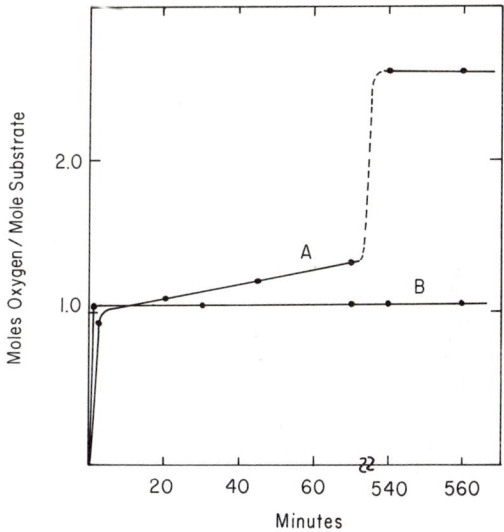

Figure 1. Oxygen absorption in dimethyl sulfoxide (80%)–tert-butyl alcohol (20%) at 27° ± 2°C.

A. Diphenylmethane
B. Fluorene
Substrate concentration 0.10M
Potassium tert-butoxide concentration 0.233M

Table I. Oxidation of Diphenylmethane in Basic Solutions[a]

Solvent[b]	Stoichiometry[c]	Initial Rate[d]
tert-Butyl alcohol	—	<0.00
Pyridine(80)–tert-BuOH(20)	—	0.02
Pyridine	1.4	0.15
DMF(80)–tert-BuOH(20)	1.0	0.37
DMSO(80)–tert-BuOH(20)	2.7	1.0
HMPA(80)–tert-BuOH(20)	1.1	1.1
HMPA	2.7	2.9

[a] 0.1M diphenylmethane (initial concentration), 0.2M potassium tert-butoxide, 749-mm. saturation oxygen pressure at 27° ± 2°.
[b] Mixtures in volume %. DMF = dimethylformamide, DMSO = dimethyl sulfoxide, HMPA = hexamethylphosphoramide.
[c] Moles of oxygen absorbed/mole of diphenylmethane at complete reaction.
[d] Moles of oxygen absorbed/ mole of diphenylmethane per minute.

a beta-E-2 elimination (Reaction 7). Reactions in HMPA hydrolyzed

$$Ar_2CHOOH + B^- \rightarrow HB + Ar_2CO + OH^- \qquad (7)$$

after the absorption of one mole of oxygen did not yield significant amounts of alcohol, while after the absorption of 1.5 moles of oxygen no diphenylmethane remained. The oxidation may involve further oxidation of the hydroperoxide (Reactions 8 to 10).

$$Ar_2CHOO^- + B^- \rightleftarrows Ar_2C(OO^-):^- \tag{8}$$

$$Ar_2C(OO^-):^- + O_2 \rightarrow Ar_2C(OO^-)_2 \tag{9}$$

$$Ar_2C(OO^-)_2 + H_2O \rightarrow Ar_2CO + 2\,OH^- + O_2 \tag{10}$$

The rate of oxidation of diphenylmethane has been determined at a number of initial diphenylmethane and base concentrations. During the initial stages, it appears reasonable that each molecule of diphenylmethane destroyed consumed exactly one molecule of oxygen and that the instantaneous concentration of diphenylmethane is given by $[(C_6H_5)_2CH_2]_0 - O_2$ absorbed. Treatment of the data in this manner yields plots for a pseudo-first-order reaction as shown in Figure 2. If $k_0[O_2][Ar_2CH:^-] \gg k_{-i}[(CH_3)_3COH][Ar_2CH:^-]$, the apparent rate constant observed will be $k_i[KOC(CH_3)_3]$. Following this assumption leads to

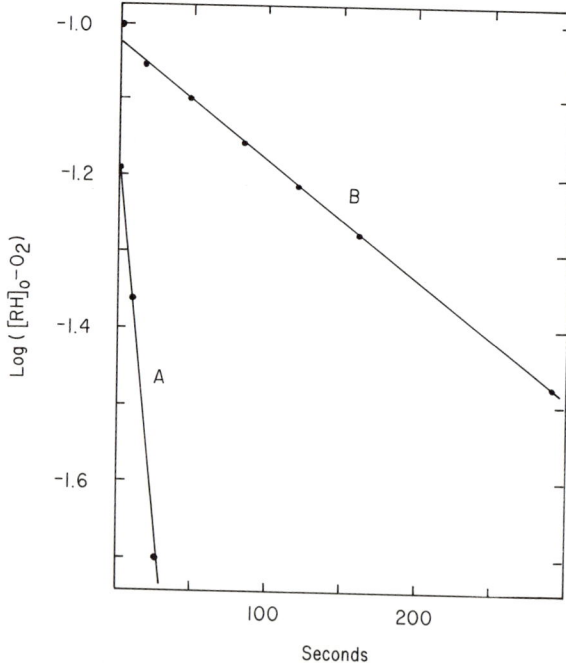

Figure 2. *Oxidation in dimethyl sulfoxide (80%)-tert-butyl alcohol (20%) at 24.5°C.*

A. Diphenylmethane
B. α-α-Dideuteriodiphenylmethane
Potassium tert-butoxide concentration 0.20M in A, 0.232M in B. Diarylmethane concentration 0.16M

the data in Table II. These data are consistent with an ionization process that is first-order in base and first-order in diphenylmethane with a value of k_i at 24.5°C. of 0.085 ± 0.015 mole^{-1} per second.

Table II. Calculated Ionization Rate Constants for Diphenylmethane in Dimethyl Sulfoxide (80%)–*tert*-Butyl Alcohol Solutions

Temp., °C.	$[(C_6H_5)_2CH_2]_0$, Mole	$[KOC(CH_3)_3]_0$, Mole	$k_i{}^a$, Mole/Sec.
31–32	0.025	0.05	0.15
31–32	0.064	0.20	0.18
31–32	0.017	0.20	0.18
25–26	0.05	0.23	0.087
25–26	0.10	0.23	0.092
25–26	0.10	0.23	0.12
24.5	0.064	0.20	0.089
24.5	0.064	0.20	0.105
24.5	0.10	0.20	0.070
24.5	0.10	0.20	0.075
24.5	0.10^b	0.23	0.015

^{a} Calculated from $-d[O_2]/dt = k_i[(C_6H_5)_2CH_2][KOC(CH_3)_3]_0$ (Figure 2). Oxidation rate measured in vigorously shaken flask containing oxygen at 750 mm.
^{b} α,α-Dideuteriodiphenylmethane.

This interpretation was proved correct by considering the oxidation of a sample of diphenylmethane that had an isotopic purity of 97.0% α,α-dideuterio and 2.7% α-deuterio by mass spectrometry. The oxidation rate observed after the initial 15-second period (*see* Figure 2), during which the undeuterated and monodeuterated material were destroyed, yielded a second-order rate constant, $k_i = 0.0148$ mole^{-1} per second. There is thus an appreciable isotope effect k_H/k_D of about 6 in the ionization of diphenylmethane by potassium *tert*-butoxide in DMSO(80%)–*tert*-butyl alcohol (20%) at 25°C. This compares with a value of k_H/k_D of 9.5 reported for the ionization of triphenylmethane (*16*). The observation of primary isotope effects of this magnitude requires that the protonation of the diphenylmethide ion by *tert*-butyl alcohol in DMSO solution does not proceed at the diffusion rate which would, by the principle of microscopic reversibility, require the absence of an isotope effect in the deprotonation step.

Final confirmation of our interpretation of the rate of oxygen absorption by diphenylmethane was provided by studying the loss of deuterium from the deuterated sample of diphenylmethane. After 1.50 and 2.70 minutes in a solution containing 0.233M potassium *tert*-butoxide and 0.10M diarylmethane, and in the absence of oxygen, the recovered diphenylmethane was found to contain dideuterio, monodeuterio, and undeuterated diphenylmethane in the ratios, 74.0:23.8:2.2 (1.5 minutes) and 58.7:34.9:6.4 (2.7 minutes). These data lead to a calculated value of k_i of 0.013 mole^{-1} per second. We consider this number to be equal to the value of k_i obtained from the rate of oxygen absorption (0.015). Thus, the rate of oxygen absorption is equal to the rate of ionization, and k_0 must be considerably greater than k_{-i} ([*tert*-BuOH] $\gg$ [O_2]). In the presence of oxygen the diphenylmethide ion once formed

$$(C_6H_5)_2CH:^- + (CH_3)_3COH \xrightarrow{k_{-i}} (C_6H_5)_2CH_2 + (CH_3)_3CO^- \quad (11)$$

is never reprotonated to yield diphenylmethane (Reaction 11). Instead, it is trapped by oxygen at a rate that must be close to the diffusion-controlled rate.

When the oxidation of α,α-dideuteriodiphenylmethane was interrupted after the absorption of 1.0 equivalent of oxygen, the product was found to be only benzhydrol and a trace of diphenylmethane (by GLPC). Mass spectroscopic analysis of the benzhydrol indicated 98.5% monodeuterated material.

Oxidation of Fluorene in Basic Solution. The oxidation rate of diphenylmethane was found to be independent of added nitroaromatics, as was previously observed to be the case with triphenylmethane (*16*), but not fluorene (*18*). These results agree for a rate-limiting ionization process for the oxidation of diphenylmethane but not for fluorene under the standard reaction conditions. We have extended the study of fluorene oxidation and now present definitive evidence that for fluorene the value of k_i [Ar_2CH_2][B^-] $\gg k_0$[O_2][Ar_2CH^-]. However, in the presence of large amounts of nitroaromatics the rate of oxygen absorption can actually become equal to the rate of ionization—*i.e.*, k[$ArNO_2$][$Ar_2CH:^-$] = k_i [Ar_2CH_2][B^-].

The oxidation of fluorene in basic solution is in sharp contrast to diphenylmethane. Figure 1 emphasizes the clean stoichiometry observed in the oxidation of fluorene (one mole of oxygen per mole of fluorene). The over-all reaction for fluorene apparently involves Reactions 1, 2, 6, and 5 (*18*).

The hydroperoxide from the more acidic fluorene prefers to react via the decomposition process (Reaction 6 or 7) rather than the reductive reaction with dimethyl sulfoxide (Reaction 3).

Table III summarizes some information on the initial rates of oxidation of fluorene in several solvents. In the alcohol-containing solvents the stoichiometry was nearly one molecule of oxygen per mole of fluorene, an observation that excludes 9-fluorenol as an intermediate. In all solvents, including HMPA, interrupted oxidations yielded only fluorenone (or the DMSO-fluorenone adduct) and fluorene. Apparently Reaction 6 or 7 occurs readily in the presence of hydroxylic solvents. In HMPA the high stoichiometry suggests further oxidation of the initially formed hydroperoxide, as was observed for diphenylmethane (Reactions 8 to 10).

Table III. Oxidation of Fluorene in Basic Solution[a]

Solvent	Stoichiometry[b]	Initial Rate[c]
tert–Butyl alcohol	1.09	0.035
DMF(80%)–tert-BuOH(20%)	1.16	2.7[d]
DMSO(80%)–tert-BuOH(20%)	1.14	2.8[d]
HMPA(80%)–tert-BuOH(20%)	1.22	2.9[d]
HMPA	2.5	3.7[d,e]

[a] $[Ar_2CH_2]_0 = 0.1M$, $[KOC(CH_3)_3] = 0.2M$, 25-27°C., saturation oxygen pressure of 750 mm.
[b] Moles of oxygen consumed per mole of fluorene.
[c] Moles of oxygen per mole of fluorene per minute.
[d] Minimum rate, may be limited by oxygen diffusion rate.
[e] 0.4M potassium tert-butoxide.

Fluorene and 9,9-dideuteriofluorene oxidized at the same rate in DMSO and in tert-butyl alcohol solution. This observation is consistent with a rapid, reversible ionization step. In tert-butyl alcohol the exchange of alpha deuterium atoms of dideuteriofluorene was measured (see experimental section) and a second-order rate constant for ionization calculated to be 0.12 ± 0.01 mole^{-1} per second. Under the conditions of this experiment the rate of oxygen absorption of undeuterated fluorene was approximately 1/50 the rate of deuterium exchange from the 9,9-dideuteriofluorene.

The rate of oxidation of fluorene was measured in the presence of various concentrations of nitrobenzene and the better electron acceptor, m-trifluoromethylnitrobenzene. It was felt that the rate of Reaction 12 should be given by the rate of oxygen absorption (Reaction 13).

$$Ar_2CH:^- + ArNO_2 \xrightarrow{k} ArNO_2 \cdot^- + Ar_2CH \cdot \quad (12)$$

$$ArNO_2 \cdot^- + O_2 \rightarrow ArNO_2 + O_2 \cdot^- \quad (13)$$

The following data would appear to substantiate this premise. At high nitroaromatic concentrations Reaction 12 should be able to compete with the reprotonation of the carbanion and the rate of ionization should become equal to the rate of oxygen absorption. Since the stoichiometry of the oxidation did not change on adding the nitroaromatic catalysts, the assumption that the absorption of only one molecule of oxygen occurred for each electron transfer step is legitimate.

To explain the constant stoichiometry Reactions 14 and 15, or 16, followed by 6, are suggested (15).

$$Ar_2CH \cdot + O_2 \rightarrow Ar_2CHOO \cdot \quad (14)$$

$$Ar_2CHOO\cdot + O_2\cdot^- \rightarrow Ar_2CHOO^- + O_2 \qquad (15)$$

$$Ar_2CH\cdot + O_2\cdot^- \rightarrow Ar_2CHOO^- \qquad (16)$$

Table IV summarizes some data observed for nitrobenzene-catalyzed reactions.

Table IV. Nitrobenzene-Catalyzed Oxidations of Fluorene in tert-Butyl Alcohol Solution at 29.5°C.

$[Fluorene]_0$, Mole	$[KOC(CH_3)_3]_0$, Mole	$[C_6H_5NO_2]$, Mole	k_i,[a] Mole^{-1} Sec.$^{-1}$	$k_i/[C_6H_5NO_2]$
0.048	0.20	—	0.0024	—
0.048	0.20	0.020	0.0169	0.85
0.048	0.20	0.039	0.035	0.99
0.048	0.20	0.078	0.062	0.80
0.048	0.20	0.118	0.091	0.77
0.023	0.024	0.376	0.155	0.41
0.022	0.023	0.725	0.328	0.45
0.022	0.023	1.05	0.396	0.38
0.020	0.021	1.63	0.560	0.34
0.023	0.024	2.73	0.852	0.31

[a] Calculated from $(-d[O_2]/dt)_0 = k_i[Ar_2CH_2]_0[KOC(CH_3)_3]$ by measurement of initial oxidation rate observed for a solution saturated with oxygen at 750 mm.

It is apparent from Table IV that with nitrobenzene as the oxidation catalyst the ionization-limited rate was not reached even at a nitrobenzene concentration of 2.7M (0.02M fluorene, 0.02M potassium tert-butoxide). The rate of oxidation at the low nitrobenzene concentrations is first-order in nitrobenzene, fluorene, and base. This is consistent with an oxidation rate determined by Reaction 12 and involving an equilibrium concentration of the fluorene anions.

Data obtained with m-trifluoromethylnitrobenzene are given in Table V. Again, at low nitroaromatic concentrations, the reaction is first-order in the nitroaromatic. However, at high nitroaromatic concentrations the oxidation rate becomes independent of nitroaromatic concentration, as expected for an ionization rate-limited process.

The catalyst in the absence of fluorene absorbed oxygen, and in Table V a correction for this process has been attempted. The corrected data indicate an ionization rate constant of fluorene of 1.2 ± 0.2 mole^{-1} per second at 29.5°C. The isotope effect in ionization of fluorene and 9,9-dideuteriofluorene is thus about 10.

Final confirmation of this interpretation was attempted by studying the electron transfer between fluorene and m-trifluoromethylnitrobenzene in basic solution monitored by ESR spectroscopy in the absence of oxygen. Table VI summarizes data yielding an ionization rate constant of 0.9

mole⁻¹ per second. The experimental uncertainties in the ESR measurement are in the neighborhood of 50%, so the agreement between the rate constants measured by oxidation (Table V) and ESR (Table VI) is excellent.

Table V. Oxidation of Fluorene Catalyzed by *m*-Trifluoromethylnitrobenzene in *tert*-Butyl Alcohol at 29.5°C.

[Fluorene]$_0$, Mole	[KOC(CH$_3$)$_3$], Mole	[m-CF$_3$C$_6$H$_4$NO$_2$], Mole	k_i,[a] Moles⁻¹ Sec.⁻¹	k_i[a,b] Moles⁻¹ Sec.⁻¹
0.048	0.20	—	0.0024	0.0024
0.024	0.025	0.015	0.15	0.15
0.024	0.025	0.030	0.38	0.34
0.024	0.025	0.15	0.67	0.63
0.023	0.024	0.29	0.85	0.74
0.022	0.023	0.68	1.36	1.03
0.022	0.022	0.81	1.38	1.05
0.020	0.021	1.25	1.64	1.19

[a] Calculated from expression $(-d[O_2]/dt) = k_i[Ar_2CH_2]_0[KOC(CH_3)_3]$ by measurement of initial oxidation rate observed for solution saturated with oxygen at 750 mm.
[b] Corrected for oxidation of catalyst observed in absence of fluorene.

Table VI. Rates of Electron Transfer from Fluorene Anion to *m*-Trifluoromethylnitrobenzene in *tert*-Butyl Alcohol Solution

Moles/Liter			$(d[O_2]/dt)_0$ (Msec.)⁻¹ × 10⁻⁴	$(d[R\cdot^-]/dt)_0$[a] (Msec.)⁻¹ × 10⁻⁴	k_i[b] Mole Sec.⁻¹
m-CF$_3$C$_6$H$_4$NO$_2$	Fluorene	KOC(CH$_3$)$_3$			
—	0.024	0.025	0.02	—	—
0.015	0.024	0.025	0.73	0.60	0.12
0.030	0.024	0.025	1.75	1.04	0.30
0.15	0.024	0.025	3.30	2.44	0.59
0.29	0.023	0.024	3.30	1.23	0.59
0.68	0.022	0.023	4.32	3.58	0.87
0.81	0.022	0.022	4.26	3.40	0.89

[a] Initial rate, moles *m*-trifluoronitromethylbenzene radical anion formed per liter second as measured by ESR spectroscopy.
[b] Calculated from expression $(d[R\cdot^-]/dt)_0 = k_i[Ar_2CH_2]_0[KOC(CH_3)_3]$.

We now feel that the experimental evidence completely proves that for diphenylmethane in basic solution the oxidation involves a rate-limiting ionization step followed by the rapid trapping of the diphenylmethide ion by oxygen. Because of the rate-limiting ionization step, catalysis by nitroaromatics cannot be observed in the oxidation of diphenylmethane. On the other hand, fluorene generates a much more stable carbanion that apparently reacts slowly with oxygen. Ionization is no longer rate-limiting except under conditions of strong catalysis.

Oxidation of Other Diarylmethanes. Table VII summarizes the products isolated in the oxidation of a number of other diarylmethanes.

Table VII. Oxidation of Diarylmethanes and Derivatives[a]

Diarylmethane, M	$KOC(CH_3)_3$, M	O_2 Absorbed[b]	Product, % Yield	
Xanthene, 0.1	0.22	1.68	Xanthone	92
Thioxanthene, 0.1	0.22	1.21	Thioxanthene	51
Anthrone, 0.1	0.22	1.02	Anthraquinone	13
			Anthraquinone-DMSO adduct	68
9,10-Dihydroanthracene, 0.1	0.22	1.91	Anthracene	69
			Anthraquinone	11
			Anthraquinone-DMSO adduct	10
Acridan, 0.1	0.22	1.94	Acridine	87
sym-Tetraphenylethane, 0.1	0.20	1.4	Tetraphenylethylene	96
1,1,2-Triphenylethane, 0.1	0.50[c]	1.5	Triphenylethylene	
Benzhydrylamine, 0.05	0.20	1.2	Benzophenone-DMSO adduct	
Benzhydryl chloride, 0.05	0.20	1.0	Benzophenone-DMSO adduct	52
Diphenylacetonitrile, 0.05	0.20	1.0	Benzophenone-DMSO adduct	26
Diphenylacetic acid, 0.1	0.20	1.0	Benzilic acid	24
4,5-Methylenephenanthrene, 0.1	0.20[d]	1.8	4,5-Carbonylphenanthrene	low
1,1-Diphenylacetone, 0.1	0.20	1.3	Benzophenone-DMSO adduct	40

[a] In dimethyl sulfoxide (80%)–tert-butyl alcohol solution at 27° ± 3°C.
[b] At complete reaction, moles oxygen absorbed per mole substrate.
[c] In hexamethylphosphoramide solution.
[d] tert-Butyl alcohol solution.

Oxidation of Benzhydrol in Basic Solution. Reaction of benzhydrol with oxygen in basic solution results in the formation of benzophenone, or in DMSO solutions the benzophenone—DMSO adduct. Table VIII summarizes data on the oxidation of benzhydrol in three solvents and in the presence of various concentrations of potassium tert-butoxide. The rates are the maximum oxidation rates, often observed after an inductive period (Figure 3).

Table VIII indicates a small effect of solvent upon the rate of the base-catalyzed oxidation. The effect of the concentration of excess base on the rate of oxidation is also small. The small variations in rate illustrated in Table VIII are not considered necessarily definitive since the stoichiometry of the reaction changes with solvent and base concentration and the rates were observed only after an induction period. The data of Table VIII provide no evidence for ionization of the α-hydrogen atom of benzhydrol to give either a carbanion or dianion. Consistent with this interpretation, nitrobenzene

$$Ar_2CHO^- \rightleftarrows {}^-{:}C(OH)Ar_2 \rightleftarrows [Ar_2C{-}O]^{-2} \qquad (17)$$

Table VIII. Oxidation of Benzhydrol in Basic Solutions at 27 ± 2°C.

		Rates of Oxygen Absorption[a]		
$[(C_6H_5)_2CHOH]_0$, M	$[KOC(CH_3)_3]_0$, M	DMSO(80%)- tert-BuOH(20%)	Pyridine(80%)- tert-BuOH(20%)	tert-BuOH
0.10	0.00	0.00	0.00	0.00
0.06	0.18	0.012	0.009	0.012
0.12	0.12	—	—	0.005
0.12	0.24	0.010	0.017	0.015
0.12	0.36	0.016	—	0.019
0.12[b]	0.36	0.0015	—	0.0007
0.12	0.48	—	—	0.023

[a] Moles oxygen/mole benzhydrol per minute.
[b] α-Deuteriobenzhydrol.

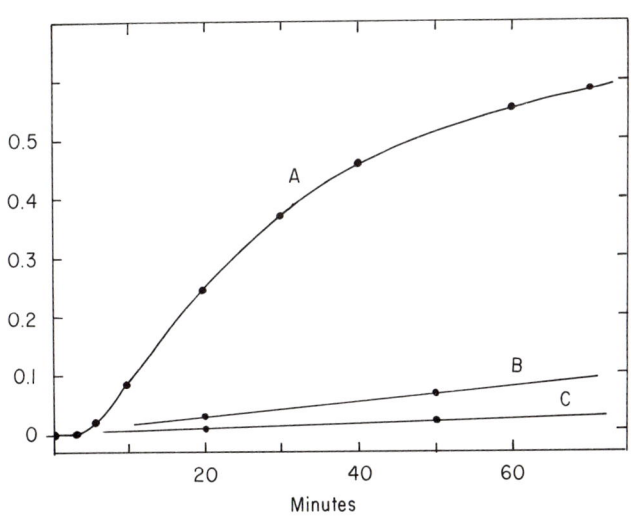

Figure 3. Oxidation of 0.1M benzhydrol in tert-butyl alcohol containing 0.4 potassium tert-butoxide

A. No additive
B. 0.002M ferric chloride
C. 0.01M arsenic trioxide

had no effect on the rate of oxidation. Moreover, α-deuteriobenzhydrol did not undergo exchange of the deuterium atom in tert-butyl alcohol containing 0.36M potassium tert-butoxide during a 24-hour period. Unreacted benzhydrol recovered from an oxidation of α-deuteriobenzhydrol in DMSO (80%)–tert-butyl alcohol (20%) showed no exchange even after an oxidation period of 6 hours.

Ferric chloride (0.002M) reduced the rate of oxidation of benzhydrol (0.15M) in the presence of 0.39M potassium tert-butoxide in tert-butyl alcohol to a rate of 0.001 mole of oxygen per mole of benzhydrol per minute, while arsenic trioxide (0.01M) reduced the rate of oxidation of 0.12M benzhydrol and 0.36M potassium tert-butoxide to 0.0001 mole of oxygen per mole of benzhydrol per minute for a 4-hour period, after which the oxidation occurred at the uninhibited rate (Figure 3). Table IX summarizes some observed stoichiometries in the oxidation of benzhydrol.

Table IX. Oxygen Stoichiometry in Oxidation of Benzhydrol[a]

$[KOC(CH_3)_3]_0$, M	Solvent	Moles Oxygen Absorbed per Mole Benzhydrol	Time, Min.	Recovered Benzhydrol, %
0.12	tert-BuOH	0.73	1100[b]	13
0.20	tert-BuOH	0.91	625	7
0.24	tert-BuOH	1.02	425	7
0.24	tert-BuOH	1.13	1100[b]	—
0.36	tert-BuOH	1.36	1200[b]	3
0.48	tert-BuOH	1.45	1400[b]	—
0.24	DMSO(80)–tert-BuOH(20)	1.33	600[b]	—
0.36	DMSO(80)–tert-BuOH(20)	1.65	275[b]	—

[a] 0.12M benzhydrol, 27° ± 3°C.
[b] Oxygen absorption had become very slow.

The stoichiometry of the oxidation appears to require the formation of potassium superoxide as one of the oxidation products, particularly at long reaction periods and high base concentrations. An oxidation of 3.00 mmoles of benzhydrol (0.12M) in the presence of 9.9 mmoles of potassium tert-butoxide (0.37M) in DMSO (80%)–tert-butyl alcohol (20%) absorbed 4.95 mmoles of oxygen in 27.7 minutes at 25°C. and yielded 2.2 mmoles of the benzophenone–DMSO adduct and 0.8 mmole of benzophenone. A precipitate formed (0.307 gram) which analyzed (23) as 103% (4.25 mmoles) potassium superoxide (KO_2).

To ascertain the possibility of the intervention of benzophenone ketyl, $(C_6H_5)_2CO \cdot ^-$, in the oxidation of benzhydrol, the oxidation of benzopinacol in basic solution was investigated. In basic solution benzopinacol is in equilibrium with the ketyl, as is easily demonstrated by ESR spectroscopy (19).

$$(C_6H_5)_2C(O^-)C(C_6H_5)_2O^- \rightleftarrows 2 \ (C_6H_5)_2CO \cdot^- \qquad (18)$$

Table X summarizes some pertinent results.

Table X. Oxidation of Pinacols[a]

Pinacol	$[KOC(CH_3)_3]_0$, M	Solvent	Moles Oxygen Absorbed[b]	Initial Rate[c]
Benzopinacol	0.18	tert-BuOH	1.04	0.16
Benzopinacol	0.36	tert-BuOH	1.34	0.32
Benzopinacol	0.18	DMSO(80)–tert-BuOH	1.50	0.23
1,1,2-Triphenylethane-1,2-diol	0.18	DMSO(80)–tert-BuOH(20)	2.20	0.93
1,2-Diphenylethane-1,2-diol	0.18	DMSO(80)–tert-BuOH(20)	2.20	0.10

[a] 0.06M, 27° ± 3°C.
[b] At complete oxidation, usually 300 minutes; moles of oxygen per mole of pinacol.
[c] Moles of oxygen per mole of pinacol per minute.

The oxidation of benzopinacol and benzhydrol have similar stoichiometries and hence may involve similar intermediates, particularly benzophenone ketyl. The ratio of KO_2/K_2O_2 formed in an oxidation of benzhydrol (3 mmoles, 0.12M) in tert-butyl alcohol containing 9 mmoles of potassium tert-butoxide (0.36M) at 27° ± 3°C. is consistent with potassium peroxides' being the precursor of potassium superoxide. The ratio of KO_2/K_2O_2 increased from 1.0 at 100 minutes (316 mmoles of oxygen absorbed) to 4.3 at 300 minutes (5.2 mmoles of oxygen absorbed). We thus conclude that under our reaction conditions, benzhydrol is first converted to benzophenone and potassium peroxide with the consumption of one mole of oxygen per mole of benzhydrol. In a secondary process a second mole of oxygen may be absorbed to yield potassium superoxide (3, 4, 8, 9).

$$(C_6H_5)_2CHO^- + O_2 + B^- \rightarrow (C_6H_5)_2CO + O_2^{-2} + HB \qquad (19)$$

$$K_2O_2 + O_2 \rightleftarrows 2 \ KO_2 \qquad (20)$$

The data apparently require a free radical chain mechanism for the oxidation of benzhydrol. Potassium superoxide filtered from a completed oxidation completely removed the induction period for a fresh oxidation. Thus, potassium superoxide must either serve as an initiation of oxidation chains or function in a chain-branching reaction. Cupric salts also removed the induction period. The inhibition by ferric ion or arsenic trioxide can be construed as arising from the decomposition of the superoxide ion or other peroxidic chain-carrying species. A specific mechanism for Reaction 19 is given in Reactions 21 and 22.

$$KO_2 + (C_6H_5)_2CHO^- \rightarrow KOOH + (C_6H_5)_2CO\cdot^- \quad (21)$$

$$(C_6H_5)_2CO\cdot^-K^+ + O_2 \rightarrow (C_6H_5)_2CO + KO_2 \quad (22)$$

Reaction 23 may be involved also.

$$(C_6H_5)_2C(O^-)OO\cdot \rightleftarrows (C_6H_5)_2CO + O_2\cdot^- \quad (23)$$

The large isotope effect (k_H/k_D = 8 to 16) appears consistent with the suggested reaction sequence.

The effect of the concentration of base requires that the formation of superoxide from peroxide is more rapid and occurs to a greater extent at the higher base concentrations. This conclusion is difficult to test, because both potassium peroxide and potassium superoxide are insoluble in the oxidation solvent and potassium superoxide precipitates from solution with as much as 35% by weight of potassium *tert*-butoxide or potassium hydroxide (which can be removed by extraction with *tert*-butyl

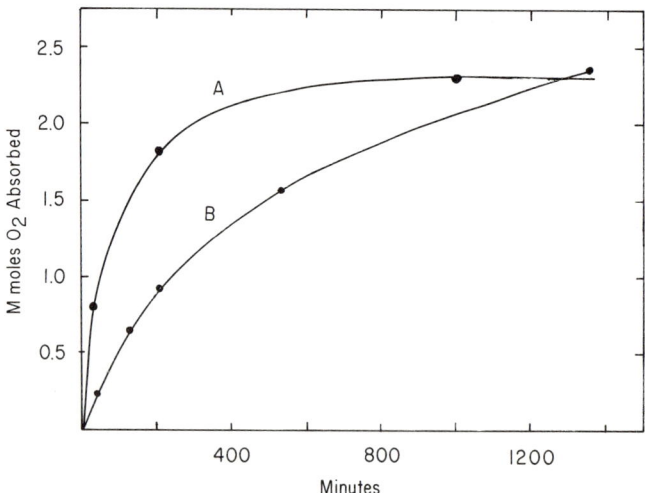

Figure 4. Oxidation of 2.5 mmoles of potassium peroxide in 25 ml. of dimethyl sulfoxide (80%)–tert-butyl alcohol (20%) solution

A. 1.15 mmoles of potassium tert-butoxide
B. 7.5 mmoles of potassium tert-butoxide
After 1500 minutes 5 mmoles of potassium superoxide isolated by filtration from Reaction A or B

alcohol). Figure 4 shows that the reaction of oxygen with potassium peroxide occurs more readily in the presence of larger amounts of excess base. Perhaps the equilibrium of Reaction 20 is displaced by the precipitation of a potassium *tert*-butoxide–potassium superoxide complex.

Oxidation of 9-Fluorenol and 9-Xanthenol in Basic Solution. Fluorenol and xanthenol react with base and oxygen to give high yields of the ketones, or in the case of fluorenol in DMSO solution, the DMSO-fluorenone adduct. The stoichiometry of the oxidation (Table XI) varies with

Table XI. Products of Oxidation of Fluorenol and Xanthenol

Substrate[a] (3 Mmoles)	Solvent	Base (9 Mmoles)	Mmoles Oxygen Absorbed (Min.)[b]	Yield of Ketone, %
Fluorenol	*tert*-BuOH	KOC(CH$_3$)$_3$	3.9 (42)	81
	Pyridine(80%)–*tert*-BuOH(20%)	KOC(CH$_3$)$_3$	6.3 (80)	—
	Pyridine(80%)–*tert*-BuOH(20%)	RbOC(CH$_3$)$_3$	4.3 (8)	71
	DMSO(80%)–*tert*-BuOH(20%)	KOC(CH$_3$)$_3$	4.2 (35)	72[c]
	DMSO(80%)–*tert*-BuOH(20%)	LiOC(CH$_3$)$_3$	2.9 (3)	71
	DMSO(80%)–*tert*-BuOH(20%)	RbOC(CH$_3$)$_3$	4.3 (20)	—
Xanthenol	*tert*-BuOH	KOC(CH$_3$)$_3$	3.2 (40)	99
	Pyridine(80%)–*tert*-BuOH(20%)	KOC(CH$_3$)$_3$	5.7 (62)	—
	Pyridine(80%)–*tert*-BuOH(20%)	RbOC(CH$_3$)$_3$	5.1 (129)	96
	DMSO(80%)–*tert*-BuOH(20%)	LiOC(CH$_3$)$_3$	3.0 (12)	87
	DMSO(80%)–*tert*-BuOH(20%)	NaOC(CH$_3$)$_3$	4.1 (11)	95
	DMSO(80%)–*tert*-BuOH(20%)	KOC(CH$_3$)$_3$	5.5 (27)	89[d]
	DMSO(80%)–*tert*-BuOH(20%)	RbOC(CH$_3$)$_3$	5.8 (43)	97

[a] 0.06M except for *tert*-butyl alcohol, where initial concentration of diarylcarbinol was 0.12M.
[b] At complete oxidation, 27° ± 3°C.
[c] 9-(Methylsulfinylmethyl)-9-hydroxyfluorene.
[d] 5.0 mmoles of KO$_2$ isolated.

the nature of the base, solvent, and substrate. The oxidations are much more rapid than for benzhydrol but further oxidation of initially formed potassium peroxide is still a complication. Using lithium *tert*-butoxide in DMSO (80%)–*tert*-butyl alcohol (20%) as the base results in the absorption of one mole of oxygen per mole of diarylcarbinol. Lithium superoxide is not stable, and the stoichiometry must follow the equation:

$$Ar_2CHOH + 2\ LiOC(CH_3)_3 + O_2 \rightarrow Ar_2CO + 2\ (CH_3)_3COH + Li_2O_2 \quad (24)$$

The oxidations involving lithium *tert*-butoxide reach completion (3 minutes for fluorenol, 12 minutes for xanthenol) much sooner than the corresponding reactions utilizing potassium *tert*-butoxide as base (35 minutes for fluorenol, 27 minutes for xanthenol). This behavior obviously involves Reaction 20 since the initial rates of oxidation were all approximately the same for the lithium- and potassium *tert*-butoxide–catalyzed reactions.

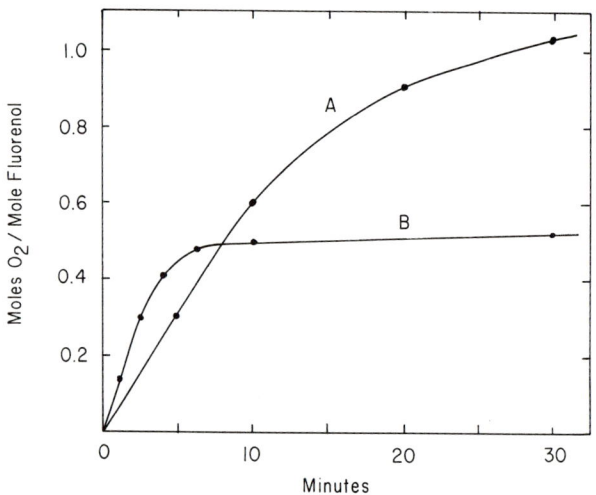

Figure 5. Oxidation of 0.15M 9-fluorenol in tert-*butyl alcohol containing 0.4M potassium* tert-*butoxide*

A. *No additive*
B. *0.002M ferric chloride*

The oxidation of xanthenol and fluorenol showed a number of differences from the oxidation of benzhydrol. No induction period was observed (Figure 5). The rate was enhanced by ferric ion, and the stoichiometry was altered to 0.5 mole of oxygen per mole of fluorenol (Figure 5), apparently because of Reaction 25.

$$KO_2, K_2O_2 \xrightarrow{Fe^{III}} K_2O + O_2 \quad (25)$$

No deuterium isotope effect was observed in the oxidations of 9-deuterio-9-fluorenol. The rates of oxidation varied considerably with solvent

(Table XII), and pronounced autocatalysis by nitroaromatics could be detected (Table XIII).

Table XII. Initial Rates of Oxidation of Fluorenol and Xanthenol[a]

Substrate, M	$KOC(CH_3)_3$, M	tert-BuOH	Initial Rate[b] Pyridine(80%)–tert-BuOH(20%)	DMSO(80%)–tert-BuOH(20%)
Fluorenol, 0.06	0.18	0.28	1.1	1.9
9-Deuteriofluorenol, 0.06	0.18	—	—	1.8
Xanthenol, 0.06	0.18	0.06[c]	0.15	0.50
Xanthenol, 0.06	0.18[d]	—	—	0.70[c]
Xanthenol, 0.06	0.18[e]	—	—	0.90[c]
Xanthenol, 0.06	0.18[f]	—	—	0.45

[a] At 27° ± 3°C.
[b] Moles of oxygen/mole of substrate per minute.
[c] Heterogeneous.
[d] $LiOC(CH_3)_3$.
[e] $NaOC(CH_3)_3$.
[f] $RbOC(CH_3)_3$.

Table XIII. Catalyzed Oxidation of Xanthenol[a]

Catalyst, 0.005M	Initial Rate[b]
None	0.15
$C_6H_5NO_2$	0.30
$p\text{-}ClC_6H_4NO_2$	0.35
$p\text{-}BrC_6H_4NO_2$	0.40
$p\text{-}CNC_6H_4NO_2$	1.0
4-Nitropyridine-N-oxide	1.2

[a] 0.10M xanthenol, 0.22M potassium tert-butoxide in pyridine (80%)–tert-butyl alcohol (20%) solution at 27° ± 3°C.
[b] Moles of oxygen/mole of substrate per minute.

The data of Tables XII and XIII appear to demand an oxidation mechanism similar to that observed for fluorene itself and involving a carbanion intermediate. The greater acidity of fluorene and xanthene relative to diphenylmethane (approximately 10 pK_a units) (28) apparently promotes ionization to yield a dianion which can react directly with oxygen or undergo a catalyzed reaction—e.g., by nitroaromatics or Fe^{III}.

$$Ar_2CHO^- \rightleftarrows Ar_2CO^{-2} \qquad (26)$$

$$Ar_2CO^{-2} + O_2 \rightleftarrows Ar_2CO^{\cdot -} + O_2^{\cdot -} \rightleftarrows Ar_2C(O^-)OO^- \rightleftarrows Ar_2CO + O_2^{-2} \qquad (27)$$

$$Ar_2CO^{-2} + ArNO_2 \rightarrow ArNO_2^{\cdot -} + Ar_2CO^{\cdot -} \qquad (28)$$

$$\text{ArNO}_2\cdot^- + \text{O}_2 \rightarrow \text{ArNO}_2 + \text{O}_2\cdot^- \tag{29}$$

$$\text{Ar}_2\text{CO}\cdot^- + \text{O}_2 \rightarrow \text{Ar}_2\text{CO} + \text{O}_2\cdot^- \tag{30}$$

Whether the ion pair $[(C_6H_5)_2\text{CO}\cdot^- \; \text{O}_2\cdot^-]$ separates to yield alkali metal superoxide (MO_2) or collapses to yield alkali metal peroxide (M_2O_2) depends upon the stability of the alkali metal superoxide. Thus, in general the yield of superoxide increases as the alkali metal is changed from lithium to sodium to potassium to rubidium, a sequence that parallels the stabilities of the superoxides. Superior yields of superoxides are observed in pyridine solution (Table XI). This is apparently connected with the ability of pyridine to stabilize the superoxide ion by complex formation (25).

Detection of Paramagnetic Products under Oxidative Conditions. Treatment of diphenylmethane in basic solution with a trace of oxygen in DMSO solutions fails to produce significant amounts of a paramagnetic product detectable by ESR spectroscopy. On the other hand, treatment of benzhydrol with traces of oxygen in basic solution can produce significant amounts of the ketyl. Pyridylthiazolylcarbinols are readily converted to the ketyls by base in alcoholic solution. (24). In pure DMSO significant amounts of the ketyl are formed whereas in *tert*-butyl alcohol or DMSO (80%)–*tert*-butyl alcohol (20%) only traces of the ketyl can be detected. These results are consistent with the formation of the ketyl under oxidative conditions by Reaction 31. Only under the most basic conditions (pure DMSO) is the dianion formed by

$$(C_6H_5)_2\text{CO} + (C_6H_5)_2\text{CO}^{-2} \rightleftarrows 2\,(C_6H_5)_2\text{CO}\cdot^- \tag{31}$$

potassium *tert*-butoxide. The equilibrium between ketone and dianion to yield ketyl either lies far to the right or else is slowly established because benzopinacol upon treatment with potassium *tert*-butoxide in *tert*-butyl alcohol or DMSO (80%)–*tert*-butyl alcohol (20%) yields significant concentrations of the ketyl (Reaction 14).

Treatment of fluorene, xanthene, thioxanthene, phenyl-4-pyridylmethane, phenyl-2-pyridylmethane, 4,5-methylenephenanthrene, tetraphenylcyclopentadiene, or phenalene with base and a trace of oxygen in DMSO (80%)–*tert*-butyl alcohol (20%) solution gave significant amounts of the ketyls—*i.e.*, Figure 6. These reactions may involve the initial formation of the ketone followed by reduction by the carbanion to yield the ketyl.

$$\text{Ar}_2\text{CO} + \text{Ar}_2\text{CH}^- \rightarrow \text{Ar}_2\text{CO}\cdot^- + \text{Ar}_2\text{CH}\cdot \tag{32}$$

This process has been demonstrated to occur in the reaction between fluorenone and the fluorene anion in the absence of oxygen in *tert*-butyl alcohol solution (20). In the case of phenalene the hydrocarbon is spontaneously converted to the ketyl in basic DMSO solutions in the absence of oxygen. Apparently DMSO can serve as the oxidizing agent.

In methanolic solutions of sodium methoxide phenalene will react with traces of oxygen to produce the phenalenyl radical, $a^H = 5.14$ (6 hydrogens), 1.47 (3 hydrogens) (12, 26), which can be converted to the ketyl with excess oxygen. [Sogo, Nakazaki, and Caluin (26) report $a^H = 7$ gauss (six hydrogens) and 2 gauss (three hydrogens) in carbon tetrachloride solution.]

Oxidation of xanthenol or fluorenol with deficient quantities of oxygen in tert-butyl alcohol produced large quantities of the ketyl, as did reaction of equal molar amounts of the ketone and alcohol in basic solution. In fact, the reaction of the pinacol of fluorenone with excess base in tert-butyl alcohol produced an essentially quantitative yield of the ketyl (19).

Interrupted oxidations of 9,10-dihydroanthracene or 9,10-dihydrophenanthrene in DMSO (80%)–tert-butyl alcohol (20%) containing potassium tert-butoxide produced the 9,10-semiquinone radical anions, apparently as a product of oxidation of the monoanion.

The major oxidation product isolated was anthracene, perhaps formed in part from the hydroperoxide (*1*). However, significant amounts of potassium superoxide accompanied the anthracene. This result suggests that the major source of anthracene involved the oxidation of the dianion. In pure DMSO in the presence of excess potassium *tert*-butoxide, a trace of oxygen converts 9,10-dihydroanthracene, 9,10-dihydrophenanthrene, or acenaphthene to the hydrocarbon radical anions. These products are apparently formed in the oxidation of the hydrocarbon dianions.

Treatment of diphenylamine with base and traces of oxygen in DMSO solution yielded a significant ESR signal that we have identified as I. The same radical is formed spontaneously from the mono-anil of *p*-benzoquinone in basic DMSO and by the oxidation of 4-hydroxydiphenylamine in this solvent (Figure 7).

Oxidation of the anion from diphenylamine apparently involves attack of oxygen at carbon rather than nitrogen.

$$(C_6H_5)_2N^- + O_2 \rightarrow C_6H_5N=\!\!\left\langle\!\!\begin{array}{c}\rule{0pt}{1ex}\\\rule{0pt}{1ex}\end{array}\!\!\right\rangle\!\!\underset{O-O^-}{\overset{H}{\diagdown\!\!\diagup}} \rightarrow C_6H_5N=\!\!\left\langle\!\!\begin{array}{c}\rule{0pt}{1ex}\\\rule{0pt}{1ex}\end{array}\!\!\right\rangle\!\!=O + OH^-$$

Discussion

Generalizations concerning the mechanism of the autoxidation of hydrocarbons in basic solution must consider the acidities of the hydrocarbons. In general the oxidizability of carbanions increases with an increase in the pK_a of the parent hydrocarbon. For di- and triphenylmethanes, pK_a's $\cong$ 30, 28 (13, 28), the carbanion is extremely reactive toward oxygen and under our reaction conditions is always trapped by oxygen before reprotonation can occur. Increasing acidity by a factor of 10 pK_a units (fluorene, pK_a = 20) (13) changes the timing of the oxidation sequence. Reprotonation of the carbanion is a common occurrence. A free radical chain oxidation of the carbanion at its equilibrium

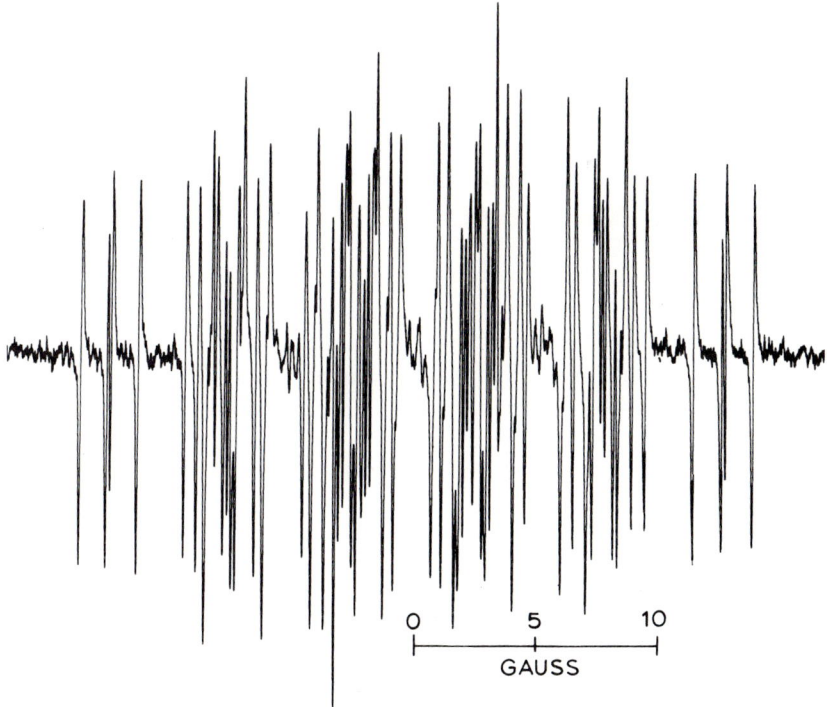

Figure 6. First derivative electron spin resonance spectrum of radical anion formed spontaneously from phenalene in dimethyl sulfoxide (80%)–tert-butyl alcohol (20%) solution

Spectrum apparently due to ketyl, because phenalen-1-one yields same spectrum

concentration can be demonstrated to occur in the presence of catalysts and may well occur in the absence of catalysts. Acidity of the α-hydrogen atom of the initially formed secondary hydroperoxides also can affect the course of the oxidation. Thus, in basic DMSO solution diphenylmethyl hydroperoxide is reduced to the alcohol, while 9-hydroperoxyfluorene undergoes a base-catalyzed E-2 elimination to yield the ketone.

The oxidation of benzhydrol and 9-fluorenol in basic solution again shows a difference in regard to mechanism that can be primarily attributed to a difference in acidity as carbon acids. In *tert*-butyl alcohol benzhydrol enters into an oxidation scheme as the mono (oxy) anion. The data strongly suggest a free radical chain. Under these conditions the more acidic fluorenol or xanthenol oxidizes via carbanions or dianions. These oxidations can be catalyzed to occur *via* a free radical chain process by one-electron acceptors, such as nitrobenzene, and a free radical chain process may well be involved in the absence of the catalyst.

Carbanions derived from hydrocarbons or substituted hydrocarbons with pK_a values in the range of 10 may—*e.g.*, 2-nitropropane—or may not (diethyl malonate) be oxidizable. When oxidation does occur, it is generally solely by the radical-chain mechanism.

The use of oxidation techniques to measure directly the rates of ionization of weak hydrocarbon acids with pK_a's in the range of 30 and indirectly for stronger acids—*e.g.*, fluorene—illustrates a unique property of oxygen. Oxygen is not usually considered to be an electrophile. It shows no tendency to interact with hydroxide or *tert*-alkoxide bases. However, as a trapping agent for carbanions oxygen can be exceedingly effective, a behavior readily explicable in terms of oxidation and reduction potentials.

The rates of oxidation of fluorenyl and diphenylmethyl anions are correlated by simple Hückel MO calculations. Loss of an electron from fluorene anion involves electron transfer from a bonding orbital ($0.18\ \beta$). On the other hand, the electron lost from the diphenylmethide ion will come from a nonbonding orbital. Thus, the transfer of an electron to oxygen will occur more readily from the diphenylmethide ion than from the fluorenyl anion.

Experimental

Oxidation Procedure, Apparatus and Reagents. The wrist-action oxidation apparatus and general oxidation procedure have been described (*18*).

Solvents and potassium *tert*-butoxide were purified or prepared as described (*18*). Rubidium-, sodium-, and lithium *tert*-butoxides were prepared by the method described for potassium *tert*-butoxide (*20*). Diphenylmethane (Eastman Kodak Co.) was vacuum-distilled. Fluorene

(Aldrich Chemical Co.) was twice recrystallized from ethanol before use. Pure grades of commercially available xanthene, thioxanthene, 9-xanthenol, benzhydrol, 4,5-methylenephenanthrene, anthrone, and 9,10-dihydroanthracene were used without further purification. 9,10-Dihydrophenanthrene was twice distilled under vacuum before use. 1,2-Dihydropyrene was obtained from Rutgerswerke Aktiengesellschaft. 9,9′-Biacridanyl (m.p. 237.5–239.5°C.) was prepared (5). 9-Fluorenol was prepared by reductions of fluoren-9-one with lithium aluminum hydride (m.p. 155–156°C.). Benzopinacol was synthesized by the method of Gomberg and Backmann (6) [m.p. 185–186°C. (dec.), lit (6) m.p. 192–194°C. (dec.)]. α,α-Dideuteriodiphenylmethane was prepared by the reaction of lithium aluminum deuteride with dichlorodiphenylmethane (29). The reaction product was pure by GLPC and showed no aliphatic hydrogen absorption by NMR and infrared. After correction for ^{13}C contributions, the mass spectrum (Atlas CH-4 spectrometer) gave (m/e, relative intensity), 168, 0.7; 169, 4.75; 170, 169.33. These values of m/e correspond to 97% d_2-, 2.72% d_1-, and 0.4% d_0-diphenylmethane.

9,9-d_2-Fluorene was made by adding deuterium oxide to a solution of fluorene and potassium *tert*-butoxide in dimethylformamide. Fluorene (6 grams), DMF (200 ml.), and potassium *tert*-butoxide (100 grams) were placed in a three-necked flask equipped with a magnetic stirring bar, nitrogen inlet, and condenser with drying tube. Prepurified nitrogen was passed through the solution for several hours before deuterium oxide (20 ml.) was added. The solution was stirred for 6 hours, another 80 ml. of deuterium oxide added, and the solution was stirred another 1/2 hour. A large excess of water was added and the fluorene filtered from the aqueous solution. The deuterated fluorene was recrystallized from aqueous ethanol. Mass spectroscopic analysis yielded (m/e, relative intensity) 166, 0.95; 167, 10.17; 168, 74.07. These values correspond to 87% d_2, 11.9% d_1, and 1.1% d_0-fluorene.

Isolation of Oxidation Products. After oxygen absorption had ceased, or reached the desired value, the oxidates were poured into water. In many cases the reaction product could be removed by filtration in high yield. In this manner xanthone (m.p. 172-174°C.), was isolated from oxidations of xanthene or xanthen-9-ol; thioxanthone (m.p. 208-210°C.), from thioxanthene; acridine (m.p. 107-109°C.), from acridan; anthracene (m.p. 216-217°C.), from 9,10-dihydroanthracene; phenanthrene (m.p. 95-99°C.), from 9,10-dihydrophenanthrene; pyrene (m.p. 151-152.5°C.) (recrystallized from benzene) from 1,2-dihydropyrene; and 4-phenanthroic acid (m.p. 169-171°C.) (recrystallized from ethanol) by chloroform extraction of the hydrolyzed and acidified oxidate of 4,5-methylenephenanthrene.

The product of the absorption of 3 mmoles of oxygen by a solution of 3 mmoles of diphenylmethane and 15 mmoles of potassium *tert*-butoxide in 20 ml. of DMSO (80%)–*tert*-butyl alcohol (20%) was poured into water. Upon standing, colorless crystals of benzhydrol (m.p. 64°C.) formed which could be recovered in 75% yield by filtration. Complete oxidation of the diphenylmethane gave an oxidate which after hydrolysis and benzene extraction yielded 2.9 mmoles of the 1,1-diphenyl-2-(methylsulfinyl)ethanol (m.p. 147-148°C.) (recrystallized from a mixture of chloroform and cyclohexane) (*14*).

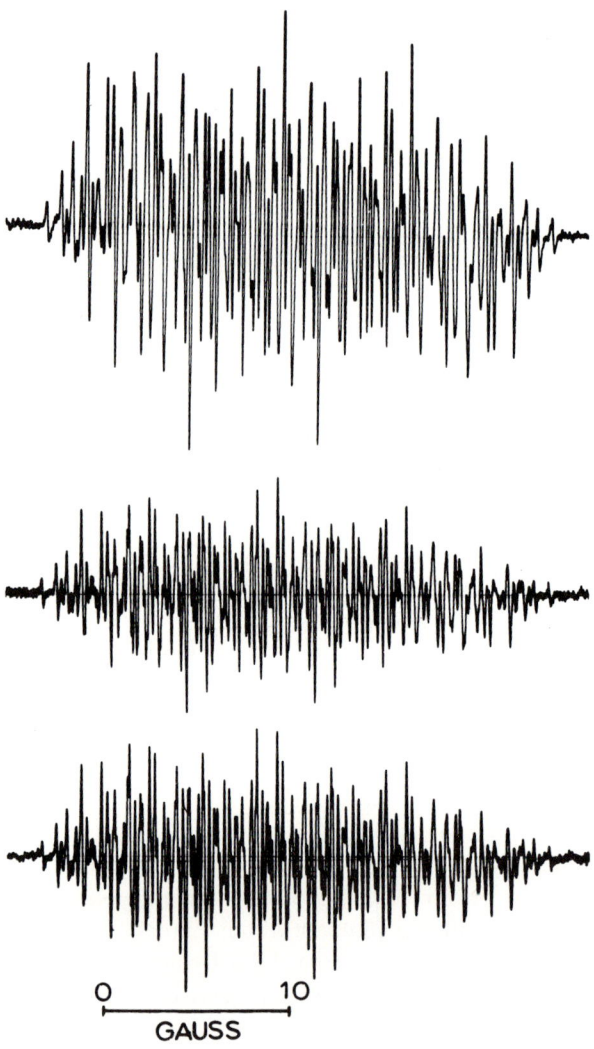

Figure 7. First derivative ESR spectra of radical anions observed in dimethyl sulfoxide solutions of potassium tert-butoxide. Top, autoxidation product of diphenylamine. No signal was observed in the absence of oxygen. Middle, autoxidation of 4-hydroxydiphenylamine. Bottom, spontaneous reduction product of mono-anil of p-benzoquinone.

Analysis Calculated C, 68.89; H, 6.12; S, 12.19. Found, C, 69.60; H, 5.95; S, 12.36.

The adduct could also be synthesized from preformed benzophenone. Thus, 5.2 mmoles of benzophenone reacted in 25 ml. of DMSO (80%)–*tert*-butyl alcohol (20%) containing 5.7 mmoles of potassium *tert*-butoxide to give a 93% yield of the adduct in 15 minutes at 25°C. The DMSO-benzophenone adduct was also isolated by benzene extraction of the hydrolyzed and acidified oxidates of benzhydryl chloride, benzhydryl bromide, benzhydryl amine, diphenylacetonitrile, and 1,1-diphenylacetone in DMSO (80%)–*tert*-butyl alcohol (20%).

Hydrolysis of the oxidate of anthrone resulted in the precipitation of anthraquinone (m.p. 263-268°C.). Extraction of the aqueous filtrate with chloroform yielded the DMSO adduct (1 to 1) of anthraquinone (m.p. 158-158.5°C.) (recrystallized from a chloroform–cyclohexane mixture).

Analysis Calculated C, 67.1; H, 4.93; S, 11.20. Found, C, 66.91; H, 5.14; S, 11.75.

Anthrone did not react with DMSO under the reaction conditions. However, 9,10-anthraquinone (2 mmoles) in 25 ml. of DMSO (80%)–*tert*-butyl alcohol (20%) containing potassium *tert*-butoxide (4 mmoles) gave a deep red solution at 25°C., from which 60% of the adduct could be isolated after 1 hour and 88% after 3 hours. This adduct was isolated from the oxidate of 9,10-dihydroanthracene (after hydrolysis, acidification, and filtrations of anthracene) by extraction of the aqueous filtrate by chloroform. Xanthone and thioxanthone failed to form isoluble adducts with DMSO in basic solution.

Oxidation of Potassium Peroxide. Determination of Potassium Superoxide. Potassium peroxide was prepared by the addition of a *tert*-butyl alcohol solution of 90% hydrogen peroxide to potassium *tert*-butoxide in DMSO or *tert*-butyl alcohol. Oxygen absorption was followed in the standard manner (20). Analysis of solid precipitates for potassium superoxide followed exactly the method of Seyb and Kleinberg (23). Potassium superoxide formed in the oxidation of benzhydrol was determined in a 15-ml. aliquot of the oxidation solution. To this aliquot 10 ml. of diethyl phthlate was added to prevent freezing of the solution. The mixture was cooled to 0°C., and 10 ml. of acetic acid–diethyl phthlate (4 to 1) added over a period of 30 minutes with stirring. The volume of the evolved oxygen was measured.

Deuterium-Hydrogen Exchange. α,α-Dideuteriodiphenylmethane $(0.1)M$ was exchanged with DMSO (80%)–*tert*-butyl alcohol (20%) in the presence of potassium *tert*-butoxide under a nitrogen atmosphere. The diphenylmethane was added to the stirred reaction flask *via* an addition sidearm, and the exchange was halted by adding a large excess of water. The hydrolyzed oxidate was extracted with ether, and the diphenylmethane was isolated by distillation, and analyzed by mass spectrometry. After hydrolysis blank experiments indicated that no further exchange took place. The sample of α,α-dideuteriodiphenylmethane described previously analyzed (m/e, relative intensity) 168, 9.0; 169, 94.7; 170, 294.5 in 1.5 minutes and 168, 4.6; 169, 24.9; 170, 42.0 in 2.7 minutes.

Fluorene-9,9-d_2 was exchanged in tert-butyl alcohol solution in a specially constructed stainless steel apparatus at 25°C. Syringes (50-ml.) filled with potassium tert-butoxide (0.0462M) in tert-butyl alcohol and the deuteriofluorene (0.050M) in tert-butyl alcohol were mechanically driven with a variable-speed syringe pump (infusion pump 975, Harvard Apparatus Co.). The flowing solutions were mixed in equal volumes at a tee and the mixed solutions allowed to flow through a 12-inch section of stainless steel tubing having a volume of 0.84 ml. from the point of mixing to the exit.

The exit stream was immediately hydrolyzed with water, and fluorene was isolated by filtration and recrystallization from ethanol. Blank experiments indicated that no further exchange occurred after hydrolysis. In one experiment using a flow rate of 4.35 ml. per minute the exchanged fluorene analyzed (m/e, relative intensity) 166, 2.25; 167, 32.68; 168, 187.93. This corresponded to 84.45% d_2, 14.6% d_1, and 1.0% d_0-fluorene. In another experiment the flow rate was 2.18 ml. per minute, and the product was analyzed (m/e, relative intensity) 166, 4.45; 167, 36.0; 168, 167.92. This corresponds to 80.7% d_2, 17.3% d_1, and 2.0% d_0-fluorene.

Detection of Radical Anion by ESR Spectroscopy. The ESR measurements of the rate of free radical formation by electron transfer from fluorene to nitroaromatics were obtained by use of the flow system and U-type mixing cells described previously (18, 20). Concentrations were estimated by comparison of the total area of overmodulated first-derivative spectra with solutions of diphenylpicrylhydrazyl under identical solvent and instrumental conditions. Relative concentrations within a given experiment are considered accurate to within a few per cent, while absolute concentrations are considered to be accurate to ±30%.

Acknowledgment

The ESR spectra of phenalenone, acridine, and benzoquinone anil radical anions were recorded by W. C. Danen, E. T. Strom, and F. A. Neugebauer.

Literature Cited

(1) Barton, D. H. R., Jones, D. W., *J. Chem. Soc.* **1965**, 3563.
(2) Cairns, T. L., McKusick, B. C., Weinmayr, V., *J. Am. Chem. Soc.* **73**, 1270 (1951).
(3) Étienne, A., Fellion, Y., *Compt. Rend.* **238**, 1429 (1954).
(4) Étienne, A., Le Berre, A., *Ibid.* **252**, 1166 (1961).

(5) Gilman, H., Towle, J. L., Ingram, R. K., *J. Am. Chem. Soc.* **76**, 2920 (1959).
(6) Gomberg, M., Bachmann, W. E., *Ibid.* **49**, 236 (1927).
(7) Hock, H., Ernst, F., *Ber.* **92**, 2732 (1959).
(8) LeBerre, A., *Bull. soc. chim. France*, **1961**, 1198, 1543; **1962**, 1682.
(9) LeBerre, A., *Compt. Rend.* **252**, 1341 (1961).
(10) Pauson, P. L., Williams, B. J., *J. Chem. Soc.* **1961**, 4153.
(11) Pratt, E. F., Trapasso, L. E., *J. Am. Chem. Soc.* **82**, 6405 (1960).
(12) Reid, D. H., *Chem. Ind.* **1956**, 1504; *Tetrahedron* **3**, 339 (1958).
(13) Ritchie, C. D., Uschold, R. E., *J. Am. Chem. Soc.* **89**, 2752 (1967).
(14) Russell, G. A., Becker, H.-D., Schoeb, J., *J. Org. Chem.* **28**, 3584 (1964).
(15) Russell, G. A., Bemis, A. G., *Inorg. Chem.* **6**, 403 (1967).
(16) Russell, G. A., Bemis, A. G., *J. Am. Chem. Soc.* **88**, 5491 (1966).
(17) Russell, G. A., Janzen, E. G., Becker, H.-D., Smentowski, F. J., *Ibid.* **84**, 2652 (1962).
(18) Russell, G. A., Janzen, E. G., Bemis, A. G., Geels, E. J., Moye, A. J., Mak, S., Strom, E. T., ADVAN CHEM. SER. **51**, 112 (1965).
(19) Russell, G. A., Janzen, E. G., Strom, E. T., *J. Am. Chem. Soc.* **84**, 4155 (1962).
(20) *Ibid.* **86**, 1807 (1964).
(21) Russell, G. A., Moye, A. J., Janzen, E. G., Mak, S., Talaty, E. R., *J. Org. Chem.* **32**, 137 (1967).
(22) Schlenk, W., Appenrodt, J., Michael, A., Thal, A., *Ber.* **47**, 473 (1914).
(23) Seyb, E., Kleinberg, J., *Anal. Chem.* **23**, 115 (1951).
(24) Shchukina, M. N., Ermolaeva, V. G., Kalmanson, A. E., *Dokl. Akad. Nauk SSSR* **158**, 436 (1964).
(25) Slough, W., *Chem. Commun.* **1965**, 194.
(26) Sogo, P. B., Nakazaki, M., Caluin, M., *J. Chem. Phys.* **26**, 1343 (1957).
(27) Sprinzak, Y., *J. Am. Chec. Soc.* **80**, 5449 (1958).
(28) Steiner, E. C., Starkey, J. D., *J. Am. Chem. Soc.* **89**, 2751 (1967).
(29) Wiberg, K. B., Evans, R. J., *Tetrahedron* **8**, 313 (1960).
(30) Wooster, C. B., *Chem. Rev.* **11**, 1 (1932).

RECEIVED October 9, 1967. Work supported by a grant from the Petroleum Research Fund administered by the American Chemical Society. Part III in a series on Oxidation of Carbanions. Part II (*21*) was published in 1967.

Discussion

H. R. Gersmann (Koninklyke Shell Laboratories, Amsterdam, Netherlands): The results obtained by Russell correlate with those obtained by Gersmann and Niewenhuis (Organic Reaction Symposium, Cork, 1964) in the study of autoxidation of esters and ketones. Here weakly acidic esters also showed rates of ionization equal to the rate of oxidation as shown by the equality of the rate of racemization of an optically active ester to the rate of oxidation.

An interesting point is still whether one can define when carbanions will react with oxygen by electron transfer, as would seem from the results of the benzhydrols. In the case of esters and ketones, even acidic substrates which were completely ionized in our basic systems always showed bimolecular kinetics and high yields of organic peroxides.

15

Base-Catalyzed Autoxidation of 9, 10-Dihydroanthracene and Related Compounds

J. O. HAWTHORNE, K. A. SCHOWALTER, A. W. SIMON, and M. H. WILT

Applied Research Laboratory, United States Steel Corp., Monroeville, Pa.

M. S. MORGAN

Mellon Institute, Carnegie-Mellon University, Pittsburgh, Pa.

> *The autoxidation mechanism by which 9,10-dihydroanthracene is converted to anthraquinone and anthracene in a basic medium was studied. Pyridine was the solvent, and benzyltrimethylammonium hydroxide was the catalyst. The effects of temperature, base concentration, solvent system, and oxygen concentration were determined. A carbanion-initiated free-radical chain mechanism that involves a single-electron transfer from the carbanion to oxygen is outlined. An intramolecular hydrogen abstraction step is proposed that appears to be more consistent with experimental observations than previously reported mechanisms that had postulated anthrone as an intermediate in the oxidation. Oxidations of several other compounds that are structurally related to 9,10-dihydroanthracene are also reported.*

A number of studies have been reported in which organic compounds with acidic hydrogens attached to carbon have been oxidized in a basic medium (5). It was not generally recognized until recently (6, 7) that a methylene hydrogen of 9,10-dihydroanthracene could dissociate in certain solvents in the presence of a base to form a carbanion, which, in contact with molecular oxygen, would yield oxidation products of dihydroanthracene. The investigation presented in this paper was conducted to establish the mechanism by which the autoxidation of 9,10-dihydroanthracene takes place under homogeneous basic conditions.

Under the conditions used here, the two major oxidation products were anthracene and anthraquinone.

Either compound could be the major oxidation product by selecting suitable solvent and base. Since anthraquinone is a more valuable product than anthracene, the major portion of this investigation was directed toward identifying the factors that are important in producing anthraquinone and minimizing the formation of anthracene.

A homogeneous reaction system was used in which benzyltrimethylammonium hydroxide was the base and pyridine the solvent. An oxidation mechanism is proposed that is consistent with observations on the reaction variables and possible oxidation intermediates of dihydroanthracene.

Experimental

Apparatus. The reaction system was a Mini-Lab assembly (Ace Glass, Inc., 50-ml. capacity) with a hollow-bore stirring rod through which oxygen or air could be introduced. The gases were metered by passing them through a flowmeter. Temperature was manually controlled by supplying heat with an electric heating mantle, or, in the event of an exothermic reaction, by directing a cold air blast against the outside of the reactor.

Procedure. The general procedure was to add a 40% pyridine solution (1.7 ml.) of the catalyst (benzyltrimethylammonium hydroxide, 3 mmoles) to a solution of 9,10-dihydroanthracene (9.0 grams, 50 mmoles) in anhydrous pyridine (50 ml.). The solution was heated to 50° to 60°C. and excess oxygen (0.29 to 1.05 liters per minute) was passed into the stirred solution. The heat of the reaction increased the temperature to 70°C., and this temperature was maintained by external cooling (air) for the first 30 minutes. (Anthraquinone began to crystallize after 10 minutes.) Heat was applied thereafter to maintain the 70°C. temperature for a total reaction time of 2 hours. The mixture was then cooled to 25°C., and the catalyst was neutralized with acetic acid (marked by a color change from red to light yellow). The solvent was evaporated under reduced pressure (bath temperature, 30°C.), and the residue was washed with water, collected on a tared filter, dried at room temperature under reduced pressure, and weighed. Tests were also conducted in which the anthraquinone that crystallized from the reaction solution was collected by filtration, washed with pyridine (15 ml.), and dried. The mother liquor was evaporated to dryness to recover material remaining in solution. The products were analyzed for anthraquinone, anthracene, and unreacted dihydroanthracene.

With the above procedure, variables such as reaction temperature, mole ratio of catalyst to dihydroanthracene, reaction solvent, and oxygen concentration were examined.

Results

Temperature Effects. The oxidation of 9,10-dihydroanthracene to anthraquinone in anhydrous pyridine solvent with benzyltrimethylammonium hydroxide as the base occurs over a wide temperature range (Table I). Some oxidation takes place at a temperature as low as −20°C., but maximum anthraquinone conversions (about 70%) occur between 50° and 70°C. Above 70°C., the conversion decreases, probably as a result of thermal decomposition of the benzyltrimethylammonium hydroxide.

Table I. Effect of Initial Reaction Temperature on Oxidation of Dihydroanthracene to Anthraquinone [a]

Temp., °C.	Conversion to Anthraquinone, Wt. %
−20	8
0	10
30	13
50	71
70	70
90	40

[a] Reaction mixture and conditions: anhydrous pyridine, 50 ml., benzyltrimethylammonium hydroxide, dihydroanthracene, 9.0 grams, 50 mmoles; reaction time, 2 hrs.

Table II. Effect of Base Concentration on Oxidation of Dihydroanthracene to Anthraquinone [a]

Mole Ratio of Base to Dihydroanthracene	Conversion to Anthraquinone, Wt. %
0.015	19
0.03	29
0.06	70
0.09	67

[a] Reaction mixture and conditions: anhydrous pyridine, 50 ml.; benzyltrimethylammonium hydroxide; dihydroanthracene, 9.0 grams; temperature, 70°C.; reaction time, 2 hrs.

Effect of Base Concentration. The effect of base concentration was studied by varying the mole ratio of base to dihydroanthracene (Table II). The maximum anthraquinone conversion is obtained with a mole ratio of catalyst to dihydroanthracene of 0.06.

Solvent Effects. The conversion of dihydroanthracene could be increased by adding water to the pyridine solvent (Table III). An 86% conversion to anthraquinone was obtained when 95% aqueous pyridine was used as the solvent. Furthermore, methanol could be substituted for the water with equivalent results. Other solvents were tried in place of pyridine (Table IV). The data indicate that 95% aqueous pyridine gave the best yields, although aniline gave nearly similar results. When acetonitrile and dimethylformamide were used, the large amounts of unreacted starting material indicate that these solvents may have deactivated the base by undergoing a hydrolysis reaction.

Table III. Effect of Water Content in Pyridine on Oxidation of Dihydroanthracene to Anthraquinone[a]

Water in Pyridine, Wt. %	Conversion to Anthraquinone, Wt. %
0	71
2	67
5	87
10	75

[a] Reaction mixture and conditions: solvent, 50 ml.; benzyltrimethylammonium hydroxide, 3 mmoles; dihydroanthracene, 9.0 grams; temperature, 70°C.; reaction time, 2 hrs.

Table IV. Effect of Various Solvents on Oxidation of Dihydroanthracene to Anthraquinone

Solvent	Conversion,[a] Wt. %		Unreacted Dihydro-Anthracene,[a] Wt. %
	To Anthraquinone	To Anthracene	
95% Aqueous pyridine	85	15	0
N,N-Dimethylaniline	31	13	60
Aniline	81	20	0
Morpholine	54	37	8
N-Methylmorpholine	53	33	14
Quinoline	15	13	71
N,N-Dimethylamino-propylamine	74	26	0
Diethylenetriamine	43	55	1
Cyclohexylamine	59	40	0
n-Hexylamine	65	35	0
Acetonitrile	15	3	81
2-Picoline	52	46	0
3-Picoline	59	39	0
4-Picoline	60	36	0
N,N-Dimethylformamide	14	3	87

[a] Totals deviate from 100% in some instances because of differences in methods of analysis.

Effect of Oxygen Concentration. The effect of oxygen concentration on the conversion to anthraquinone and anthracene was also determined. As the oxygen partial pressure was increased, the ratio of anthraquinone to anthracene formed increased significantly (Table V). Thus, the data indicate that higher oxygen concentrations favor anthraquinone formation.

Table V. Effect of Oxygen Concentration on Oxidation of Dihydroanthracene[a]

Oxygen Pressure, atm.	Conversion, Wt. %		Anthraquinone-Anthracene Ratio
	To Anthraquinone	To Anthracene	
0.2[b]	60	40	1.5
1.0	72	28	2.6
4.0	90	10	9.0

[a] Reaction mixture and conditions: anhydrous pyridine, 50 ml.; benzyltrimethylammonium hydroxide, 3 mmoles; dihydroanthracene, 9.0 grams; temperature, 70°C.; reaction time, 2 hrs.
[b] Carbon dioxide–free air at 1 atm.

Table VI. Oxidation of Various Compounds[a] in Pyridine with Benzyltrimethylammonium Hydroxide Catalyst

Compound	Product (Conversion, %)
9,10-Dihydroanthracene	Anthraquinone (70), anthracene (30)
Xanthene	Xanthone (79)
5,10-Dihydrotetracene	5,10-Tetracenequinone (39), tetracene (61)
Acridan	Acridine (47), acridone (0)
1-Methyl-9,10-dihydroanthracene	1-Methylanthraquinone (30)
2-Methyl-9,10-dihydroanthracene	2-Methylanthraquinone (35)
Anthrone	Anthraquinone (40)

[a] Reaction mixture and conditions: anhydrous pyridine, 50 ml.; benzyltrimethylammonium hydroxide, 3 mmoles; substrate, 50 mmoles; temperature, 70°C.; reaction time, 2 hrs.

Oxidation of Related Compounds. Several other compounds related to dihydroanthracene in structure were oxidized in pyridine solvent (Table VI). No attempt was made to optimize the yields in any instance except with dihydroanthracene. It was surprising that anthrone reacted much more slowly than dihydroanthracene and that only a 40% yield of anthraquinone was obtained.

Discussion

Mechanism for Base-Catalyzed Autoxidation of 9,10-Dihydroanthracene. The autoxidation of 9,10-dihydroanthracene in pyridine as the solvent and in the presence of benzyltrimethylammonium hydroxide, a strong base, is believed to involve the reaction of a carbanion and molecular oxygen. Indirect evidence of the existence of the carbanion of dihydroanthracene in pyridine solution comes from the color that forms in the presence of the base. When dihydroanthracene is added to a pyridine solution of the base, a deep blood-red color develops immediately. This color is not completely attributable to carbanions since a trace of anthraquinone alone will produce it. However, under an inert atmosphere (nitrogen) in which no anthraquinone can be formed, a deep red color is also formed.

Background and Possible Intermediates. Accepting the premise of carbanion formation in the basic media, the mode of reaction with molecular oxygen can now be considered. Sprinzak (8) reported that the autoxidation of fluorene in basic media proceeds by direct reaction of the fluorenyl carbanion with oxygen to form initially the hydroperoxide, which decomposes to yield 9-fluorenone, as depicted below.

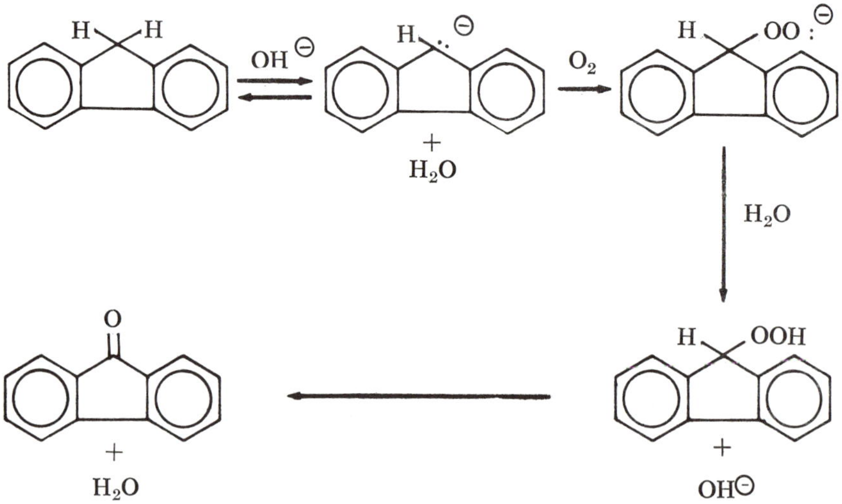

If the 9,10-dihydro-9-anthranyl carbanion were to react directly with oxygen, 9,10-dihydro-9-anthranylhydroperoxide would be formed. This could decompose to give anthrone and/or anthracene. Anthrone, which would exist mainly as anthranol in a basic medium, generally is oxidized easily to anthraquinone. The following equations illustrate this reaction.

However, when anthrone was oxidized under the same conditions as the dihydroanthracene, the conversion to anthraquinone was estimated to be only 40%, and that value was probably high because of interference by unreacted anthrone during analysis. Anthrone, then, was not readily oxidized, contrary to expectation if it were the intermediate to the quinone.

A sample of the monohydroperoxide, previously reported by Bickel and Kooyman (2), was obtained by autoxidation of 9,10-dihydroanthracene in benzene under ultraviolet irradiation. When this compound was treated under nitrogen with benzyltrimethylammonium hydroxide, it decomposed to give a mixture of anthracene and anthrone. (Under acidic conditions, it decomposed entirely to anthracene.) A fresh sample of the hydroperoxide was then oxidized. The physical appearance of the reaction mixture was similar to that in the oxidation of anthrone. The product was analyzed, and the conversion to anthraquinone was only 59%. Again, other oxidation products or anthrone may have contributed to the anthraquinone estimate.

Both Russell (5) and Barton (1) have examined the oxidation of dihydroanthracene in a solvent system consisting of 80% dimethyl sulfoxide and 20% *tert*-butyl alcohol and with potassium *tert*-butoxide as the base. In both studies, a large excess of base was used, so that there is a possibility of dicarbanion formation. In the present investigation, only catalytic amounts of base were used, which makes it unlikely that a dicarbanion could be formed. Barton and Russell both proposed reaction schemes in which a hydroperoxide is formed. The hydroperoxide is converted to anthrone, which is then oxidized to anthraquinone. However, Russell's rate data (5), which agree with our observations, indicate that anthrone oxidizes at a significantly slower rate than dihydroanthracene itself. Since both the hydroperoxide and anthrone oxidize under our reaction conditions at a slower rate than dihydroanthracene, it does not seem likely that they are intermediates in the oxidation. Attempts were made to identify oxidation intermediates by quenching the oxidation products of dihydroanthracene before the reaction had gone to completion, followed by an examination of the reaction products by gas chromatography. There was no evidence of anthrone or any other oxidation intermediates. Only unreacted dihydroanthracene, anthraquinone, and anthracene were found.

The direct reaction of oxygen with the carbanion from dihydroanthracene does not seem likely. Russell (5) has indicated a preference for a one-electron transfer process to convert the carbanion to a free radical, which then reacts with oxygen to form an oxygenated species. Therefore, we considered a mechanism involving one-electron transfer to form a free radical from the carbanion, which would lead to the formation of anthraquinone and anthracene without having either the hydroperoxide or anthrone as an intermediate.

Postulated Mechanism. The first phase of the oxidation of 9,10-dihydroanthracene involving a free-radical process would be the following chain initiation.

$$\text{dihydroanthracene} + OH^{\ominus} \rightleftharpoons \text{monocarbanion} + H_2O \tag{1}$$

$$\text{monocarbanion} + O_2 \longrightarrow \text{radical} + O_2^{\cdot\ominus} \tag{2}$$

Reaction 1 is again the abstraction of a proton by a base in a reversible reaction to give the monocarbanion. The carbanion reacts with oxygen (Reaction 2) by a one-electron transfer to give the free radical and the charged oxygen molecule, which can react again to become a peroxide ion. [A similar step has been presented for the autoxidation of *p*-nitrotoluene under basic conditions (*4*).] The chain propagation is expressed as follows.

$$\text{DHA}^\cdot \xrightarrow{O_2} \text{DHA-OO}^\cdot \quad (3)$$

$$\text{DHA-OO}^\cdot \longrightarrow \text{DHA(OOH)}^\cdot \quad (4)$$

With the formation of free radicals having been initiated, these radicals react with oxygen (Reaction 3) to begin the propagation of the radical chains in forming a peroxy radical. The peroxy radical then attacks the 10-carbon-hydrogen bond to form the hydroperoxide radical (Reaction 4). [The possibility of such an intramolecular attack has been demonstrated in an aliphatic system where two reactive hydrogen atoms are located in the favorable 1,4-positions (*9*)].

$$\text{DHA(OOH)}^\cdot \xrightarrow{O_2} \text{DHA(OOH)(OO}^\cdot) \quad (5)$$

$$\text{DHA(OOH)(OO}^\cdot) + \text{DHA} \longrightarrow \text{DHA(OOH)}_2 + \text{DHA}^\cdot \quad (6)$$

The hydroperoxide radical reacts with another molecule of oxygen (Reaction 5) to give the hydroperoxide-peroxy radical. This radical in turn reacts with a molecule of dihydroanthracene (Reaction 6), to give the dihydroperoxide and generate a radical to propagate the chain. However, the hydroperoxide radical formed in Reaction 4 may be decomposed by a carbanion to the anthracene diradical (Reaction 7). [An example of the decomposition of an unstable hydroperoxide by reaction with an anion is found in the basic autoxidation of 2-nitropropane (*3*).]

$$\text{(cyclohexadiene-OOH)} + \text{(cyclohexadienyl anion)} \longrightarrow \text{(benzene ring)} + \text{HOO}^{\ominus} + \text{(cyclohexadiene)} \quad (7)$$

The diradical can decay to form anthracene with termination of a chain.

$$\text{(diradical)} \longrightarrow \text{(anthracene)} \quad (8)$$

In Reaction 6, one of the products was the dihydroperoxide, which is decomposed by the base to give anthraquinone (Reaction 9).

$$\text{(9,10-dihydroperoxide)} \longrightarrow \text{(anthraquinone)} + 2\,H_2O \quad (9)$$

The reactions are summarized as follows, where R equals 9,10-dihydro-9,10-anthrylene:

1. $H-R-H + OH^- \rightleftarrows H-R{:}^- + H_2O$
2. $H-R{:}^- + O_2 \rightarrow H-R\cdot + O_2\cdot^-$
3. $H-R\cdot + O_2 \rightarrow H-ROO\cdot$
4. $H-ROO\cdot \rightarrow \cdot ROOH$
5. $\cdot ROOH + O_2 \rightarrow \cdot OOROOH$
6. $\cdot OOROOH + H-R-H \rightarrow HOOROOH + H-R\cdot$
7. $\cdot ROOH + H-R{:}^- \rightarrow \cdot R\cdot + HOO{:}^- + H-R\cdot$
8. $\cdot R\cdot \rightarrow$ Anthracene
9. $HOOROOH \rightarrow$ Anthraquinone $+ 2\,H_2O$

Supporting Evidence. Reactions 5 and 7 indicate that the hydroperoxide radical can react with either oxygen or a carbanion to give anthraquinone or anthracene, respectively. Thus, high oxygen and low carbanion concentrations would favor the formation of anthraquinone;

the reverse order would favor a greater proportion of anthracene. Experimentally, the ratio of anthraquinone to anthracene in the product was directly related to the oxygen concentration (Table V).

As Equation 1 indicates, adding water would decrease the concentration of carbanions by shifting the equilibrium to the left. Experimentally, the ratio of anthraquinone to anthracene was 2.6 to 1 in anhydrous pyridine with oxygen at 1 atm., but was increased to 7.2 to 1 in 95 volume % aqueous pyridine. Further addition of water to 10% decreased the over-all reaction rate. Water had no effect at higher oxygen concentrations (oxygen pressure of 4 atm.).

The stoichiometry of the reaction was examined by measuring the amount of oxygen consumed. If one disregards the small amount of oxygen which reacted in the initiation step (Reaction 2), the reactions involved are 10 and 11.

$$\text{DHA} + 2 O_2 \longrightarrow \text{anthraquinone} + 2 H_2O \qquad (10)$$

$$\text{DHA} + O_2 \longrightarrow \text{anthracene} + H_2O_2 \qquad (11)$$

Under basic conditions, the hydrogen peroxide would be decomposed as follows:

$$H_2O_2 \rightarrow 1/2\, O_2 + H_2O$$

Oxidation of the dihydroanthracene (50 mmoles) by oxygen at 4 atm. consumed 1.80 molecular equivalents (90 mmoles) of oxygen. This amount of oxygen corresponds to an 87% conversion to anthraquinone and a 13% conversion to anthracene. Analysis of the product gave corresponding values of 90 and 10%. The difference between calculated and experimental conversions may well be within experimental error.

An attempt was made to determine the amount of hydrogen peroxide formed and correlate it with the amount of anthracene. An experiment was made with oxygen at atmospheric pressure and a reaction temperature of −20°C., so that any hydrogen peroxide formed would be less likely to decompose. The solid product (88% recovery of dihydroanthracene) was isolated and found to contain 1 mmole of anthracene. The aqueous filtrate from the isolation of the product contained 1.2 mmoles of hydrogen peroxide as determined by active oxygen content. Within probable experimental error, the stoichiometry of the oxidation to anthracene follows Reaction 11.

Conclusions

This study indicates that the oxidation of dihydroanthracene in a basic medium involves the formation of a monocarbanion, which is then converted to a free radical by a one-electron transfer step. It is postulated that the free radical reacts with oxygen to form a peroxy free radical, which then attacks a hydrogen atom at the 10-position by an intramolecular reaction. The reaction then proceeds by a free-radical chain mechanism. This mechanism has been used as a basis for optimizing the yield of anthraquinone and minimizing the formation of anthracene.

Literature Cited

(1) Barton, D. H. R., Jones, D. W., *J. Chem. Soc.* **1965**, 3563.
(2) Bickel, A., Kooyman, E., *Ibid.*, **1956**, 2215.
(3) Russell, G. A., *J. Am. Chem. Soc.* **76**, 1595 (1954).
(4) Russell, G. A., "Oxidation to Produce Petrochemicals," Symposium Preprints, Division of Petroleum Chemistry, 137th Meeting, American Chemical Society, Cleveland, Ohio, April 1960, Vol. 5, No. 2-C, p. C-25.
(5) Russell, G. A., Janzen, E. G., Bemis, A. G., Geels, E. J., Moye, A. J., Mak, S., Strom, E. T., ADVAN. CHEM. SER., No. **51**, 112-171 (1965).
(6) Russell, G. A., Janzen, E. G., Becker, H-D., Smentowski, F. J., *J. Am. Chem. Soc.* **84**, 2652 (1962).
(7) Simon, A. W., Morgan, M. S., Wilt, M. H., U. S. Patent **3,163,657** (Dec. 29, 1964).
(8) Sprinzak, Y., *J. Am. Chem. Soc.* **80**, 5449 (1958).
(9) Wibaut, J., Strang, A., *Koninkl. Ned. Akad. Wetenschap. Proc.* **B54**, 102 (1951).

RECEIVED October 9, 1967.

Discussion

G. A. Russell (Iowa State University, Ames, Iowa): I find it difficult to accept the intramolecular hydrogen transfer. Why do you think this occurs?

K. A. Schowalter: Our data indicate that both the hydroperoxide and anthrone oxidize at a slower rate than dihydroanthracene in our system. In the experiment where we quenched the reaction prematurely, neither the hydroperoxide nor anthrone was found among the reaction products. Based on this, we concluded that anthrone was not a reaction intermediate and the intramolecular hydrogen transfer appeared to provide best for a mechanism that would not proceed through the hydroperoxide or anthrone.

Dr. Russell: Is it possible that the oxidation of 9,10-dihydroanthracene is more exothermic than that of anthrone and that the heat generated could have accounted for the high yield of anthraquinone in the 2-hour reaction?

Dr. Schowalter: The reaction was carefully conducted at 70°C. in both cases; therefore, the difference in rate could not have been the result of temperature difference.

A. J. Moye (California State College, Los Angeles, Calif.): An alternative mechanism to the intramolecular hydrogen transfer could be the transfer of a single electron to give a radical-anion intermediate in place of the radical-peroxide of Reaction 4. Have you considered this possibility?

Dr. Schowalter: It appears that this is another mechanism that could account for the oxidation of dihydroanthracene without intermediate anthrone formation.

16

Liquid-Phase Oxidation of Thiols to Disulfides

J. D. HOPTON, C. J. SWAN, and D. L. TRIMM

Department of Chemical Engineering and Chemical Technology, Imperial College, London, S.W.7, England

> *The oxidation of a series of alkyl and aryl thiols in aqueous alkaline solution has been studied in the presence of various metal ions. Quantitative amounts of disulfide were produced in all cases. The oxidation rate of thiols has been found to be affected by the geometric size and electron-directing properties of substituent groups in the organic chains of the thiols. The best three catalysts, when added as simple salts, have been found to be copper, cobalt, and nickel. The dependence of the rates of oxidation on the concentrations of reactants have been investigated in some detail.*

The oxidation of thiols to disulfides (3, 10) or to sulfonic acids (11, 12) in the absence and presence of catalysts has been the subject of several recent investigations. Attention has been focused primarily on the range of products which may be produced as a function of the reaction conditions. In general, the reaction seems to lead to the almost exclusive production of disulfides when oxidation occurs in aqueous solution, and to a variety of disulfides and sulfenic, sulfinic, and sulfonic acids when carried out in nonaqueous solutions (2, 12). In both cases the reaction has been suggested (10, 12) to involve the initial production of disulfide, followed by hydrolysis of this compound to produce sulfur-containing acids

$$RSH + B = RS^- + BH^+ \quad (1)$$

$$RS^- + O_2 = RS\cdot + O_2^- \quad (2)$$

$$RS^- + O_2^- = RS\cdot + O_2^{2-} \quad (3)$$

$$2RS\cdot = RSSR \quad (4)$$

$$O_2^{2-} + H_2O = \tfrac{1}{2}O_2 + 2OH^- \quad (5)$$

$$RSSR + OH^- = RS^- + RSOH \quad (6)$$

$$RSOH + OH^- = RSO^- + H_2O \quad (6)$$

$$3RSO^- = RSO_3^- + RSSR \quad (6)$$

Alternatively, Berger (*2*) has suggested that disulfide, on the one hand, and sulfinic and sulfonic acids, on the other hand, are alternative products formed from the thiol molecule and anion, respectively. Although no unequivocal evidence for the mechanism involving attack on disulfides by hydroxyl ions has been reported, support for this concept has been obtained from experiments involving potassium hydroxide-hexamethyl phosphoramide-water solvents (*11*), where the yield of disulfide relative to higher acids increases with the water content of the solvent mixture. Owing to the lack of further supporting evidence, the conclusions of Berger must be considered somewhat doubtful.

The importance of the electron transfer reaction between RS^- and an electron acceptor (Reactions 2 and 3) has been amply confirmed by the observation that the least acidic thiols are least resistant to oxidation (*2*), and by the enormously enhanced rate of reaction in the presence of redox catalysts, such as transition metal ions (*13*) or organic redox additives (*14*). In these latter cases, reactions of the type below become important,

$$RS^- + X^{2+} = RS\cdot + X^+ \quad (7)$$

the ion X^+ being reoxidized by molecular oxygen.

Although the oxidation of thiols to disulfides in the presence of a catalyst is a reaction of commercial interest, it is only comparatively recently that the marked effects of impurities on the system has been realized. Wallace and co-workers (*13, 14*) have studied the metal-catalyzed oxidation of some thiols in the presence of a few metal ions and complexes under comparable conditions, and they have suggested a general mechanism for the reaction, based on Reactions 1, 4, 5, 6, and 7. The rate of reaction was found to depend on the chemical nature and the physical state of the catalyst. The reaction was suggested to involve metal complexes in the solid state (*13*).

As a preliminary to the detailed investigation of the kinetics and mechanism of the oxidation of thiols in the presence of metal-containing catalysts (*8*), the present paper describes a survey of the rates and end products of oxidation of a series of alkyl and aryl thiols under comparable conditions. The reaction in the absence and presence of various metal-containing catalysts has been studied under conditions of minimal impurity concentrations.

Experimental

Materials. Since the oxidation of thiols is strongly catalyzed by traces of metal ions, all experimental techniques were designed to prevent the introduction of extraneous metallic impurities. Preparation and storage of reagents were completed in acid-washed, steam-cleaned glassware. Deionized water was used to prepare all solutions.

The purest available commercial samples of thiols were carefully distilled, and the middle fractions were stored under nitrogen. Solutions of sodium hydroxide and of simple metal salts were prepared from Analar grade chemicals. Oxygen and nitrogen, from cylinders, were purified by passage through traps cooled to $-198°C.$, the oxygen being condensed and fractionally distilled before use.

Apparatus. The course of reaction was followed by periodically measuring the volume of oxygen absorbed at constant pressure, and by analysis. The apparatus and techniques for measuring oxygen uptakes have already been described (3). Spectrophotometric investigation of complexes in solution was carried out using a Unicam SP800 spectrophotometer.

Analysis. The titration of thiols and of disulfide by iodine and bromine has been described (3). Control experiments involving the titration of solutions containing known amounts of thiol and disulfide showed the methods to be accurate to within $\pm 2\%$.

Samples of metal complexes isolated from the final solutions were subjected to microanalysis (for carbon, hydrogen, oxygen, and sulfur). Metals were determined colorimetrically by the following methods—copper: as the complex formed with sodium diethyl dithiocarbamate (6); cobalt: as the nitroso-R salt complex (7); nickel: as the dimethylglyoxime complex (4).

Results

Although the investigation of the oxidation of thiols in the absence of added metal catalysts has been reported elsewhere (3), it is necessary to compare some results with experiments in the presence of catalysts. The oxidation of thiols to disulfides was studied under standard conditions in both cases which, unless otherwise stated, involve 50-ml. samples of solutions containing sodium hydroxide $(2M)$, metallic catalyst (copper, $10^{-5}M$; other metals, $10^{-3}M$) and thiol $(0.5M)$ maintained at $30°C.$ under a constant pressure of oxygen (750 mm. Hg). Reaction rates determined by measuring oxygen uptake as a function of time were confirmed by the regular analyses of thiol and disulfide; the stoichiometry agreed with the over-all equation

$$4RSH + O_2 = 2RSSR + 2H_2O \qquad (8)$$

Experiments were carried out on the oxidation of ethanethiol using a wide variety of metal catalysts. Some typical results are shown in Table I from which it is evident that adding metal salts always results in the quantitative production of disulfides and that copper, cobalt, and nickel are by far the most effective of the simple metal salts added to the solutions.

Table I. Oxidation of Ethanethiol Catalyzed by Metal Ions at Standard Conditions

Metal	Added as	Time for Completion[a] of 90% Reaction, Hrs.[b]	RSSR in Final Solutions, %
Ce	$(NH_4)_2Ce(NO_3)_6$	7	100
U	$UO_2(NO_3)_2 6H_2O$	8	98
V	$VOSO_4$	$7\frac{1}{2}$	98
Cr	$Cr_2(SO_4)_3 K_2SO_4 24H_2O$	8	—
Mo	$(NH_4)_6Mo_7O_{24} \cdot 4H_2O$	8	100
W	$Na_2WO_4 \cdot 2H_2O$	7	99
Mn	$MnSO_4 \cdot 4H_2O$	$6\frac{1}{2}$	100
Fe	$FeSO_4 \cdot 7H_2O$	7	100
Fe	Haemin	1	100
Co	$CoSO_4 \cdot 7H_2O$	$4\frac{1}{2}$	100
Ni	$NiSO_4 + aq$	4	99
Pd	$PdCl_2$	8	—
Pt	$PtCl_4$	8	105
Cu	$CuSO_4 \cdot 5H_2O$	1	100
Ag	$AgNO_3$	8	—
Zn	$ZnSO_4 \cdot 7H_2O$	6	99
Cd	$3CdSO_4 \cdot 8H_2O$	$7\frac{1}{2}$	103
Hg	$HgCl_2$	8	—
Al	$K_2SO_4 \cdot Al_2(SO_4)_3 \cdot 24H_2O$	8	—
Tl	Tl_2SO_4	8	—
Sn	$SnCl_2 \cdot 2H_2O$	8	—

[a] Expressed in terms of the reaction $4 RSH + O_2 = 2 RSSR + 2 H_2O$.
[b] To the nearest 30 minutes.

The oxidation of various thiols by these three catalysts was investigated under standard conditions. At concentrations of copper of $10^{-3}M$, the rate of reaction was controlled by the rate of shaking of the reaction vessel (Table II) and depended on the oxygen pressure above the solutions. At copper concentrations of $10^{-5}M$ the reaction rate was independent of the rate of shaking. This difference was attributed to the rate of diffusion of oxygen into the solutions being rate determining at $10^{-3}M$ [Cu]. Under such conditions, increased shake rates result in better mixing of the reactants in the liquid phase and in a greater surface area of contact between gas and liquid, thereby enhancing the diffusion of oxygen into and throughout the solutions. If the rate of oxygen diffusion is the slowest step in the sequence of reactions, then the over-all reaction rate should increase with shake rate, as observed at $10^{-3}M$ [Cu].

Table II. Effect of Shake Rate on the Kinetics of Oxidation in the Copper-Catalyzed Reaction[c]

[Cu], M	Shake Rate,[a]	Rate of Oxygen Uptake[b]
10^{-3}	360	0.80
10^{-3}	380	1.23
10^{-4}	310	0.74
	360	0.84
	400	1.57
10^{-5}	310	0.60
	400	0.60

[a] Cycles per minute of the shaker.
[b] Initial rates, expressed as percentage of final uptake/min.
[c] Standard conditions, using ethanethiol.

Table III. Oxidation of Various Thiols in the Presence of Copper, Cobalt, and Nickel Salts[a]

Thiol	Copper (10^{-5}M)		Cobalt (10^{-3}M)		Nickel (10^{-3}M)	
	Time to Complete 90% of Reaction, Hrs.	Final RSSR, %	Time to Complete 90% of Reaction, Hrs.	Final RSSR, %	Time to Complete 90% of Reaction, Hrs.	Final RSSR, %
EtSH	1	100	$4\frac{1}{2}$	101	4	96
Bu^{n}SH	$1\frac{1}{2}$	101	>10	—	15	102
Bu^{i}SH	$1\frac{1}{2}$	102	6	100	12	99
Bu^{s}SH	2	100	8	98	>10	—
Bu^{t}SH	>10	—	>10	—	>10	—
Hex^{n}SH	$1\frac{1}{2}$	104	5	101	$4\frac{1}{2}$	101
PhSH	>10	—	>10	—	>10	—
PhCH$_2$SH.	3	98	>10	—	>10	—

[a] Standard conditions; metals added as simple salts.

Studies of the copper-catalyzed system were completed, then, at a copper concentration of $10^{-5}M$. Results for all three systems, averaged over a large number of determinations, are presented in Table III.

In all cases, the oxidation rate was smallest for experiments involving thiophenol and *tert*-butanethiol. The oxygen uptake *vs.* time curves for cobalt-catalyzed reactions showed an initial high slope followed by a decrease in slope after *ca.* 30% reaction to a final steady value.

Reaction systems containing cobalt and nickel were characterized by the production of flocculent precipitates of compounds other than hydroxides in the presence of all thiols, except for thiophenol and *tert*-butanethiol. Samples of these complexes produced from the ethanethiol system were washed, dried, and subjected to microanalysis. For nickel, the precipitate could be separated into two fractions by extracting with chloroform. The results of analysis of the cobalt complex agreed with the empirical formula $Co(SEt)_{3.1}$, and of the nickel complexes with the formulae, $Ni(SEt)_2$ (insoluble fraction), $Ni(SEt)_3(OH)$ (soluble fraction). It would seem that these complexes contain coordinated thiyl entities, although coordinated disulfide may be involved in some cases (1).

Some investigation was made of the oxidation rates of ethanethiol after filtering these precipitates from the reaction solution. The oxidation rate under standard conditions (Case 1) was compared with the rate after filtering the solutions before adding thiols, but after adding all other reagents (Case 2), with the oxidation rate of solutions filtered immediately after adding all reagents (Case 3), and with the rate where metals had been added to the systems as the complex produced in the reaction (Case 4). The results are reported in Table IV.

Using the ethanethiol system as a model, we investigated the dependence of the oxidation rate on the concentration of thiol, of oxygen, and of hydroxide ion. The results for the copper- ($10^{-5}M$), cobalt- ($10^{-3}M$), and nickel- ($10^{-3}M$) catalyzed oxidations, together with the comparable system in the absence of added catalysts are recorded in Table V.

Discussion

The contention that disulfides are the major products of reaction when thiols are oxidized in aqueous alkaline solution has been amply confirmed by the present investigation. Thus, in the absence and in the presence of various metal ions (Table I) and in the oxidation of various simple alkyl and aryl thiols (Table III) disulfide has always been produced quantitatively. Under these circumstances, experiments have been designed to investigate the kinetics and mechanism of the reaction as a basis for further detailed studies.

Although it is unrewarding to compare the kinetics of oxidation of various thiols (Table III) in detail since the systems are complicated by such factors as the differing degrees of ionization of thiols and differing partition of various thiols between the alkaline solutions and the product disulfide, it is instructive to consider reaction trends. Comparing the oxidation rates of *n*-, iso-, *sec*-, and *tert*-butane thiols and noting the low oxidation rate of *tert*-butanethiol and of thiophenol (*cf*. benzylthiol) shows that the geometric configuration of the organic chain of the thiol must play an important part in controlling the rate of oxidation. On the other hand, Wallace *et al.* (13) have suggested that the electron transfer reaction (Reaction 7) may control the over-all reaction rate—a conclusion supported in general by the relative rates of oxidation of *n*-alkanethiols

Table IV. Oxidation of Ethanethiol by Soluble

Metal	Added Conc., M	Case 1		Case 2	
		Rate, Mole/Liter/Sec.	Active[b] Metal, M	Rate, Mole/Liter/Sec.	Active[b] Metal, M
Cu	10^{-5}	13.2×10^{-6}	1.0×10^{-5}	13.2×10^{-6}	1.0×10^{-5}
Co	10^{-3}	10.3×10^{-6}	1.0×10^{-3}	7.6×10^{-6}	8.9×10^{-5}
Ni	10^{-3}	15.1×10^{-6}	1.0×10^{-3}	3.4×10^{-6}	1.3×10^{-5}

[a] Case 1, no filtration; Case 2, reactant solutions filtered before adding thiol; Case 3,
[b] Amount of metal present in reacting solutions, by analysis.

observed in the present study (Table 3). No correlation could be observed, however, between the catalytic activities of metal ions and their redox potentials, presumably as a result of the fact that metal ions must be present largely as insoluble metal hydroxides.

Metal catalytic activity may be expected to be a function of the solubility of the active species and/or the ease of electron transfer to the catalyst. The results given in Table IV show conclusively that the suggestion that catalysis occurs at a gas-solid interface (13) does not hold in these systems. Preliminary experiments showed that copper ion- and haemin-catalyzed systems oxidized rapidly with no trace of solid precipitation, and that cobalt and nickel catalysis were characterized by the production of colored solutions and precipitates. Filtration experiments showed these precipitates played only a small part in catalysis (Table IV).

Filtration in Case 2 removed insoluble hydroxides from solution, and the resulting rate is much lower than the standard experiment (Case 1). The close agreement between results for Cases 1 and 3 shows that insoluble, colored precipitates are not involved in the reaction, the effective catalysis depending on the amount of metal in solution. On the other hand, there seems to be a saturation value of complex in solution since the results for Case 3 (metal originally present as hydroxide) and Case 4 (metal originally present as complex) are quite similar.

The coordination atmosphere of the metal ion in solution can also be expected to affect the reaction rate. Microanalytical results indicate that the active catalysts in cobalt and nickel systems could well be metal thiolic species produced *in situ*. However, these complexes are appreciably more soluble in the alkaline solutions than are metal hydroxides (see, for example, the analysis results reported in Table IV), and it is not possible on the present evidence to differentiate between catalysis as a result of increased solubility (comparing metal hydroxides and metal thiolic complexes), and catalysis as a result of differences in the allowed ease of electron transfer. It is apparent, however, that most of the metals investigated (Table I) are poor catalysts because they form only the insoluble hydroxide complexes.

Metal Salts at Standard Conditions[a]

	Case 3		Case 4
Rate, Mole/Liter/Sec.	Active[b] Metal, M	Rate, Mole/Liter/Sec.	Active[b] Metal, M
13.2×10^{-6}	1.0×10^{-5}	—	—
9.9×10^{-6}	6.4×10^{-4}	10.2×10^{-6}	1.0×10^{-3}
14.6×10^{-6}	5.3×10^{-4}	14.8×10^{-6}	0.99×10^{-3}

reactant solution filtered after adding thiol; Case 4, metal added as thiol complex.

The dependence of the oxidation rate of ethanethiol on the concentrations of reactants other than metal has also been investigated (Table V). The dependence of rate upon the concentration of reactants was obtained from plots of oxygen uptake *vs.* time and of disulfide production *vs.* time (3). No significant differences in results between analytical methods were observed once oxygen-uptake figures had been corrected for the stoichiometry of the reaction (*see* Reaction 8). The reaction rates increased with alkali concentration, but this was caused by the increasing solubility of oxygen in the solutions of higher alkali concentration. In all cases the reaction rate increased linearly with oxygen pressure in the system, and in most cases the reaction rate was independent of the amount of thiol present. Some unusual results were found at very low alkali concentrations and high thiol concentrations, owing to the fact that thiol was not completely ionized under these conditions.

Table V. Kinetic Parameters for the Metal-Catalyzed Oxidation of Ethanethiol

Metal	Order of Dependence of Rate on Concentration of			Initial Rate Constant (30°C.)	Activation Energy, Kcal./Mole
	[EtSH]	[O_2[a]]	[NaOH[a]]		
—	1	1	0	4.9×10^{-2} liter mole^{-1} sec.$^{-1}$	16.4
Cu	0	1	0	2.3×10^{-1} sec.$^{-1}$	4.3
Co	0	1	0	2.1×10^{-1} sec.$^{-1}$	7.5
Ni	1	1	0	3.6×10^{-1} liter mole^{-1} sec.$^{-1}$	8.0

[a] Corrected for solubility or oxygen in NaOH solutions of varying concentration.

Oxidation in the presence of copper ions may well be controlled by the rate of reoxidation of cuprous ions since the reaction between cupric ions and thiols is known to be almost instantaneous (5).

The over-all process is, however, rapid, and the rate of reoxidation of copper ions must also be fast. The kinetic results tend to support the contention that this reaction is rate controlling since the oxidation of ethanethiol (Table V) depends on oxygen concentration and is independent of thiol concentration. Again, if the reoxidation of copper is rate controlling, the over-all oxidation rate of thiols in the presence of copper should be approximately the same except where the reaction between copper ions and thiols is slow—*e.g.*, when the reaction is sterically hindered by the geometry of the thiol molecule. The results in Table III bear out these suggestions.

In summary, the oxidation of thiols to disulfides is quantitative in aqueous alkaline solution and may best be effected at high oxygen pressures in the presence of a catalyst. The catalyst should dissolve in the alkaline solutions, and of the simple metal salts, the addition of copper, cobalt, and nickel results in the most effective catalysis.

Acknowledgement

The authors wish to express their gratitude to the British Petroleum Company for the award of Research Bursaries to two of them (J.D.H. and C.J.S.).

Literature Cited

(1) Akerfeld, S., Lovgren, G., *Anal. Biochem.* **8**, 223 (1964).
(2) Berger, H., *Rec. Trav. Chim.* **82**, 773 (1963).
(3) Cullis, C. F., Hopton, J. D., Trimm, D. L., unpublished data.
(4) Mitchell, A. M., Mellon, M. G., *Ind. Eng. Chem., Anal. Ed.* **17**, 380 (1945).
(5) Reid, E. E., "Organic Chemistry of Bivalent Sulfur," Vol. 1, Chemical Publishing Co., New York, 1958.
(6) Sandell, E. B., "Colorimetric Determinations of Traces of Metals," p. 444, Interscience, New York, 1959.
(7) Shipham, W. H., Lai, J. R., *Anal. Chem.* **28**, 1151 (1956).
(8) Swan, C. J., Trimm, D. L., ADVAN. CHEM. SER. **76**, 182 (1968).
(9) Tarbell, D. S., "Organic Sulfur Compounds," Pergamon Press, New York, 1961.
(10) Wallace, T. J., Schriesheim, A., *J. Org. Chem.* **27**, 1514 (1962).
(11) Wallace, T. J., Schriesheim, A., *Tetrahedron* **21**, 2271 (1965).
(12) Wallace, T. J., Schriesheim, A., *Tetrahedron Letters* **1967**, 1131.
(13) Wallace, T. J., Schriesheim, A., Hurwitz, H., Glaser, M. B., *Ind. Eng. Chem., Process Design Develop.* **3**, 237 (1964).
(14) Wallace, T. J., Miller, J. M., Pobiner, H., Schriesheim, A., *Proc. Chem. Soc.* **1962**, 384.

RECEIVED October 23, 1967.

17

Oxidation and Chemiluminescence of Tetrakis(dimethylamino)ethylene

Decay Rates of the Chemiluminescent Intermediate

CARL A. HELLER

Mail Code 6059, Michelson Laboratory, China Lake, Calif. 93555

When oxygen is removed from a reaction solution of tetrakis-(dimethylamino)ethylene (TMAE), the chemiluminescence decays slowly enough to permit rate studies. The decay rate constant is pseudo-first-order and depends upon TMAE and 1-octanol concentrations. The kinetics of decay fit the mechanism proposed earlier for the steady-state reaction. The elementary rate constant for the dimerization of TMAE with $TMAE^{2+}$ is obtained. This dimerization catalyzes the decomposition of the autoxidation intermediate.

When tetrakis(dimethylamino)ethylene (TMAE) is autoxidized, it chemiluminesces with the reaction catalyzed by proton donors (12). The 1-octanol–catalyzed reaction has been studied in detail at 30°C. (2, 3, 8). The over-all oxidation (11) can be written as the sum of two reactions:

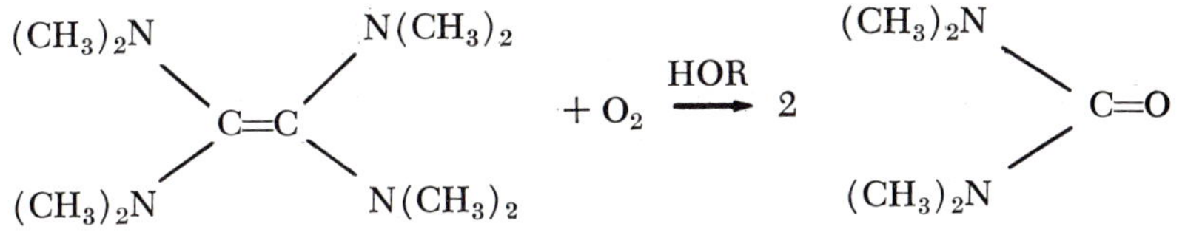

$$(CH_3)_2N\!\!>\!\!C\!\!=\!\!C\!\!<\!\!N(CH_3)_2 \;+\; O_2 \;\xrightarrow{HOR}\; 2\; (CH_3)_2N\!\!>\!\!C\!\!=\!\!O$$
$$(CH_3)_2N \qquad\qquad\qquad\qquad\qquad\quad (CH_3)_2N$$

(TMAE)   (TMU)

$$TMAE \;+\; O_2 \;\xrightarrow{HOR}\; (CH_3)_2N\!\!>\!\!\underset{O}{\overset{\|}{C}}\!\!-\!\!\underset{O}{\overset{\|}{C}}\!\!<\!\!N(CH_3)_2 \;+\; 2\; CH_3\!\!>\!\!N\cdot$$
$$\qquad\qquad\qquad\qquad\qquad\qquad\qquad\qquad\qquad\qquad\qquad\quad CH_3$$

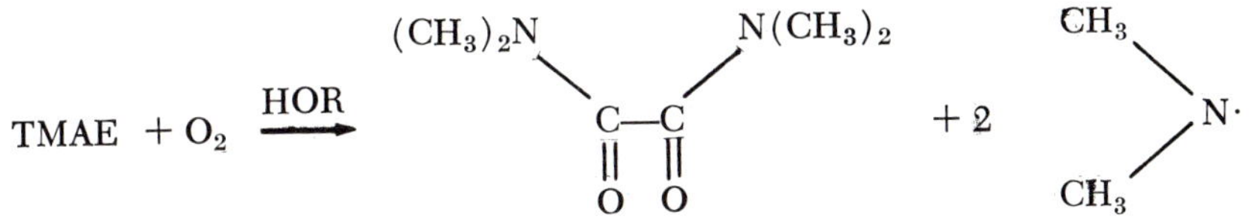

(TMOA)

The amino radical undergoes a series of reactions to give mainly $(CH_3)_2N-N(CH_3)_2$ plus minor products, including $(CH_3)_2NH$ and $(CH_3)_2NCH_2N(CH_3)_2$ *(10)*.

The chemiluminescent kinetics are second-order in TMAE and can be written

$$2E + O_2 \xrightarrow{HOR} E^* + \text{products}$$

$$E^* \rightarrow E + \text{photon}$$

where E is TMAE, and E^* is electronically excited TMAE.

The relationship between the two sets of reactions has been discussed *(3)*, and the following mechanism was proposed to explain the kinetics and products:

$$E + O_2 \rightleftarrows E \cdot O_2 \quad (1)$$

$$E \cdot O_2 + HA \rightarrow (E^{2+} O_2H^- A^-)_{\text{solvated}} \quad (2)$$

$$(E^{2+} O_2H^- A^-)_s + E \rightleftarrows (^+E - E^+ O_2H^- A^-)_s \quad (3)$$

$$(E^{2+} O_2H^- A^-)_s \rightarrow HA + TMOA + 2(CH_3)_2N \cdot \quad (4)$$

$$(^+E - E^+ O_2H^- A^-)_s \rightarrow HA + 2TMU + E^* \quad (5)$$

where HA is a proton donor, and TMOA and TMU are tetramethyl oxamide and tetramethylurea. E^{2+} is the dication of TMAE and $^+E - E^+$ is a dimer *(3)* of the radical cation, $E \cdot ^+$.

Analysis of the kinetic data from the steady-state reactions permitted some factoring and evaluation of rate constants *(3)*. The oxidation rate equation is

$$-\frac{dE}{dt} = K_1 k_{2i} (E)(HA)(O_2) \quad (6)$$

There is a different value of the observed constant $K_1 k_{2i}$ for each proton donor in the system. These donors in our system were the glass wall, 1-octanol monomer, and 1-octanol tetramer.

The equation for light intensity *(3)* is

$$I = \phi_e \times \frac{\phi_5 E}{E + C} \times \left(-\frac{dE}{dt}\right) \quad (7)$$

From the data we obtained the 30°C. values:

$$\phi_e = \frac{0.35}{1 + 48(HA)_1 + 1040(HA)_2 + 6400\, O_2}$$

$$\phi_5 = 0.104 \quad (7a)$$

$$C = \frac{k_4}{k_3 k_5}(k_{-3} + k_5) = 0.087$$

ϕ_e has the same form as for fluorescence quenching. The constants are larger, indicating alcohol and O_2 in solvation shells around the ionic intermediate (5).

ϕ_5 is the quantum yield of the elementary Reaction 5. The value of about 10% is not surprising. About 90% of the TMAE must be formed in the ground electronic state. Studies by Thrush and coworkers (6) have shown that more than one electronic level may be accessible for reactions of small molecules in the gas phase. In the case of TMAE some of the inefficiency may also be caused by the formation of a second product in an electronically excited state. We need more information to tell whether this actually can or does occur with the nonfluorescent TMU or 1-octanol.

As can be seen, we do not have much additional information about Reactions 3 and reverse Reactions 3, 4, or 5. The rate constants of these reactions are buried in C, which is independent of TMAE concentration by definition, and appears to be independent of 1-octanol. The experimental data on this latter point were weak since C was difficult to evaluate; there is as yet no measure of its temperature dependence.

It would be of interest to measure the values of k_5, such as activation energy, entropy, etc. Measurements of C from the steady-state studies are slow and not highly accurate.

However, the decay reactions may be studied by another method. If oxygen is removed from a reacting solution, the light decays relatively slowly (3). Presumably this is a measure of the decay rate of the intermediates.

We have measured the rates of the decay curves. Mathematical analysis of the proposed mechanism permits some comparison of observed rates and elementary rate constants of the proposed mechanism.

Experimental

Chemicals. Purification of n-decane, TMAE, and 1-octanol has been described (2, 3, 8). Oxygen was passed through Drierite to remove water vapor.

Solutions were prepared in an inert atmosphere box. Volumetric measurements based upon density were used. Stock solutions appeared stable for several months in the box. Hypodermic syringes and small volumetric flasks were used to measure and store the solutions.

Apparatus. A conventional vacuum apparatus was connected to the reaction vessel. A Pace gage could be used to measure oxygen uptake during runs.

The reaction vessels as shown in Figure 1 were cylindrical borosilicate glass vessels, with i.d., 25 mm. and internal height, 36 mm. A standard-taper joint permitted connection to a stopcock. These two sections could be filled in a nitrogen atmosphere box. Then the sealed vessel could be connected to the vacuum system at the ball joint.

A Teflon-coated iron stirring bar was kept in the vessel. Usually a bar 3 mm. in diameter and 8 mm. long was used.

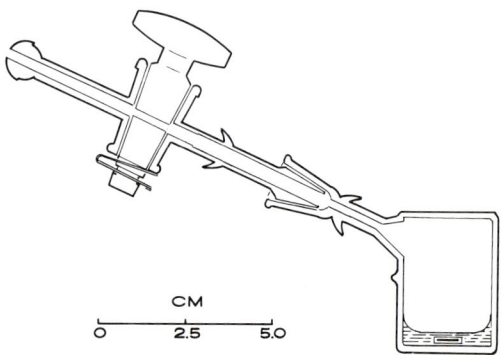

Figure 1. Reaction vessel
Borosilicate glass = 25 mm. i.d., filled in nitrogen atmosphere and attached to vacuum apparatus by ball and socket joint

Various bath systems were used. In general the reaction vessel was immersed for half its height in the bath. The bath was stirred by a large stirring bar which acted upon a small bar. Temperature was controlled by thermal inertia of the bath. Thus, only the 0° and 25°C. runs had reasonably good control. For intermediate temperatures the temperature changed slowly, and an average reading was used.

Light detection was by a Photovolt photometer which used an RCA 1P21 photomultiplier. The detector box was placed above the reaction vessel, and the system made light-tight by a black collar and cloth.

Both Pace gage and photometer output were recorded on a strip chart recorder.

Procedure. The reaction vessel was filled with a desired amount of a solution and connected to the vacuum system. Nitrogen (+ 3% hydrogen) was removed by pumping with stirring at 0°C. The absence of light was taken as evidence that there were no leaks. Then the oxygen was added quickly, and the stopcock to the differential Pace gage was closed. Both light and pressure change were measured until ready to evacuate. Oxygen could be removed in about 15 seconds. The low pressure limit was the vapor pressure of n-decane. There was some problem owing to the n-decane vaporization at 26°C., but not at 0°C.

Results

Order and Rate Constants. An example of light *vs.* time for stationary-state and decay runs was given previously (3). Figure 2 shows a semilog plot of typical decay data, replotted from pen records on the strip chart recorder. Each order of magnitude represents a separate decade on the photovolt amplifier.

Obviously the data follow a rate equation of the form:

$$dI/dt = -k_o I \text{ or } \ln I = -k_o t + \text{constant} \tag{8}$$

The results show the behavior of k_o with changes of concentrations of TMAE and alcohol and of temperature between 0° and 26°C.

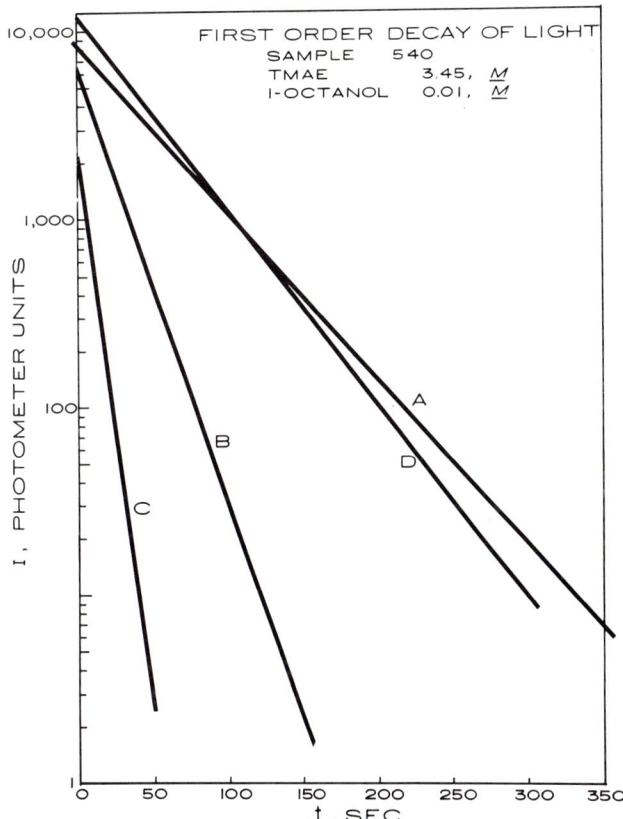

Figure 2. Decay curves plotted semilog
Smooth recorder curves replotted using points at reasonable intervals. Runs A and D at 0°C., line B at 12.4°C., and line C at 25.2°C.

Figure 2 is a good representation of almost all the 112 runs made with 1-octanol. There was curvature on only a few runs, undoubtedly caused by experimental error since they could not be reproduced. This feature was checked carefully after mathematical analysis indicated reasons to expect curvature. The conditions of reactions and values of k_o have been tabulated (Tables I and II). Since k_o depends upon 1-octanol and TMAE, it is called the pseudo-first-order rate constant.

Table I. Rate Constants

Run	Solution	[TMAE], M	[1-Octanol], M	Init. P, Torr	T, 0°C.	$k_o \times 10^3$ Sec.$^{-1}$
395[a]	126g	0.01	0.1	—	23	33
396[a]	126g	0.01	0.1	—	0	3.8
409[a]	128	0.01	0.1	—	0	3.0
410[a]	128	0.01	0.1	—	24	24
411[a]	128a	0.01	0.1	—	0	4.2
412[a]	128a	0.01	0.1	—	24	29
433	176	0.01	0.1	—	0	3.3
434A	177	0.1	0.1	—	0	4.07
434B				—	0	4.0
435	178	0.01	0.05	—	0	3.2 → 3.7
440	178	0.01	0.05	506	0	2.6
439	176	0.01	0.1	—	26.4	56
445A	183	4.25	0.1	—	0.2	26
445C				—	25	175
447A	187	3.97	0.5	—	0	51
447G				—	25	340
448	186	0.21	0.1	—	0	4.1
449A	185	2.1	0.1	—	27	122
449C	185	2.1	0.1	—	0	14.5
450	186	0.21	0.1	—	0	4.2
451	186	0.21	0.1	—	30	82
452A	177	0.1	0.1	—	25	37
452B	177	0.1	0.1	—	25	37
453A	185	2.1	0.1	—	27.0	127
453B	185	2.1	0.1	—	27.0	127
453C	185	2.1	0.1	—	0	14.5
453E	185	2.1	0.1	—	12.7	42.7
453F	185	2.1	0.1	—	17.6	87.9
454A	187	3.97	0.5	—	0	51
454F	187	3.97	0.5	—	0	51
454C	187	3.97	0.5	—	14	166
454D	187	3.97	0.5	—	27.B	308
454G	187	3.97	0.5	—	25.1	298
455A	188	0.43	0.1	62	0	11.6
455B	188	0.43	0.1	86	23.5	49
455C	188	0.43	0.1	70	23.4	61
456B	189	0.43	5.67	400	0	202
456D	189	0.43	5.67	380	17	—
457A	188	0.43	0.1	99	9	13.2
457B	188	0.43	0.1	106	30	85
457C	188	0.43	0.1	—	0	4.9
458	185	2.1	0.1	16	0	14
459A	185	2.1	0.1	15	0	13.8
459B	185	2.1	0.1	20	25.5	112
459C	185	2.1	0.1	15	11.5	35
460A	188	0.43	0.1	50	24.2	58
460B	188	0.43	0.1	50	12.4	19

Table I. (Continued)

Run	Solution	[TMAE], M	[1-Octanol], M	Init. P, Torr	T, 0°C.	$k_o \times 10^3$ Sec.$^{-1}$
460C	188	0.43	0.1	50	0	5.0
461A	191	0.01	0.02	438	0	1.4
462	192	0.1	0.5	497	0	14.1
463	193	0.1	0.02	500	0	1.6
464	194	0.1	0.5	499	0	11.2
465	191	0.01	0.02	600	25	20
467	193	0.1	0.02	498	25	22
468	194	0.1	0.5	497	25.5	150
476	191	0.01	0.02	599	0	1.4
477	192	0.01	0.05	501	0	15.8
478	193	0.1	0.02	500	0	2.1
479	194	0.1	0.5	—	0	12.0
480	181	0.01	0.02	598	0	1.7
481	192	0.01	0.5	—	0	13.4
482	193	0.1	0.02	—	0	1.7
483	194	0.1	0.5	—	0	10.5
512A		0.43	0.02	—	0	2.4
512B		0.43	0.02	150	18	16
512C		0.43	0.02	—	0	2.5
513		0.43	0.02	150	0	2.7
513B		0.43	0.02	150	23	27.5
514A	204	0.43	0.04	154	0	3.2
514B	204	0.43	0.04	151	23	21.9
514C	204	0.43	0.04	—	0	3.0
515A	205	0.43	0.2	152	0	6.9
515B	205	0.43	0.2	156	24	—
516A	204	0.43	0.04		0	3.1
516B	204	0.43	0.04		15.5	14.6
516C	204	0.43	0.04		26	36
517	206	0.01	0.03		0	1.8
518	206	0.01	0.03		12	8.4
519	206	0.01	0.03		24	—
520	207	0.01	0.04	601	0	2.1
521	207	0.01	0.04		0	2.2
522	207	0.01	0.04		17.2	8.1
523	207	0.01	0.04		25	29
524	207	0.01	0.04	506	$14^1/_2$	—
525	208	0.01	0.08	500	0	2.7
526	208	0.01	0.08	502	11	11.3
527	208	0.01	0.08	504	24	30.8
528	209	0.01	0.2	502	0	5.7
530	209	0.01	0.02	503	24	52
529	209	0.01	0.2	—	11.2	23
531	191	0.01	0.02	501	25	21
533A	210	0.43	0	201	0	1.8
533B	210	0.43	0	202	$16^1/_2$	9.8
533C	210	0.43	0	203	28	27

Table I. (Continued)

Run	Solution	[TMAE], M	[1-Octanol], M	Init. P, Torr	T, 0°C.	$k_o \times 10^3$ Sec.$^{-1}$
534A	TMAE	4.3	0	201	0	9.9
534B	TMAE	4.3	0	205	8.2	20
534C	TMAE	4.3	0	207	27	Curve
535	213	0.86	0.1	213	0	7.3
536	213	0.86	0.1	153	13.5	23
537	213	0.86	0.1	100	24	65
538A	212	0.43	0.1	151	0	5.4
538B	212	0.43	0.1	153	12$^1/_2$	17
538C	212	0.43	0.1	182	27.5	86
539A	214	1.72	0.1	162	0	10.4
539B	214	1.72	0.1	157	12$^1/_2$	30.3
539C	214	1.72	0.1	151	26.2	89
540A	215	3.45	0.1	103	0	20
540B	215	3.45	0.1	101	12.4	54
540C	215	3.45	0.1	101	25.2	148
540D	215	3.45	0.1	101	0	19
547	1 ml.	4.3	0	106	0	Curve
548	5 ml.	4.3	0	107	0	1.2
550	233	0.01	0.01	601	0	0.31
551	233	0.01	0.01	600	0	0.18
552	233	0.01	0.01	102	0	0.19
554	234	0.01	0.02	601	0	1.78
555	234	0.01	0.02	600	0	1.98
556	234	0.01	0.02	601	0	1.27
587A	292	0.1	0.5	159	0	11.9
587B	292	0.1	0.5	482	0	10.4
588A	292	0.1	0.5	140	0	8.8
588B	292	0.1	0.5	521	0	9.8
589A	293	0.43	0.5	202	0	12.1
589B	293	0.43	0.5	202	0	11.5
590A	293	0.43	0.5	200	0	12.8
590B	293	0.43	0.5	309	0	11.9
591A	294	1.0	0.5	200	0	15.2
591B	294	1.0	0.5	201	0	—
592A	294	1.0	0.5	145	0	15.4
592B	294	1.0	0.5	103	0	14.8
593A	295	2.0	0.5	99	0	21.8
593B	295	2.0	0.5	70	0	22.2
594A	295	2.0	0.5	56	0	22.3
594B	295	2.0	0.5	35	0	31.8
595A	296	3.0	0.5	32.5	0	30.3
595B	296	3.0	0.5	17.2	0	29.8
596A	296	3.0	0.5	21.5	0	29.5
596B	296	3.0	0.5	27.3	0	29.2

[a] Mixture of 1-octanol and 2-octanol.

Table II. Rate Constants

Run	Solution	[TMAE], M	Catalyst	Concn., M	P_i, Torr	T, 0°	k_o, × 10^3, Sec.$^{-1}$
469	2-phase	4.3	H_2O	1 ml./1 ml.		25	No good
470	2-phase	4.3	H_2O	1 ml./1 ml.		25	No good
474A	TMAE	4.3	H O	Satd		0	No good
474B	TMAE	4.3	H O	Satd		22	No good
474C	TMAE	4.3	H O	Satd		0	8.5
475	TMAE	4.3	H O	Satd		0	4.5
438	180	0.1	Ethanol	0.1	104	0	1.2
439	180	0.1	Ethanol	0.1	109	26.5	11.5
441	180	0.1	Ethanol	0.1	505	0	0.92
442	181	0.01	Ethanol	0.05	448	0	0.87
444	182	0.5	Ethanol	0.5	111	0	9.5
446A	184	4.29	Ethanol	0.1	111	30	Curve
446B	184	4.29	Ethanol	0.1	—	0	3.8
446C	184	4.29	Ethanol	0.1	—	$13^1/_2$	4.15
446D	184	4.29	Ethanol	0.1	—	23.2	7.3
541A	216	3.9	Silanol	0.2^a	151	0	—
541B	216	3.9	Silanol	0.2	100	$12^1/_2$	88
541C	216	3.9	Silanol	0.2	103	24.5	Curved
541D	216	3.9	Silanol	0.2	103	0	28
544A	217	0.43	Silanol	0.1	200	0	7.9
544B	217	0.43	Silanol	0.1	205	$11^1/_2$	19.5
*544C	217	0.43	Silanol	0.1	205	25	65
544D	217	0.43	Silanol	0.1	—	0	4.4

^{a} Calculated concentrations of OH. Solutions 216 and 217 in ratio indicated, although exact concentrations are doubtful.

Samples at TMAE concentrations above 0.43M (10% by volume) could be run more than once. This was convenient for measuring temperature effects. We usually began and ended our series with a 0°C. run (Figure 2). As many as seven runs had no apparent effect on k_o. No series was run to see when an effect became apparent. Samples at lower concentrations were run only once. Although each such run gave straight decay lines, there was much scatter in the slopes, probably caused by impurity or wall effects.

The initial oxygen pressure did not affect k_o. It did affect initial I and was controlled so that most decay runs started at about the same light intensity.

Some observations of the solutions also help confirm the earlier identification of TMAE as the emitting species. TMAE fluorescence is green from liquid n-decane, but becomes blue in frozen n-decane. The same sudden shift occurs in chemiluminescence.

A glowing solution from which O_2 had been removed was frozen with dry ice. The frozen portion glowed but changed color, and a sharp color boundary could be seen between the liquid and solid. The completely frozen solution stopped glowing as its temperature approached −78°C. Upon warming, a blue glow appeared, and upon partial melting we again saw two colors. The warming and cooling could be repeated until the light level became too low to see.

Table III. Decay Rate Constant vs. TMAE at 0°C.

A_o [a]	TMAE, M	$k_o \times 10^3$, Sec.$^{-1}$	Data Points
0.02	0.01	1.6	6
	0.43	2.5	3
0.04	0.01	2.1	2
	0.43	3.1	3
0.1	0.01	3.3	2
	0.1	4.7	1
	0.21	4.2	1
	0.43	5.2	5
	0.86	7.2	1
	1.72	10.4	1
	2.1	14	4
	3.45	20	2
	4.25	26	1
0.2	0.01	5.7	1
	0.43	6.9	1
0.5	0.01	13.6	4
	0.1	10.9	3
	3.97 [b]	51	3
	0.1 [b]	10.2	4
	0.43 [b]	12.1	4
	1.0 [b]	15.1	3
	2.0 [b]	22.1	3
	3.0 [b]	29.7	4

[a] Formal concentration ignoring association (4) of 1-octanol.
[b] Data obtained subsequent to plot of Figure 3.

Effect of TMAE. Table III gives the values of k_o found at 0°C. Figure 3 shows the 0° data plotted against TMAE concentration. Evidently the data fit a linear equation of the form

$$k_o = aE + b \tag{9}$$

Similar plots can be drawn at other temperatures. The constant b definitely depends upon alcohol concentration. The data are less certain, but a seems to show little dependence upon alcohol concentration.

Effect of Alcohol. Table IV and Figure 4 show the effect of 1-octanol. It is evident that the alcohol does affect k_o. The steady-state parameter,

C, gave no indication of alcohol dependence, so the present results are somewhat surprising.

The effect might be caused by the dielectric constant, which is 10 for 1-octanol and 2 for n-decane, or a hydrogen bonding to the peroxy anion may be involved.

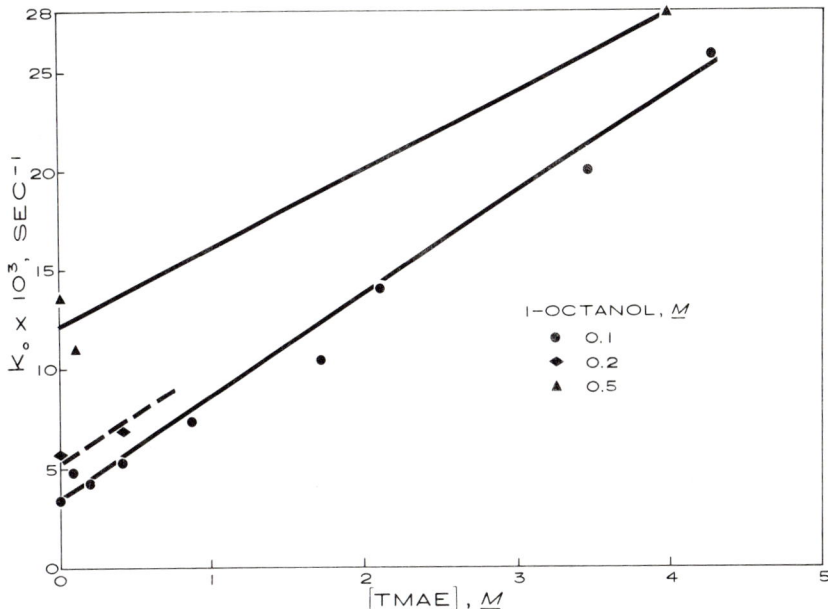

Figure 3. Effect of TMAE concentration on k_o at 0°C. and different 1-octanol concentrations
Lines drawn by eye to fit data

Attempts were made to obtain values of k_o for water, ethanol, and a silanol. Water and ethanol were too volatile and were partially lost during oxygen pumpoff. Silanol gave poorly reproducible results for unknown reasons. The results obtained for these catalysts are given in Table II.

Temperature Effects. Runs made at temperatures above 0°C., when plotted on Arrhenius graphs, gave fairly straight lines over the 25° to 30°C. interval (Figure 5). Table V shows activation energies calculated from the slopes, including some solutions for which only two temperatures were used.

The activation energies of k_o are independent of 1-octanol but depend on TMAE (Table V). Figure 6 shows the change of activation energy with TMAE. This change confirms the suspicion that k_o is not an elementary reaction rate constant. Also, k_o must be a sum of elementary constants rather than a product. This fits with the form of k_o in Equation 9.

Table IV. Decay Rate Constant *vs.* 1-Octanol at 0°C.

TMAE	A_o, M[a]	$k_o \times 10^3$, Sec.$^{-1}$	E, M
0.01	0.02	1.1	
	0.02	1.35	
	0.02	1.4	
	0.04	2.1	
	0.04	2.2	
	0.05	2.6	
	0.05	3.4	
	0.08	2.7	
	0.1	3.3	
	0.2	5.6	
	0.5	12.4	
	0.5	15.8	
	0.5	14.0	
0.43	0.0	1.8	
	0.02	2.4	
	0.02	2.7	
	0.04	3.2	
	0.04	3.0	
	0.04	3.0	
	0.1	11.6	
	0.1	4.9	
	0.1	5.0	
	0.2	6.9	
	5.67	202	
4.1 ± 0.2	0.0	9.9	4.32
	0.1	26	4.25
	0.5	51	3.97
	0.5	50.6	3.97

[a] Formal concentration ignoring association (4) of 1-octanol.

The activation energies found from the plots can have real meaning only at the extremes of TMAE concentration. The Arrhenius plots at intermediate concentrations must be curved—the sum of rates with the two different activation energies as slopes. However, the Arrhenius plots can be useful for interpolating values of k_o at particular temperatures—for example, Table VI shows k_o vs. TMAE at 25°C.—the values of k_o taken from Arrhenius plots. These values can be plotted as was done for the 0°C. values in Figure 3. The probable significance of these plots is discussed below.

Discussion

Mechanism. We wish to analyze the decay data in terms of the mechanism proposed for the steady-state reaction (3). Obviously the two reactions are related in fact. If the steady-state mechanism can be analyzed to give the observed decay kinetics, we will have support for the mechanism. We may also obtain rate constants for some of the elementary reactions.

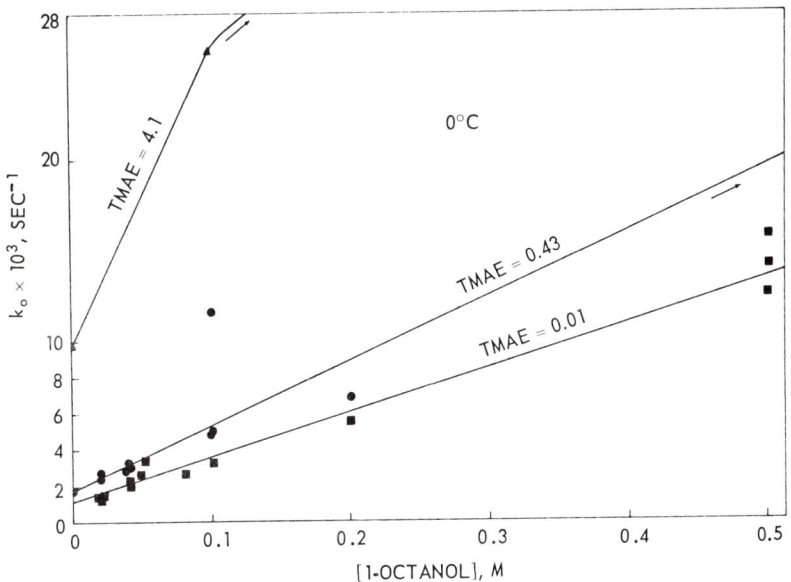

Figure 4. Rate constant vs. 1-octanol concentration
A few points are off the graph, as indicated by arrows

The mechanism is shown in Reactions 1 to 5. When oxygen is removed, Reactions 1 and 2 no longer occur, and only 3 to 5 are operative. The stability of the intermediates to back reaction has been discussed (3).

This stability of the ions to decompose to O_2 is probably caused by solvation by alcohol and O_2 molecules. This would give more stability than expected from reaction energetics alone.

One feature of the mechanism is obvious. Alcohol does not enter into any of the decay reactions as written. To explain the alcohol effect we must either introduce new reactions or call upon solvent effects.

Thus, the mathematical solution can explain only the TMAE effects upon k_o and the activation energy results.

Mathematical Analysis. Reactions 3 to 5 are first-order or pseudo-first-order reactions. Thus, the pseudo-first-order constant for Reaction 3 is k_3E. For brevity we rewrite the two intermediates $(E^{2+} O_2H^-A^-)_2$ and $(^+E{-}E^+ O_2H^-A^-)_s$ as C_1 and C_2, respectively. We assume that the light intensity is proportional to C_2 during any one run. Thus, $dI/dt = dC_2/dt$ and the calculated decay of C_2 can be related to k_o.

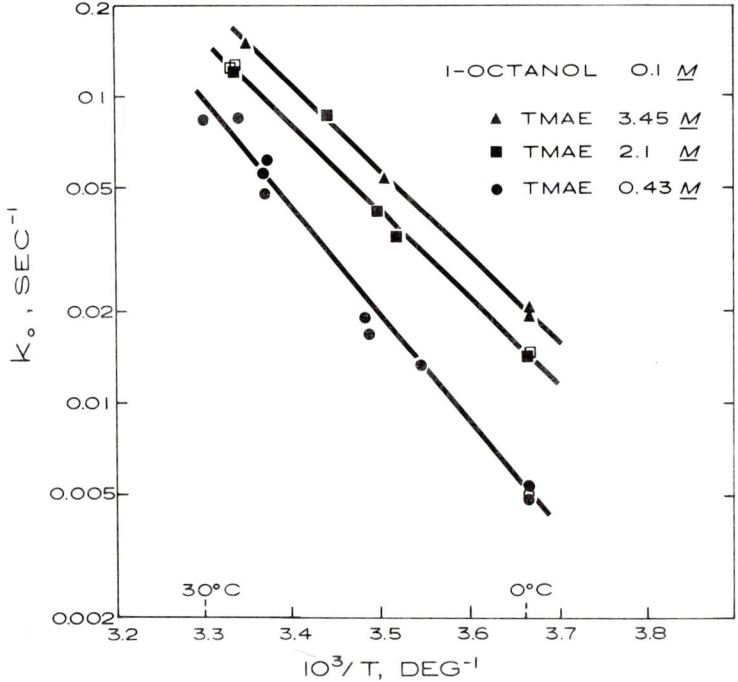

Figure 5. Data at three TMAE concentrations showing typical Arrhenius plots
3.45M data taken from Figure 2

Benson (*1*) shows a method for solving such a set of linear rate equations. The general solution will be

$$C_m = \sum_{j=1}^{s} a_{mj} e^{-\lambda_j t} + \theta_m$$

where C_m is concentration of the chemical species, λ is a function of the rate constants, while A_{mj} and θ_m depend upon the λ and initial or final concentrations. The various parameters may be evaluated as follows: For C_2 we get the solution:

$$C_2 = a_{23}\, e^{-\lambda_3 t} + a_{24}\, e^{-\lambda_4 t} + \theta_m$$

The values of λ arise from a quadratic equation:

$$\frac{k_3 E + k_{-3} + k_4 + k_5 \pm \sqrt{(k_3 E + k_{-3} + k_4 + k_5)^2 - 4(k_3 E k_5 + k_3 k_4 + k_4 k_5)}}{2}$$

With λ_3 and λ_4 as the two roots, we let the root with positive sign be λ_3. Both must be real since the mechanism cannot give periodic behavior for the decay of C_1 or C_2. However, $\lambda_3 > \lambda_4$, and the difference can be large.

The general form of the decay of C_2 is shown in Figure 7. This could probably reduce to only one line, as found in our results. It is not immediately obvious whether we are measuring λ_3 or λ_4. This will depend upon the magnitudes of a_{23} and a_{24}. We must examine the behavior of this solution further.

Table V. Activation Energies

[TMAE]	Alcohol[a]	E_A, Kcal./Mole	Data Points
0.43	0.10	17	5
0.01	0.03	20	2
0.01	0.04	17	5
0.01	0.08	18	3
0.01	0.10[b]	14	7
0.01	0.20	18.1	3
0.1	0.02	16.4	4
0.1	0.1	15	2
0.1	0.5	17.6	4
0.21	0.1	16.4	2
0.43	0.00	16.1	3
0.43	0.02	16.7	5
0.43	0.04	14.7 ± 1	6
0.43	0.1	17.0	12
0.43	0.16	14.3 ± 1	4
0.86	0.1	14.7	3
1.72	0.1	13.4	3
2.1	0.1	13.2	7
3.45	0.1	12.9	4
3.97	0.50	12.8	9
4.25	0.1	12.4	6
4.31	0.00	13.2	2
0.43	Silanol	17	4

[a] 1-Octanol except as noted.
[b] Mixture of 1-octanol and 2-octanol.

The radicand in λ can be further factored to give

$$r = k_3^2 E^2 + 2k_3 E (k_{-3} + k_4 - k_5) + (k_{-3} + k_5 - k_4)^2$$

Using this we can calculate the behavior of λ_3 and λ_4 as E changes. For λ_3 when E is zero

$$2\lambda_3 = k_3 + k_4 + k_5 + \sqrt{(k_{-3} + k_5 - k_4)^2}$$

$$\lambda_3 = k_{-3} + k_5$$

As E becomes large, $k_3 E$ predominates over the other constants and we get

$$\lambda_3 = k_3 E + b$$

Actually a plot of λ_3 would not be linear in E but would have the intercept $k_{-3} + k_5$ and soon become almost a straight line with slope k_3.

This fits the behavior found for k_o in Figure 3. Thus, we can get values for k_3 from the slope. To get $k_{-3} + k_5$ precisely we would need data which showed the curvature at low E. However, an approximate value can be obtained from a straight line. In fact, the nearly straight line shown suggests that k_4 is small at 0°C. and 0.1M 1-octanol.

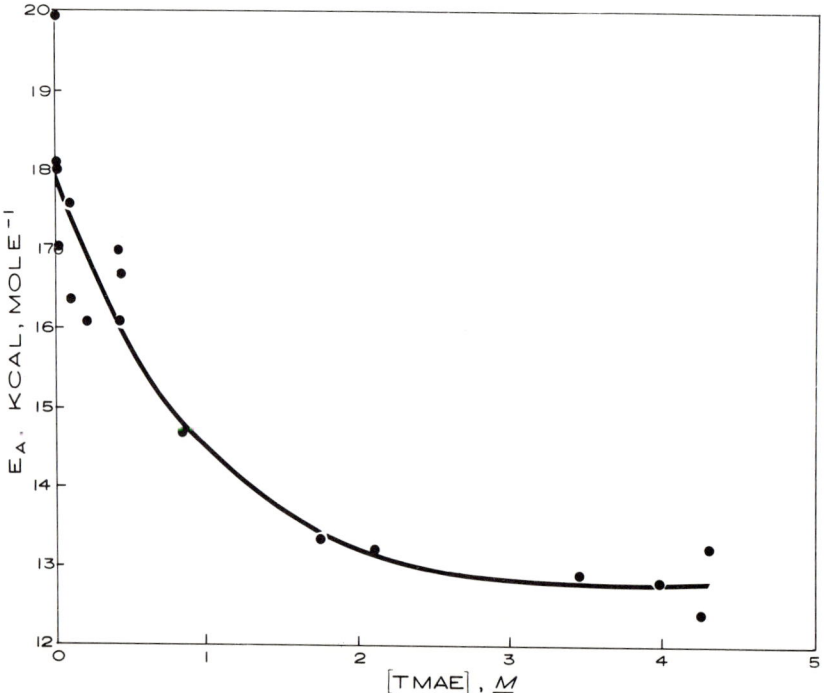

Figure 6. Experimental activation energies for k_o *vs. TMAE concentration*

Parameter θ_m must be zero since $C_2 = \theta_m$ at $t = \infty$, and C_2 must go to zero. If another line should appear in a decay run, its slope would be λ_4. When $E = 0$, $\lambda_4 = k_4$ and we could measure this constant.

The mathematical possibility that one term had a negative coefficient could account for the jump in light when the run is started by removing oxygen. However, this term would need to have the larger eigenvalue, λ_3, since it decreases more rapidly. This contradicts the observations about the behavior of the decay with respect to TMAE. Therefore, we are brought back to the view that both terms are positive, with λ_4 too small to see.

The coefficients could be obtained in terms of initial concentrations of C_1 and C_2, but these are not known. The use of light to measure C_2 would require difficult absolute measurements. Absorbance measurements would also be difficult since the concentrations are so low. Thus, the only information that can be used is k_o, the decay constant.

Rate Constants. The behavior of k_o fits that of λ_3 with respect to TMAE concentration. We can now evaluate the k_3 and $(k_{-3} + k_5)$ at $A_o = 0.1 M$.

The slope of Figure 3 is k_3. We have made a similar plot of k_o at 25°C. using interpolations on the Arrhenius plots. The values obtained

Table VI. Decay Rate Constant[a] vs. TMAE at 25°C.
$A_o = 0.1$

TMAE, M	$k_o \times 10^3$, Sec.$^{-1}$
0.01	35 ± 8
0.1	47
0.21	53
0.43	68
0.86	73
1.72	83
2.1	110
3.45	154
4.25	175

[a] Values of k_o taken from Arrhenius plots.

At 0°C. $k_3 = 5.1 \times 10^{-3} M^{-1}$ sec.$^{-1}$

At 25°C. $k_3 = 33 \times 10^{-3} M^{-1}$ sec.$^{-1}$

give an activation energy of 13 kcal. per mole, as was to be expected since the slope is based mainly upon Arrhenius plots at high TMAE concentration.

We can find only one sample of a rate constant for a similar reaction. Anthracene dianion (A^{2-}) reacts with 1,1-diphenylethylene (D) as follows:

$$2Na^+A^{2-} + D \xrightarrow{k_6} 2Na^+A^-\text{---}D^-$$

The rate constant is estimated (9) as $k_6 = 10^3 M^{-1}$ sec.$^{-1}$ at 26°C. with no activation energy measured. These two dimerization reactions thus differ by a factor of 10^5.

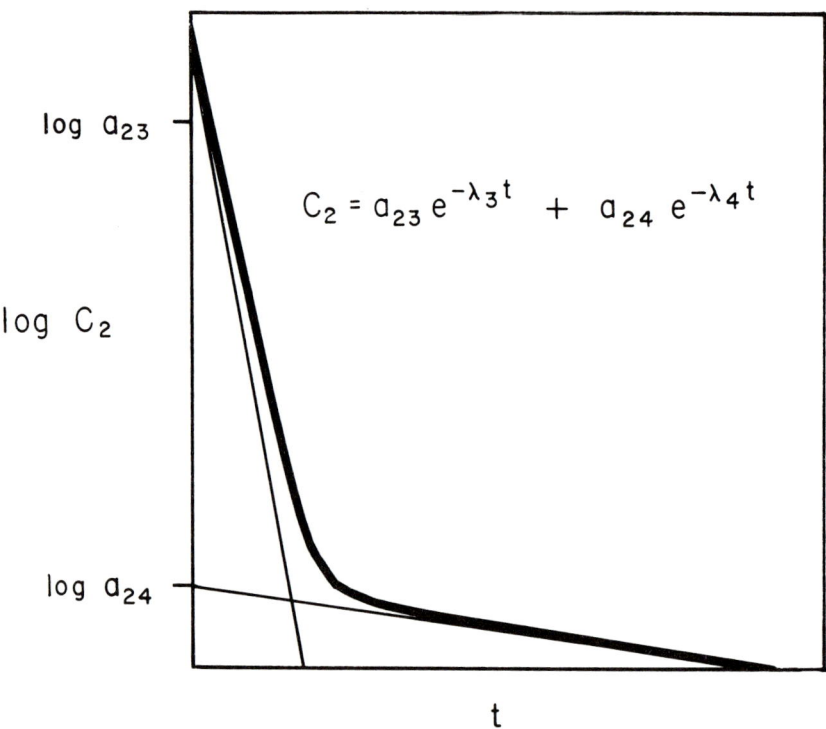

Figure 7. Form of solution of differential equations arising from decay mechanism

If we look back to Equation 7a, we notice that the value for C includes the factors $(k_{-3} + k_5)$ and k_3. We can use our Arrhenius plots to get k_3 and $(k_{-3} + k_5)$ at 30°C. by a slight extrapolation, then substitute these values into Equation 7a to get:

$$\frac{k_4}{k_5} = 0.087 \; \frac{80 \times 10^{-3}}{43 \times 10^{-3}} = 0.05$$

This ratio is seen to be that of the two paths of oxidation. We have assumed that the value of C is roughly independent of 1-octanol and can be used with the rate constants at 0.1M. The rate constant data obtained are summarized in Table VII.

The actual values of the sum $(k_{-3} + k_5)$ are shown in Table VII. The sum has an activation energy of 16 kcal. per mole. It does not seem possible with the present data to make a clearcut assignment between k_{-3} and k_5.

Table VII. Rate Constants and Activation Energies

T, °C.	1-Octanol, M	$k_3 \times 10^3$, M^{-1} Sec.$^{-1}$	$(k_{-3} \times k_5) \times 10^3$, Sec.$^{-1}$
0	0.0	—	1
0	0.1	5.1	4.5
25	0.1	33	50
30	0.1	43	80
0	0.5	6.4	11
—	0.1	$E_3 = 13$ kcal./mole	E_{-3} or $E_5 = 16$ kcal./mole
30	~0.1	$k_5/k_4 = 20$	

Reactions

The effects of TMAE and 1-octanol on k_o are similar. One is attributed to a reaction, and the other to a change in dielectric constant. This treatment is justified as follows. Previous work shows that TMAE is a reactant. TMAE should not have a dielectric constant as large as 1-octanol. Furthermore, it affects the absorbed activation energy in a predictable fashion—predictable from the mechanism. The octanol has a dielectric constant considerably higher than n-decane and is a strongly hydrogen-bonding material. The reactions of the intermediates involve a proton shift from the HOO$^-$ to the RO$^-$. The 1-octanol might be expected to facilitate this shift. In fact, it seems possible that an equilibrium reaction may occur before the decomposition.

$$(^+E - E^+ \text{ OOH}^- \text{ OR}^-)_s \leftrightarrows (^+E - E^+ \text{ OO}^{2-} \text{ HOR})_s$$

$$(^+E - E^+ \text{ O}-\text{O}^{2-} \text{ HOR})_s \rightarrow E^* + 2\text{TMU} + \text{HOR}$$

HOO$^-$ and RO$^-$ are not necessarily adjacent to each other in the intermediate ion triplet. Thus, the solvation shell of 1-octanol could help transfer the proton.

The higher rate constant of k_5 relative to k_4 means that TMAE is a catalyst for the decomposition of the first intermediate. We need temperature data to tell whether this catalysis is caused by the steric factor or activation energy.

The major data concern Reaction 3, the dimerization reaction. Here we are pushing together a TMAE molecule and a TMAE^{2+} dication.

Reaction presumably occurs when the electrons of TMAE sufficiently overlap the vacant orbitals of TMAE^{2+}. The methyl groups project out of the plane of the C_2N_4 and cause the activation energy. The same effect has been noticed in donor-acceptor complexes of TMAE, and a detailed explanation has been calculated (7).

The much larger rate constant (9) for naphthalene dianion plus diphenylethylene undoubtedly reflects the greater ease of overlap for the ring orbitals. Part of the effect also may be caused by the fact that anions will have larger orbitals than cations.

The effect of 1-octanol upon k_3 is small and our data are uncertain. Certainly the TMAE dication will be solvated, and this may affect the rate.

Acknowledgment

Edith Kirk performed much of the experimental work. Aaron Fletcher and Wayne Carpenter helped with purification of material and discussion of results. Richard Knipe, Alvin Gordon, Roy Leipnik, and William Ware discussed the chemical and mathematical possibilities.

Literature Cited

(1) Benson, S. W., "The Foundations of Chemical Kinetics," p. 39, McGraw-Hill, New York, 1960.
(2) Fletcher, A. N., Heller, C. A., *J. Catalysis* **6**, 263 (1966).
(3) Fletcher, A. N., Heller, C. A., *J. Phys. Chem.* **71**, 1507 (1967).
(4) *Ibid.*, **71**, 3742 (1967).
(5) Fletcher, A. N., Heller, C. A., *Photochem. Photobiol.* **4**, 1051 (1965).
(6) Halstead, C. J., Thrush, B. A., *Photochem. Photobiol.* **4**, 1007 (1965).
(7) Hammond, P. R., Knipe, R. H., *J. Am. Chem. Soc.* **89**, 6063 (1967).
(8) Heller, C. A., Fletcher, A. N., *J. Phys. Chem.* **69**, 3313 (1965).
(9) Jagur-Grodzinski, J., Szwarc, M., *Proc. Roy. Soc. (London)* **A288**, 224 (1965).
(10) Urry, W. H., personal communication, 1967.
(11) Urry, W. H., Sheeto, J., *Photochem. Photobiol.* **4**, 1067 (1965).
(12) Winberg, H. E., Downing, J. R., Coffman, D. D., *J. Am. Chem. Soc.* **87**, 2050 (1965).

RECEIVED October 9, 1967.

18

Oxidation of Ozonization Products

DENNIS G. M. DIAPER

Royal Military College of Canada, Kingston, Ontario, Canada

Ozonides and alkoxyhydroperoxides from 1-octene and ethyl 10-undecenoate were isolated by column chromatography and oxidized to acids, RCOOH, using (a) O_2 at 95°C., (b) O_2 at room temperature, catalyzed by reduced PtO_2, (c) Ce^{IV}, or (d) HCO_3H. Degradation by-product yields (ester and lower acid) were compared, and (b) was superior to (a). Alkoxyhydroperoxides at 0°C. or below were oxidized rapidly and exothermically to diperoxides, $R^1O.CHR.OO.CHR.OR^1$, by alcoholic hexanitratoammonium cerate, evolving O_2. The stoichiometry, determined by isothermal calorimetry, was one equivalent of Ce^{IV} per mole. Diperoxides were slowly hydrolyzed to $RCOOR^1$ and $RCHO$. Acids were also obtained from the ozonization products from elaidic acid and from 1-methylcyclopentene.

Although reductive methods for decomposing ozonides are by far the most usual, oxidation to carboxylic acids has considerable potential industrial and analytical importance. Azelaic acid has been prepared commercially for many years by oxidizing ozonized oleic acid, and a similar preparation of brassylic acid from erucic acid has been suggested (6). Such α,ω-dicarboxylic acids of the aliphatic series are of interest in polymer technology. Analysis of natural olefins, particularly unsaturated fats, by ozonization would be greatly facilitated by a quantitative procedure for converting the ozonide to some stable end product. Performic acid oxidation is nearly quantitative (1), but by-product ester formation, with accompanying loss of one carbon atom of the chain, reaches 2 to 5% of the total yield. Up to 30% yields of by-product esters have been found in the uncatalyzed autoxidation of oleic acid ozonization products at 95°C. This high yield of by-product esters was unaffected by wide variations of experimental conditions, including change of terminal function (*20, 21, 22, 23, 24*). The mechanism of their formation is thought to be (*15, 20*):

$$RCHO + O_2 \rightarrow RCO_3H$$
$$2\ RCO_3H \rightarrow (RCO)_2O_2 + H_2O_2$$
$$(RCO)_2O_2 \rightarrow CO_2 + RCOOR$$

Another method for oxidizing olefin ozonization products in one step in aqueous alkaline emulsion (*12*) and a study of alkylcycloalkene oxidative ozonization (*9, 11*) also found chain degradation. In the present study an attempt was made to find conditions for converting ozonides to acids in high yields without chain degradation. Cerium(IV) oxidation was tried because the known intermediate of by-product ester formation —the aldehyde—is relatively insensitive to this reagent (*30*). In the form of hexanitratoammonium cerate, it may be used in various organic solvents and forms deeply colored complexes with many of them, losing this color on reduction (*10*). The high redox potential (1.6 volts) of the ceric-cerous system recommends it for use in difficult oxidations. Although carbonyl compounds having α-hydrogen atoms have been thought to be attacked by such reagents in the enol form, recent studies (*17*) indicate that the carbonyl form is the reactive species, and we hoped that the substances under study, all of which possess α-hydrogens, would be little affected at positions other than the oxygenated carbon atoms. An alternative approach to the problem of controlled low-temperature oxidation of ozonides was a catalytic study, stimulated by the success of Heyns (*13*) in the sugar series and Sneeden (*29*) in steroid oxidation.

Experimental

Materials. 10-Undecenoic acid (m.p., 23–24°) contained 10 to 15% of undecanoic acid and a trace of 9-undecenoic acid, determined by reducing the total ozonization product with excess of LiAlH$_4$, treatment with (CF$_3$CO)$_2$O, and gas chromatographic (GLC) recognition of the trifluoroacetates of undecanol, 1,10-decanediol, and 1,9-nonanediol. Its ethyl ester after fractionation (b.p., 136-144°/14 mm.; n_D^{20}, 1.4390), contained negligible amounts of ethyl 9-undecenoate, gave a single GLC peak, and was ozonized without further purification.

1-Octene had negligible amounts of 2-octene impurity (GLC on AgNO$_3$) and was ozonized without further purification.

Elaidic acid (m.p., 41-43°), prepared by nitrogen oxides elaidinization of oleic acid, gave an ethyl ester (m.p., 3-5°C.; n_D^{25}, 1.4476).

1-Methylcyclopentene (b.p., 74-77°C.), prepared from 1-methylcyclopentanol, contained 1% of the exocyclic isomer (*11, 26*).

Ceric ammonium nitrate (hexanitratoammonium cerate, G. Frederick Smith Chemical Co.) was made up by weighing in 1.0N HNO$_3$ and was standardized against Fe^{II}. Solutions were generally 0.2–0.8N.

Platinum oxide (2) was reduced as required. Four batches were used, with colors ranging from yellow-brown to almost black.

Procedures. CHROMATOGRAPHIC PURIFICATION OF OZONIZATION PRODUCTS. Ozonization products from ethyl 10-undecenoate and 1-octene were chromatographed on silica gel columns (Baker) and eluted with 15 or 25% ether in petroleum ether (b.p., 30°–60°). Fractions were examined by thin-layer chromatography (TLC) on silica gel G Chromagram sheet eluted with 40% ether in petroleum ether. For development of ozonide and peroxide spots, 3% KI in 1% aqueous acetic acid spray was better than iodine. The spots (of iodine) faded, but a permanent record was made by Xerox copying. Color of the spots varied from light brown (ozonide) to purple-brown (hydroperoxide), and the rate of development of this color was related to structure (diperoxide > hydroperoxide > ozonide). 2,4-Dinitrophenylhydrazine spray revealed aldehyde spots and also reacted with ozonides and hydroperoxides. Fractions were evaporated at room temperature or below in a rotary evaporator.

The following chromatographically pure substances were obtained in preparative quantities from terminal olefins: ethyl 10-undecenoate ozonide (m.p., 2-3°C.), the methoxyhydroperoxide from ethyl 10-undecenoate ozonolysis at 0°C. in methanol (10–20% solutions, ozonized to completion as shown by the bromine test), the ozonide from 1-octene, and the following alkoxyhydroperoxides from 1-octene: 1-methyl-, 1-ethyl-, and 1-*tert*-butylheptane-1-hydroperoxide. R_f values for the first two compounds, separated by TLC as above, were, respectively, 0.670 and 0.553. Spots of lower R_f values were attributed to polymeric peroxides. The chromatographed products were stable in storage at $-10°C.$ for several weeks. 1-Methylcyclopentene similarly gave a chromatographically pure ozonide and methoxyhydroperoxide. Ethyl elaidate ozonization products were more complex mixtures, and chromatographic purification in quantity was impractical.

ISOTHERMAL CALORIMETRY. An all-glass ice calorimeter, similar to that described by Vallee (32), was used to determine heats of reaction between hexanitratoammonium cerate samples and alkoxyhydroperoxides. Melting of an ice mantle surrounding the reaction vessel causes a mercury meniscus to move along a graduated horizontal capillary, calibrated electrically to correspond to 1.23 cal. per cm. Initially, the cerate solution (25-35 ml.) was pipetted into the vessel and cooled to 0°C. A solution of alkoxyhydroperoxide of known concentration was sealed into a weighed fragile glass bulb and introduced. A steady drift (foredrift) of approximately 0.1 cal. per minute was observed. The bulb was broken with a stirrer, and heat evolved rapidly. After 30–40 minutes the rate of drift (afterdrift) was usually about the same as the foredrift. The heat change was taken as the difference between the extrapolated foredrift and afterdrift lines at the time of mixing. Blanks (0.3–0.5 cal.) were determined and subtracted from the heat measured (5–30 cal. per determination).

OXIDATION OF OZONOLYSIS PRODUCTS FROM 1-OCTENE AND 1-METHYLCYCLOPENTENE. Uncatalyzed autoxidation was performed in a small vessel provided with a reflux condenser and a fritted disk for introducing oxygen. The sample was raised to 95°C. during 2 hours and held at that temperature, with oxygen passing, until peroxide and aldehyde could no longer be detected (24–48 hours). Catalyzed autoxidation was performed in a vessel connected to an oxygen buret and provided with a magnetic stirrer. Completion of the reaction was assumed when no more oxygen uptake occurred. Cerate oxidations were performed at 0°C. in the solvent indicated. Oxygen evolution occurred, and the color was instantaneously discharged on mixing less than one equivalent of cerate solution with the alkoxyhydroperoxides; reaction with the ozonide was much slower. Up to 4 equivalents of cerate were used, and reduction was always complete in 3–5 hours at room temperature. The solution was diluted with an equal volume of water and extracted with petroleum ether. Performic acid oxidation of the alkoxyhydroperoxide or ozonide was performed in ether (5 grams in 50 ml.) by adding formic acid (10 ml.) and 70% hydrogen peroxide in excess (10 ml.). Heat was evolved. The mixture was allowed to stand at room temperature for 3 days, after which aldehyde could no longer be detected.

OXIDATION OF ELAIDIC ACID OZONIZATION PRODUCTS. Elaidic acid was ozonized in chloroform solution (5.0 grams in 100 ml.) at 0°C., and the solvent was removed by rotatory evaporation at room temperature. Aliquots were oxidized catalytically and noncatalytically as described above.

SAPONIFICATION TITRATIONS. Weighed aliquots of oxidation products (0.1–0.5 grams) were titrated with ethanolic potassium hydroxide (0.2N) in aldehyde-free ethanol to a phenolphthalein end point while nitrogen was passed in. A measured excess of alkali was then added, and the solution was boiled for 15 minutes with nitrogen slowly passing through. Back-titration with standard HCl gave the amount of alkali consumed by saponification.

PRODUCT IDENTIFICATION BY GLC. When helium through a neopentylglycol sebacate column at 75°C. was used, retention data for authentic samples of heptanal, methyl heptanoate, and ethyl heptanoate were, respectively, 5.0, 10.7, and 17.2 cm. Products from ozonization were identified by co-chromatography with these authentic products. Acids, aldehydes, and esters of the same chain length produced in oxidative ozonolysis are conveniently converted to a single product for GLC analysis by treating the substance in ether with excess $LiAlH_4$. Ethyl acetate is added to destroy the hydride, and the organic layer is washed with water and dried. Trifluoroacetic anhydride is added, and the mixture is diluted with petroleum ether, shaken with water and sodium bicarbonate aqueous solution, dried, and concentrated.

The trifluoroacetates of 1-hexanol and 1-heptanol had retention times of 3.0 and 5.1 minutes, respectively, on Carbowax at 80°C. The trifluoroacetates of 1-undecanol, nonane-1,9-diol, and decane-1,10-diol had retention times of 4.0, 5.1, and 7.6 minutes on Apiezon at 170°C. Peaks were well formed and did not have the characteristic "tail" of alcohol peaks. The reduction-trifluoroacetylation treatment of 5-ketohexanoic acid gave a major product, retention time 4.3 minutes on neopentyl glycol sebacate at 75°C., and a by-product, not identified, having a retention time of 6.6 minutes and giving a peak whose area was 15–20% of that of the major product (1,5-hexanediol bistrifluoroacetate). After saponification of autoxidized (or catalytically autoxidized) elaidic acid ozonization products, the dry ether solution was treated with diazomethane and four GLC peaks were identified, corresponding to 1-octanol (1.3 minutes), methyl nonanoate (1.7), dimethyl azelate (10.7), and methyl 8-hydroxyoctanoate (11.9). The presence of hexanoic acid in impure heptanoic acid was measured by the ratio of the peak areas caused by their methyl esters. Retention times on neopentyl glycol sebacate were 4.9 and 10.7 minutes. Retention times on Apiezon L of 9-carbethoxynonanal, methyl ethyl sebacate, and diethyl sebacate were 4.0, 4.8, and 5.8 minutes at 128°C.

Results

Isothermal Calorimetry of Hexanitratoammonium Cerate Oxidation of Products from 1-Octene and 10-Undecenoic Acid. The heat developed in the oxidation of ethyl 10-ethoxydecanoate 10-hydroperoxide in ethanol is shown in Figure 1. Samples of 10% solutions of peroxide in ethanol were used with 5-ml. aliquots of 0.1465N cerate in 25 ml. of ethanol. The intersection of the two lines shows a ratio of 1.04 moles of peroxide per equivalent of cerium and maximum heat evolution of 42 kcal. per equivalent of cerium. Similar plots were made for the reaction of the corresponding methoxyhydroperoxide in ethanol (1.10 equivalents, 47 kcal.) and in methanol (1.08 equivalents, 45 kcal.). 1-Ethoxyheptane-1-hydroperoxide was oxidized in acetone (0.98 equivalent, 36 kcal.), in

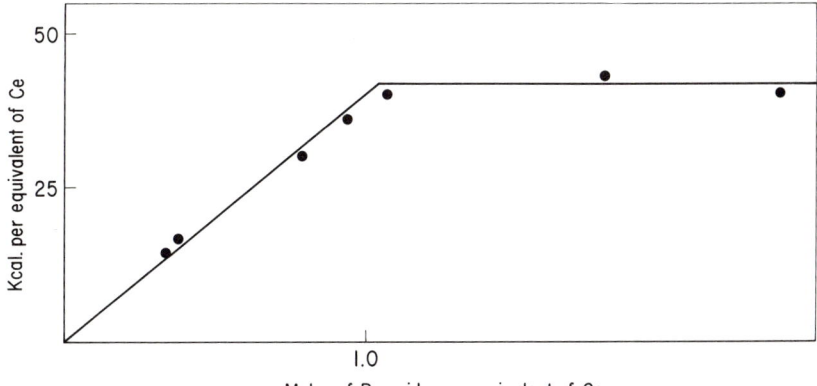

Figure 1. Heat developed as a function of reactant ratio for ceric ammonium nitrate oxidation in ethanol of ethyl 10-ethoxydecanoate-10-hydroperoxide

ethanol (1.10 equivalents, 47 kcal.), in methanol (1.08 equivalents, 43 kcal.), and in 60% aqueous *tert*-butyl alcohol (1.13 equivalents, 47 kcal.). The corresponding *tert*-butoxyhydroperoxide in aqueous *tert*-butyl alcohol gave 43 kcal. per equivalent of cerium with the consumption of 1.06 moles of peroxide, while in methanol the figures were 45 kcal. and 1.01 equivalents. Aqueous *tert*-butyl alcohol was used rather than the alcohol itself because the absorption or evolution of the latent heat of freezing of solvent interferes with the measurement of heats of reaction. The ozonides of 1-octene and ethyl 10-undecenoate reacted so slowly with hexanitratoammonium cerate at 0°C. that no measurement could be made.

Chain Degradation during Oxidation. In Table I, the substances indicated were oxidized by four different procedures. Method A was 48-hour uncatalyzed oxidation with oxygen at 95°C. Catalytic oxidation (B) was performed in acetone (10–15 ml.) over reduced platinum oxide (50 mg.) using 0.5- to 1.0-gram samples. Oxygen uptake was at such a rate that half of the final uptake occurred at room temperature in 2–9 hours, and the volume absorbed was 50–70% of the theoretical value. In method C, the oxidant was methanolic ceric ammonium nitrate (4.0 equivalents, 0.135N), and the substrate (0.5–1.0 gram) was dissolved in 10-20 ml. of methanol. The solutions were cooled by adding ice while they were mixed, and the mixture was then stirred for 24 hours, becoming colorless. Method D, performic acid oxidation, was described above. The first column gives hexanoic acid content of the isolated product measured as the peak area caused by methyl hexanoate, expressed as a percentage of the methyl heptanoate peak. Similarly, the third column gives the hexyl trifluoroacetate peak area expressed as a percentage of the heptyl trifluoroacetate peak. The ester yields in the second column are calculated from saponification titrations and expressed as percentage conversion of the starting material, assuming that one ester function is formed for each mole of substrate.

Oxidation of Elaidic Acid Ozonization Products. Aliquots of the unseparated ozonization products from elaidic acid were autoxidized at 95°C. uncatalyzed and in acetone over reduced platinum oxide as before. Total yields of acids and esters were determined by titration and were found to be 74.6 and 19.2%, respectively, in the catalyzed reaction with uptake of 63% of the theoretical volume of oxygen. Time required for uptake of half this volume was 4 hours at 21°C. Uncatalyzed oxidation at 95°C. of the other fraction gave 27.4% yield of esters and 74.5% yield of acids, calculated on the assumption that one original olefinic linkage can produce one ester function or two acid functions. When elaidic acid was ozonized in methyl acetate and the catalyzed oxidation performed in the same solvent, acid yield was 80.8%, and ester yield was 7.3% with a half-uptake time of 5.6 hours and 88% of the theoretical quantity of oxygen absorbed. After ozonization in chloroform, the uncatalyzed oxidation of the products gave 29.2% yield of esters and 68.9% yield of acids. The catalytically oxidized portion (in acetone) had 23.8% yield of esters and 71.2% yield of acids. Besides the four saponification products identified above, lower acids were found in the products of uncatalyzed autoxidation, but the GLC trace of the products from catalyzed autoxidation showed that lower acids were absent.

Table I. Ester By-products and Chain Degradation in Ozonization Product Oxidation

Substrate	Method	Lower Acid, %	Ester, %	Trifluoroacetate, %
1-Octene ozonide	A	4.3	17.02	18
	B	0.6	17.91	16.7
	C	11.2	28.4	10.1
	D	0	4.93	3.1
1-Methylcyclopentene ozonide	A	—	8.13	12
	B	—	8.01	6.7
	C	—	21.2	10.7
	D	—	4.42	3.6
MeO\\ /CH(CH$_2$)$_5$CH$_3$ HOO/	A	2.9	15.1	9.3
	C	15.3	31.8	12.2

Diperoxides from 1-Alkoxyheptane-1-hydroperoxide Oxidation. From the oxidation of 0.74 grams of 1-methoxyheptyl-1-hydroperoxide, by 1 equivalent of hexanitratoammonium cerate at 0°C., 0.23 grams of the diperoxide was isolated by column chromatography on silica with 20% ether in petroleum ether as eluting solvent. Although the NMR spectrum showed this to be a mixture of *dl* and meso isomers, a single spot reacting with KI was obtained on thin-layer chromatography. It was a rather viscous water-white liquid of characteristic, biting odor. (Found: C, 66.32; H, 11.84%. $C_{16}H_{34}O_4$ requires C, 66.16; H, 11.80%). Its infrared spectrum had no hydroxyl peak. It liberated iodine from potassium iodide and did not dissolve in aqueous sodium bicarbonate. Its R_f value on TLC analysis was not much different from that of the starting material, but the corresponding product (C, 70.76; H, 12.41% $C_{22}H_{46}O_4$ requires C, 70.54; H, 12.38%) from 1-*tert*-butoxyheptane-1-hydroperoxide oxidation had an R_f value of 0.562 compared with 0.765 for the starting material. When the product from the methoxyhydroperoxide was reduced by LiAlH$_4$ and trifluoroacetylated as described, a single GLC peak corresponding to 1-heptanol was obtained. Decomposition of the ethoxyhydroperoxide by boiling for a few minutes with dilute H_2SO_4 gave a liquid product whose GLC showed two peaks of approximately equal area corresponding to heptaldehyde and ethyl heptanoate. This liquid had the characteristic odor of heptaldehyde and a strong broad carbonyl peak in the infrared and gave a positive Schiff test.

The NMR spectra of 1-methoxyheptyl-1-hydroperoxide and of the derived diperoxide were examined in $CDCl_3$ with TMS internal standard. The hydroperoxide had a triplet (δ = 4.77 p.p.m., O—CH—O), a singlet (δ = 3.52 p.p.m., CH_3O), a multiplet (δ = about 1.3 p.p.m., CH_2), and a distorted triplet (δ = 0.88 p.p.m., CH_3—C). A separate peak ascribable to OH was not observed, but the presence of exchangeable protons was confirmed by appearance of a HDO signal after treatment with D_2O. The diperoxide showed no exchangeable protons, and the spectrum was similar to that of the hydroperoxide. The methoxyl signals, however, now appeared as two singlets, δ = 3.31 and 3.66 p.p.m., of approximately equal intensity. It was concluded that the diperoxide was a mixture of *dl* and meso forms.

Liberation of methanol during decomposition of 1-methoxy-heptyl-1-hydroperoxide was demonstrated by holding a hot copper wire in the vapor. The odor of formaldehyde was detected. From the solution, the oxime of heptaldehyde was obtained (m.p., 54-55.5°C.) undepressed in admixture with an authentic sample. (Found: C, 65.16; H, 11.55; N, 10.83%. $C_7H_{15}ON$ requires C, 65.07; H, 11.70; N, 10.84%.) Another sample of the hydroperoxide (0.73 gram) was boiled for a few minutes with dilute H_2SO_4. The solution was cooled, excess of sodium hydroxide was added, and the mixture was boiled under reflux for 1.5 hours, then acidified and steam-distilled. The ether extract of the distillate was separated into neutral and acid (0.071-gram) fractions. From the latter, the amide of heptoic acid (m.p. 92-94°C.) was obtained.

Heptoic Acid from 1-Octene. The olefin (1.8 grams) in methanol (50 ml.) was ozonized at 0°C. to completion. Ice was added, plus 35 grams (3.98 equivalents) of ceric ammonium nitrate and 10 ml. of 2N HNO_3. There was a brisk effervescence, and the solution was stirred for 8 hours, at the end of which time it was colorless. It was extracted with petroleum ether and the extract was shaken with water and divided into acidic and neutral fractions with sodium bicarbonate. The acid fraction (1.32 grams, 63%) was divided into two portions. One was treated with ethereal diazomethane and found to give GLC peaks corresponding to methyl heptanoate and methyl hexanoate (12% of the area of the former); the other was converted to the amide (m.p., 92–94°C.). The neutral fraction (0.83 gram) gave a negative KI test and a negative Schiff test. Saponification titration of a weighed aliquot showed that 5.1 meq. of ester were present.

Transesterification. The product obtained from cerate oxidation of the methoxyhydroperoxide from 1-octene in ethanol contained methyl heptanoate but no ethyl heptanoate, and the oxidation product of the ethoxyhydroperoxide in methanol contained ethyl heptanoate but no methyl heptanoate.

Ethyl 10-undecenoate ozonide was autoxidized at 95% in the absence of a catalyst. Diethyl sebacate, methyl ethyl sebacate, and dimethyl sebacate were detected after diazomethane treatment. When the ozonide was oxidized in acetone over platinum black and similarly treated with diazomethane, dimethyl sebacate could not be detected.

Oxidation of Ethyl 10-Undecenoate Ozonide. The ozonide (1.214 grams, chromatographically pure) was treated with ceric ammonium nitrate (10.2 grams, 4.0 equivalents) in 50 ml. of methanol and 10 ml. of $2N$ HNO_3. After stirring for 3.5 hours at room temperature, the solution, now colorless, was diluted with water and extracted three times with petroleum ether. The extract was washed with water and divided into acid and neutral fractions. The latter (0.872 gram after evaporation) gave a positive Schiff test and did not liberate iodine from potassium iodide. A portion was reduced and trifluoroacetylated in the usual manner, giving two major GLC peaks corresponding to the trifluoroacetates of decanediol and nonanediol, the latter having 8.1% of the area of the former. The acid fraction (0.441 gram, 41%) similarly showed 9.5% of degraded acid. A further portion of the acid fraction, treated with diazomethane, gave a GLC peak corresponding to dimethyl sebacate as well as the major peak corresponding to methyl ethyl sebacate. Its area was 11.0% of the area of the major peak.

The color of a methanolic solution of ceric ammonium nitrate gradually fades at room temperature, several days being required for complete discharge of the Ce^{IV} color. Because the cerate reagent clearly attacks methanol, a solvent inert to the reagent was sought. It was found that aqueous acidic cerate solutions in trifluoroacetic acid did not lose their color in a week at room temperature. The ozonide (0.714 gram) in trifluoroacetic acid (10 ml.) was stirred at room temperature with hexanitratoammonium cerate (5.6 grams) in 10 ml. of $2N$ HNO_3 for 2.5 hours, becoming colorless. The acid fraction (0.294 gram) recovered as before, was reduced and trifluoroacetylated, giving peaks as follows: C_{10}, area 782 units; C_9, 307; C_8, 69. Under these conditions chain degradation occurred.

The ozonide (4.3 grams) in ethyl acetate (25 ml.) was shaken with reduced platinum oxide (approximately 50 mg.) in a Burgess-Parr apparatus containing oxygen at 50 p.s.i.g. for 24 hours. The solution was aldehyde-free and gave a negative peroxide test. There was no detectable nonvolatile nonacidic fraction, and diazomethane treatment showed less than 1% of sebacic acid.

Discussion

Alkyl hydroperoxides are known to be oxidized to alkylperoxy radicals by cupric, cobaltic, and manganic salts. Although Ce^{IV} has not yet been reported to oxidize hydroperoxides and although mechanisms for ionic oxidation of alkoxy derivatives have not been put forward, Reaction 1 is a possible first step.

$$Ce^{IV} + R-\underset{\underset{OMe}{|}}{\overset{\overset{H}{|}}{C}}-OOH \rightarrow Ce^{III} + R-\underset{\underset{OMe}{|}}{\overset{\overset{H}{|}}{C}}-OO^{\cdot} + H^{+} \quad (1)$$

The *tert*-butylperoxy radical is known to decompose by various routes, of which Reaction 2 is more important than Reaction 3 by a factor of 3 to 1 (8).

$$2 \text{ } tert\text{-BuOO}^{\cdot} \rightarrow tert\text{-BuOO-}tert\text{-Bu} + O_2 \quad (2)$$

$$tert\text{-BuOOH} + tert\text{-BuOO}^{\cdot} \rightarrow tert\text{-BuOH} + tert\text{-BuO}^{\cdot} + O_2 \quad (3)$$

By analogy with Reaction 2 oxygen should be evolved from alkoxy hydroperoxide oxidation and a nonacidic diperoxide produced (Reaction 4)

$$2 \text{ } R-\underset{\underset{OMe}{|}}{\overset{\overset{H}{|}}{C}}-OO^{\cdot} \rightarrow O_2 + R-\underset{\underset{OMe}{|}}{\overset{\overset{H}{|}}{C}}-O-O-\underset{\underset{OMe}{|}}{\overset{\overset{H}{|}}{C}}-R \quad (4)$$

Protonation in the acid medium will favor decomposition of the diperoxide as shown by Reaction 5.

$$R-\underset{\underset{OMe}{|}}{\overset{\overset{H}{|}}{C}}-O-O-\underset{\underset{OMe}{|}}{\overset{\overset{\overset{+}{H}}{|}}{C}}-R \rightarrow H^{+} + RC\overset{\diagup O}{\underset{\diagdown OMe}{}} + \underset{MeO\diagup}{\overset{HO\diagdown}{}}CHR \quad (5)$$

Formation of ester and aldehyde is thus explained since the latter is produced in an acidic environment by decomposition of a hemiacetal such as that produced in Reaction 5. The formation of carbonyl groups from peroxides by nucleophilic displacement has been discussed by Waters (25) and by Kharasch (14). Production of the corresponding carboxylic acid by excess of ceric salt was accompanied by chain degradation and attack of the solvent by oxidizing agent (27, 28).

Decomposition of the peroxy radical to give equimolecular amounts of ester and aldehyde can be explained without invoking a diperoxide intermediate. Dimerization giving an unstable tetroxide would lead to an alkoxyalkoxy radical as shown in Reaction 6 (19).

$$2 \; R\!-\!\underset{\underset{OMe}{|}}{\overset{\overset{H}{|}}{C}}\!-\!OO^{\cdot} \longrightarrow \left[R\!-\!\underset{\underset{OMe}{|}}{\overset{\overset{H}{|}}{C}}\!-\!OOOO\!-\!\underset{\underset{OMe}{|}}{\overset{\overset{H}{|}}{C}}\!-\!R \right] \longrightarrow O_2 + 2 \; R\!-\!\underset{\underset{OMe}{|}}{\overset{\overset{H}{|}}{C}}\!-\!O^{\cdot} \quad (6)$$

Such a radical would disproportionate (4, 31) to aldehyde and hemiacetal (Reaction 7).

$$2 \; R\!-\!\underset{\underset{OMe}{|}}{\overset{\overset{H}{|}}{C}}\!-\!O^{\cdot} \longrightarrow RC\!\!\overset{O}{\underset{OMe}{\diagdown}\!\!\!\!\diagup} + RCH\!\!\overset{OH}{\underset{OMe}{\diagdown}\!\!\!\!\diagup} \quad (7)$$

Isolation of the diperoxide (as mixed dl and meso forms) by column chromatography shows that in this case the peroxy radical decomposes significantly by Routes 4 and 5, although participation by Routes 6 and 7 cannot be excluded.

Observed stoichiometry of the cerium-alkoxyhydroperoxide reaction was a little more than 1 mole of peroxide per equivalent of Ce. This agrees well with Reaction 4's predominating over a bimolecular decomposition analogous to Reaction 3. Because the observed stoichiometry was so close to 1:1, chain reaction decomposition initiated by metal ions appears of little importance in this case.

Catalytic oxidation of ozonides over platinum appears to be accompanied by the same ester by-product disadvantage found in the thermal process. Chain degradation by other reactions is less serious, however, and transesterification does not occur. The method can therefore be used to prepare a half-ester of a dicarboxylic acid from an ester of a suitable unsaturated acid. If ozonide autoxidation occurs by the route, ozonide → aldehyde → peracid, with the latter acting as precursor of both acid and ester products (20–24), it is interesting to compare reaction rates observed in the present study with the rate of uptake of oxygen by decanal (7), nonanal (16), heptanal (3, 18), and butanal (5), which in all cases was slower. While no comparative study of reaction rates under controlled conditions of initiator concentration, etc., was made, it was demonstrated that reduced PtO_2 is an effective catalyst for ozonolysis product autoxidation.

Acknowledgment

S. S. Barton generously assisted with the isothermal calorimetry. Financial support of Defence Research Board of Canada Grant 9530-17 is gratefully acknowledged.

Literature Cited

(1) Ackman, R. G., Retson, M. E., Gallay, L. R., Vandenheuvel, F. A., *Can. J. Chem.* **39**, 1956 (1961).
(2) Adams, R., "Organic Syntheses," Vol. I, p. 463, Wiley, New York, 1947.
(3) Backstrom, H. L. J., *J. Am. Chem. Soc.* **49**, 1460 (1927).
(4) Bartlett, P. D., Traylor, T. G., *J. Am. Chem. Soc.* **85**, 2407 (1963).
(5) Briner, E., Papazian, G., *Helv. Chim. Acta* **23**, 497 (1940).
(6) *Chem. Eng. News* **42** (48), 31 (1964).
(7) Cooper, H. R., Melville, H. W., *J. Chem. Soc.* **1951**, 1984.
(8) Dean, M. H., Skirrow, G., *Trans. Faraday Soc.* **54**, 849 (1958).
(9) Diaper, D. G. M., *Can. J. Chem.* **33**, 1720 (1955).
(10) *Ibid.* **34**, 1835 (1956).
(11) *Ibid.* **44**, 2819 (1966).
(12) Fremery, M. I., Fields, E. K., *J. Org. Chem.* **28**, 2537 (1963).
(13) Heyns, K., Paulsen, H., *Angew. Chem.* **69**, 600 (1957).
(14) Kharasch, M. S., Fono, A., Nudenberg, W., *J. Org. Chem.* **15**, 748 (1950).
(15) Lefort, D., Tempier, D., Sorba, J., *Bull. Soc. Chim. France* **1960**, 442.
(16) Lillard, D. A., Day, E. A., *J. Am. Oil Chemists, Soc.* **41**, 549 (1964).
(17) Littler, J. S., *J. Chem. Soc.* **1962**, 832.
(18) McNesby, J. R., Ph.D. Thesis, New York University, 1951.
(19) Milas, N. A., Djokic, S. M., *Chem. Ind.* **1962**, 405.
(20) Pasero, J., Ph.D. Thesis, Marseille, 1963.
(21) Pasero, J., Chouteau, J. Naudet, M., *Bull. Soc. Chim. France* **1960**, 1717.
(22) Pasero, J., Comeau, L., Naudet, M., *Bull. Soc. Chim. France* **1965**, 493.
(23) Pasero, J., Naudet, M., *Rev. Franc., Corps Gras* **7**, 189 (1960).
(24) *Ibid.* **10**, 453 (1963).
(25) Robertson, A., Waters, W. A., *J. Chem. Soc.* **1948**, 1574.
(26) Shabtai, J., Gil-Av, E., *Tetrahedron Letters* **1964**, 467.
(27) Shorter, J., *J. Chem. Soc.* **1950**, 3425.
(28) Shorter, J., Hinshelwood, C. N., *J. Chem. Soc.* **1950**, 3276.
(29) Sneeden, R. P., Turner, R. B., *J. Am. Chem. Soc.* **77**, 190 (1955).
(30) Trahanovsky, W. S., Young, L. B., *J. Chem. Soc.* **1965**, 5777.
(31) Traylor, T. G., Russell, C. A., *J. Am. Chem. Soc.* **87**, 3698 (1965).
(32) Vallee, R. E., *Rev. Sci. Instr.* **33**, 856 (1963).

RECEIVED October 9, 1967.

Radical Initiation and Interactions

T. G. TRAYLOR
Session Chairman

19

Determination of Rate Constants for the Self-Reactions of Peroxy Radicals by Electron Spin Resonance Spectroscopy

J. R. THOMAS and K. U. INGOLD

Chevron Research Co., Richmond, Calif.

> *Rate constants for the self-reactions of a number of tertiary and secondary peroxy radicals have been determined by electron spin resonance spectroscopy. The pre-exponential factors for these reactions are in the normal range for bimolecular radical-radical reactions (10^9 to 10^{11} M^{-1} $sec.^{-1}$). Differences in the rate constants for different peroxy radicals arise primarily from differences in the activation energies of their self reactions. These activation energies can be large for some tertiary peroxy radicals ($\sim$10 kcal. per mole). The significance of these results as they relate to the mechanism of the self reactions of tertiary and secondary peroxy radicals is discussed. Rate constants for chain termination in oxidizing hydrocarbons are summarized.*

Peroxy radicals have been detected by electron spin resonance (ESR) spectroscopy in many systems in which hydroperoxides are being decomposed and/or organic materials are being oxidized by molecular oxygen. However, ESR spectroscopy has only occasionally been applied to the determination of the rate constants for the self-reactions of peroxy radicals. Bielski and Saito (10) have measured the bimolecular rate constant for the disappearance of hydroperoxy radicals generated by the reaction of acidified ceric sulfate with a large excess of hydrogen peroxide. It has also been reported (33, 45) that the cumylperoxy radical concentration in the azo-bisisobutyronitrile (AIBN) and dicyclohexyl peroxydicarbonate–initiated oxidation of cumene at moderate temperatures (40° to 90°C.) is compatible with the known rates of chain initiation

(R_i) and the (corrected) chain termination rate constant ($2k_t$) reported by Melville and Richards (37) i.e.,

$$[CO_2^{\cdot}] \approx (R_i/2k_t)^{1/2}$$

A wealth of evidence (6, 7, 13, 19, 23, 46, 48) now supports the idea originally introduced by Blanchard (12) that tertiary peroxy radicals undergo both terminating and nonterminating interactions.

$$RO_2^{\cdot} + RO_2^{\cdot} \overset{K}{\rightleftarrows} (ROOOOR) \overset{2k'}{\rightarrow} \boxed{RO^{\cdot} + O_2 + {}^{\cdot}OR} \overset{2k_1}{\longrightarrow} 2RO^{\cdot} + O_2$$

cage (nonterminating)

$$\searrow 2k_2 \downarrow$$
?

ROOR + O_2 (terminating)

Thomas (46) has used ESR techniques to measure the rate constants for terminating ($2k_2$) and nonterminating ($2k_1$) reactions of *tert*-butylperoxy radicals (38) over a range of temperature. Both reactions have large activation energies, values for $E_1 \sim 15.5$, $E_2 \sim 10.2$, and E_1-E_2 ~ 5.3 kcal. per mole being obtained (46). In contrast, preliminary results suggested that cumylperoxy radicals interacted with one another without significant activation energy. A more complete ESR study of the self-reactions of peroxy radicals generated from their hydroperoxides has now been undertaken. The results for *tert*-butylperoxy, cumylperoxy, and several other peroxy radicals shows that an appreciable activation energy is a fairly general feature of these reactions. In particular, the results suggest that differences in the rate constants for the self-reactions of different peroxy radicals can be mainly assigned to differences in the activation energies of these reactions rather than to differences in the activation entropies. In certain cases, the ESR rate constants correspond extremely well with the rate constants for chain termination in the autoxidation of the appropriate hydrocarbon as determined by the rotating sector technique.

Experimental

Radical Generation. The ESR spectrometer, flow system, and general procedure have been described (46). The apparatus was calibrated with freshly prepared diphenyl picrylhydrazyl (DPPH) solutions. The peroxy radical concentrations were determined by double integration of derivative spectra. A standard coal sample in the dual cavity allowed corrections to be made for changes in cavity Q. The rates of decay of the less reactive radicals were determined by stopped-flow techniques with manually or electrically operated valves. The decay was recorded on a Tektronix 564 storage oscilloscope (46). For the more reactive radicals this procedure was unsatisfactory and decay rates were obtained by measuring the steady-state radical concentration at various flow rates. The transit times from mixer to cavity were calculated from the known volume of the system and the measured flow rates.

Radicals were generated from the hydroperoxides in two ways:

(1) By oxidizing 3×10^{-3} to $1 \times 10^{-5} M$ hydroperoxide with a large excess of ceric ammonium nitrate in methanol. The ceric and hydroperoxide solutions were mixed in an efficient mixer and then flowed through a highly fluted tube of 0.5-cc. volume before entering the ESR cavity.

The ceric concentrations were generally much higher than those used in the previous work (~ 0.3 to $0.7 M$ initial Ce^{4+} compared with $0.02 M$). This was necessitated by the fact that the higher molecular weight hydroperoxides reacted more slowly with ceric ion than the lower molecular weight materials. Rate constants for decay of their radicals therefore could not be obtained at low ceric concentrations because the radicals were still being formed as the reactant stream entered the ESR cavity. Even with the highest ceric concentrations it is possible that some hydroperoxides were not completely decomposed before entering the cavity. At high ceric concentrations and with flow rates above a certain minimum value the concentration of the less reactive peroxy radicals—e.g., tert-butylperoxy—became independent of the flow rate. This implies that the hydroperoxide ($\sim 10^{-3}$ to $10^{-4} M$) had been quantitatively converted to peroxy radicals and that no radicals were lost before entering the cavity. The peroxy radical concentration calculated with this assumption was in excellent agreement with the value based on the calibration with DPPH.

(2) By photolysis of 2.0 to $0.1 M$ hydroperoxide in benzene or methanol. A flow system proved most satisfactory. The hydroperoxide solution was photolyzed before entering the cavity and rate constants were obtained by the stopped-flow technique.

This procedure could be applied only to hydroperoxides which were available in large quantity. Direct photolysis of static systems in the ESR cavity tended to yield irreproducible data, with the rate constants exhibiting some dependence on the initial hydroperoxide concentration, presumably as a result of the nonuniform generation of radicals at the higher hydroperoxide concentrations. The results of the static photolysis are not considered reliable in comparison with the results of the flow photolysis. One interesting observation with tert-butylperoxy and cumylperoxy radicals in benzene with static photolysis at low temperatures was that the apparent rate constant for decay increased when the benzene froze. This must have been caused by the fact that reaction occurred in the liquid regions present in the frozen solution (41, 42). The liquid regions would contain all the hydroperoxide, and the true peroxy radical concentration in these liquid regions would be much higher than the apparent concentration based on the entire reaction volume.

Temperature Control. In the flow system the reservoirs containing the reactants were heated or cooled to approximately the required temperature. The flow was started and continued until the temperature of the liquid issuing from the cavity became constant before any rate measurements were made. A temperature range from $-10°$ to $50°C$. could be readily covered in this way. The Varian variable temperature accessory was employed in the static photolysis experiments.

ESR Signal. The peroxy radicals all gave a single line with no detectable fine structure at $g = 2.015 \pm 0.001$ (29, 47). The line widths decreased with increasing mass of the radical and, for any one radical, the line width decreased with decreasing temperature (47). The peak-to-peak signal height cannot be used as a relative measure of radical concentrations except for measurements which are made at the same temperature.

Materials. Chemically pure solvents and reagent grade ceric ammonium nitrate were used as received. Cumene hydroperoxide was purified via the sodium salt. Lucidol *tert*-butyl hydroperoxide was purified by low temperature crystallization. Tetralin hydroperoxide, cyclohexenyl hydroperoxide, and 2-phenylbutyl-2-hydroperoxide were prepared by hydrocarbon oxidation and purified by the usual means. 1,1-Diphenylethyl hydroperoxide and triphenylmethyl hydroperoxide were prepared from the alcohols by the acid-catalyzed reaction with hydrogen peroxide (10).

Results

The results obtained in this work are summarized in Table I. The rate constants were obtained from good second-order decays in all cases. The absolute values of the rate constants given in Table I are probably correct to within a factor of 2 to 3 in most cases. The ceric results should be somewhat more reliable than the photolytic results because more time was devoted to their study. It seems likely that more accurate rate constants will be obtained by both methods as more experience is gained in the use of the ESR technique.

For the ceric ion oxidation the decay occurs in the absence of hydroperoxide, whereas, for photolysis, decay occurs in the presence of hydroperoxide. The ceric rate constant should represent the sum of the rate constants for all processes in which peroxy radicals are destroyed—*i.e.*, it should represent both terminating and nonterminating interactions $(2k_1 + 2k_2)$. In contrast, photolytic generation of the radicals in the presence of a large excess of hydroperoxide should yield the rate constant for the conversion of peroxy radicals to nonradical products—*i.e.*, for terminating interactions $(2k_2)$—because the alkoxy radicals formed in nonterminating reactions (and their radical decomposition products) will react with the hydroperoxide to regenerate peroxy radicals

$$RO \cdot + ROOH \rightarrow ROH + RO_2 \cdot$$

We were unable to detect any signal which could be assigned to an alkoxy radical in the titanous ion reduction (17, 38) of *tert*-butyl hydroperoxide from $+25°$ to $-60°C$., nor could an alkoxy radical signal be

Table I. Second-Order Decay Constants and Ceric Oxidation[a]

Peroxy Radical	$2k \times 10^{-4}$, M^{-1} sec.$^{-1}$ 30°C.	$\log_{10} A$, M^{-1} sec.$^{-1}$	E, Kcal./Mole
tert-Butyl	0.35 ± 0.02[d]	10.4	9.5 ± 2
1,1,3,3-Tetramethylbutyl	0.83	11.0	9.8 ± 1
From dihydroperoxide[f]	2.8	10.9	9.0 ± 2
Cumyl	12 ± 2[g]	10.7	7.8 ± 2
2-Phenylbutyl-2-	29	9.4	5.5 ± 1
1,1-Diphenylethyl	~100[h]	—	—
Cyclohexenyl	280	10.8	6.0 ± 3
α-Tetralyl	520	10.0	4.6 ± 1
Cyclopentenyl	~1500	—	—
n-Butyl	~3000[i]	—	—
Hydro[j]	530[k]	9.8	4.7

[a] In methanol, stopped flow for first five radicals, variable flow for remainder.
[b] Flow photolysis in benzene (B) and methanol (M) unless otherwise noted.
[c] RH, rotating sector on hydrocarbon. ROOH, rotating sector on hydrocarbon in presence of hydroperoxide.
[d] Average of 23 determinations over a range of initial radical concentrations from 2.6×10^{-3} to $2.0 \times 10^{-4} M$.
[e] Static photolysis in benzene.

detected in the reduction of cumyl hydroperoxide, 1,1-diphenylethyl hydroperoxide, triphenylmethyl hydroperoxide, and α-tetralyl hydroperoxide at room temperature. Alkoxy radicals therefore do not contribute to the peroxy radical signal and do not interfere with the decay measurements.

Some indication of a trend toward first-order decay kinetics was observed previously (46) at high tert-butylperoxy radical concentrations ($\sim 1.7 \times 10^{-3} M$). At the high ceric ion concentrations used in the present work there was no evidence of any significant first-order process over a range of initial tert-butylperoxy radical concentrations from $4.0 \times 10^{-3} M$ to $4.0 \times 10^{-5} M$, nor for cumylperoxy radicals over a range of initial concentrations from $3.2 \times 10^{-4} M$ to $5.0 \times 10^{-5} M$. The first-order process was previously attributed (46) to a rapid equilibrium between peroxy radicals and a tetroxide, existing as a true intermediate, the slow decomposition of the tetroxide determining the kinetics of radical disappearance. No further evidence for this process was obtained in the present work over the temperature and concentration ranges investigated. It appears likely that the previous observation may have arisen from incom-

Kinetic Parameters for Peroxy Radicals (ESR)

Photolysis[b]			Rotating Sector (24–27), $2k \times 10^{-4}$, M^{-1} sec.$^{-1}$, 30°C.	
$2k \times 10^{-4}$, M^{-1} sec.$^{-1}$ 30°C.	$\log_{10} A$, M^{-1} sec.$^{-1}$	E, Kcal./Mole	RH[c]	ROOH[c]
0.13 (B)	6.4	4.5 ± 1	—	~0.1
0.50 (M)	8.2	6.3 ± 1		
(0.23)[e]	(7.3)[e]	(5.4 ± 2)[e]		
1.5 (M)	9.2	7.0 ± 2	—	0.6
1.8 (B)	12.7	11.0 ± 3	—	—
(1.0)[e]	(13.0)[e]	(12.5 ± 3)[e]		
3.7 (B)	8.8	5.8 ± 1	1.5	0.6
(3.2)[e] (4.4)[k]	(9.6)[e]	(7.1 ± 2)[e]		
—	—	—	18	4
(220)[e]	(8.1)[e]	(2.5 ± 1)[e]	9.4	4
—	—	—	560	—
—	—	—	760	760
—	—	—	620	—
—	—	—	—	—
—	—	—	—	—

[f] 2,5-Dimethylhexyl-2,5-dihydroperoxide (see text).
[g] Average of 15 determinations over a range of initial radical concentrations from 3.2×10^{-4} to 5×10^{-5} M.
[h] Hydroperoxide probably reacted too slowly with ceric ion (see text).
[i] At 7°C.
[j] Neat hydroperoxide (50).
[k] In water.

plete oxidation before entering the cavity of the more concentrated hydroperoxide solutions. The equilibrium constant between radicals and tetroxide must be considerably smaller than the previous estimate of $158 M^{-1}$.

The biperoxy radical produced by the ceric ion oxidation of 2,5-dimethylhexane-2,5-dihydroperoxide decays rapidly with first-order kinetics $[k = 10^{10.6} \exp(-11,500 \pm 1000)/RT$ sec.$^{-1} = 180$ sec.$^{-1}$ at 30°C. (30)]. After the first-order decay has run to completion, there is a residual radical concentration (~4% of the initial hydroperoxide concentration) which decays much more slowly by a second-order process. The residual second-order reaction cannot be eliminated or changed even by repeated recrystallization of the dihydroperoxide. This suggests that a small fraction of the biperoxy radicals react intermolecularly rather than by an intramolecular process and thus produce monoperoxy radicals. The bimolecular decay constant for this residual species of peroxy radical is similar to that found for the structurally similar radical from 1,1,3,3-tetramethylbutyl hydroperoxide. Photolysis of the dihydroperoxide gave radicals with second-order decay kinetics which are presumed to be 2,5-hydroperoxyhexyl-5-peroxy radicals.

Triphenylmethyl hydroperoxide appeared to react relatively slowly with ceric ion and the peroxy radicals decayed extremely rapidly, probably by the unimolecular decomposition $\phi_3COO\cdot \rightarrow \phi_3C\cdot + O_2$ (3, 21, 28). Only traces of peroxy radicals could be detected even at high flow rates, and so the decay kinetics could not be examined.

Discussion

The ESR results in Table I and the rotating sector results given in Table II show that the self-reactions of peroxy radicals (particularly tertiary peroxy radicals) may involve significant activation energies. [Activation energies of 3 kcal. per mole or less are close to the activation energies for solvent diffusion which are the minimum possible activation energies for simple bimolecular processes taking place in solution (40). Some of the termination processes involving the polymeric secondary peroxy radicals may be diffusion controlled (40).] Differences in the rate constants for the different peroxy radicals appear to be caused primarily by differences in the activation energies of their self-reactions.

Table II. Rate Constants and Kinetic Parameters for Chain Termination in Autoxidation of Hydrocarbons as Determined with the Rotating Sector (25, 26, 27, 28) (Neat Hydrocarbon or Hydrocarbon Diluted with Chlorobenzene)

Peroxy Radical	$2k_t \times 10^{-4}$, M^{-1} sec.$^{-1}$ 30°C.	$\log_{10} A$, M^{-1} sec.$^{-1}$	E, Kcal./Mole
Cumyl	1.5	11.0	~9.5
Poly(α-methylstyrylperoxy)	60	8.5	3.7 ± 1.6
α-Tetralyl	760	9.9	4.3 ± 1.5
Poly(α-deuterostyrylperoxy)	2100	10.0	3.7 ± 1.2
Poly(styrylperoxy)	4200	9.0	1.8 ± 1.2
Cyclohexanolperoxy[a]	13	6.7	2.2

[a] Photochemical aftereffect (2).

The measured rate constants show some inconsistencies in relation to other work. The most noticeable is the low ratio of $k_{ceric}/k_{photolysis}$ at 30°C. for *tert*-butyl hydroperoxide and cumene hydroperoxide compared with estimates, ~5 to 10 for k_1/k_2, obtained from studies of the induced decomposition of these hydroperoxides (22, 46, 48). The photolytic rate constant for cumene hydroperoxide is considerably larger than the termination constant for the oxidation of cumene containing cumene hydroperoxide as determined by the rotating sector (25, 26, 27, 28). It is not clear whether these differences represent some unappreciated features of the reaction or are caused by errors in the ESR determination of peroxy radical concentrations. The rate constants should be accurate to within a factor of 2 to 3, as noted above, and most of the inconsistencies can be resolved within these limits of error.

There appears to be some disagreement in the literature as to whether or not the nonterminating and terminating reactions of tertiary peroxy radicals—*i.e.*, Processes 1 and 2—have the same activation energy (*6, 18, 46*). A study of the AIBN-induced decomposition of *tert*-butyl and cumyl hydroperoxides has shown that the ratio of the oxygen evolution rate to the chain initiation rate probably increases with increasing temperature (*46*). (This conclusion rests on the assumption that the efficiency of radical generation from AIBN in these systems has a negligible temperature coefficient.) This implies that $E_1 > E_2$. The results for *tert*-butylperoxy, tetramethylbutylperoxy, and cumylperoxy radicals given in Table I provide some support for an activation energy difference of ~2 to 5 kcal. per mole between the two reactions. A difference of this magnitude might well arise from the temperature coefficient of fluidity of the solvent. As an indication of the sensitivity of the rate of combination of alkoxy radicals to their ease of diffusion it might be noted that the fraction of the *tert*-butoxy radicals formed by decomposition of di-*tert*-butylperoxyoxalate, which combine, increases rapidly as the viscosity of the solvent increases (*23*). The sensitivity to viscosity arises from the high rate constant [~$2 \times 10^8 M^{-1}$ sec.$^{-1}$ (*14*)] for the combination of *tert*-butoxy radicals.

Although the pre-exponential factors are in the expected range for bimolecular free radical reactions, they are somewhat inconsistent with the high activation energies obtained with the tertiary peroxy radicals. (We are indebted to S. W. Benson for drawing our attention to this point.) This inconsistency is most easily resolved by assuming that the tetroxide is formed reversibly and that the ceric measured rate constants represent $K2k'$—*i.e.*, $K(2k_1 + 2k_2)$. Tetroxide formation should have a negative entropy of activation of about 24 to 25 e.u. and should be exothermic by about 5 kcal. per mole, according to Benson's estimates (*8*)—*i.e.*, $K \sim 10^{-5}\, e^{5000/RT}$ liters/mole. Hence, the unimolecular decomposition of the tetroxide to two alkoxy radicals and oxygen (or RO· + ROOO·) can be represented by $2k' = 2k_{\text{measured}}/K$, which for the *tert*-butyl hydroperoxide–ceric reaction, for example, gives $2k' \approx 10^{10.4}\, e^{-9500/RT}/10^{-5}\, e^{5000/RT} \approx 10^{15.4}\, e^{-14500/RT}$ per second. The pre-exponential factor is more or less in line with the values found for other unimolecular decompositions—*i.e.*, typically $10^{14.5 \pm 0.5}$ per second.

At 30°C. both k_1 and k_2 increase with increasing size of the *tert*-alkyl and *tert*-aralkyl groups. Apparently, the elimination of oxygen from the presumed tetroxide intermediate is accelerated by an increase in the size of R. A similar steric acceleration occurs in nitrogen evolution from azonitriles RNNR—*e.g.*, [*tert*-$C_4H_9C(CH_3)CN$]$_2N_2$ decomposes about 100 times faster than [$(CH_3)_2CCN$]$_2N_2$ (*44*).

Rust and Youngman (*44*) have shown that peroxy radicals which can form intramolecular hydrogen bonds with hydroxyl groups are less reactive in hydrogen abstraction reactions than structurally related unbonded peroxy radicals. The similarity of the photolytic decay rate constants in benzene for 2,5-dimethyl-2-hydroperoxyhexyl-5-peroxy and 1,1,3,3-tetramethylbutylperoxy radicals indicates that if there is internal hydrogen bonding in the first-named radical, it has little effect on the rate of the self reaction. The photolytic results for *tert*-butyl hydroperoxide in methanol and benzene suggest that any intermolecular hydrogen bonding also has comparatively little effect on the rate of the terminating self-reaction of *tert*-butylperoxy radicals. These results are not unexpected in view of the generally small effects of solvent polarity on hydrocarbon oxidation rates (*20, 24*) and chain termination constants (*25, 26, 27, 28*). By way of contrast to organic peroxy radicals (C_4 and up), the rate constant for the self-reaction of hydroperoxy radicals is much smaller in water (*1, 10, 16*) and acetonitrile (*27*) than in less polar solvents such as *n*-decane, CCl_4, and chlorobenzene (*27*).

There is excellent agreement between the decay constants obtained by ceric ion oxidation of secondary hydroperoxides and the rate constants for chain termination in hydrocarbon autoxidation determined by the rotating sector. The agreement suggests that secondary peroxy radicals do not undergo many nonterminating interactions, so that most self-reactions of secondary peroxy radicals must be chain terminating.

Since the pre-exponential factors for the self-reactions of secondary and tertiary peroxy radicals appear to be rather similar (Tables I and II), the self-reaction of secondary peroxy radicals by way of a highly oriented intermediate (*43*) might seem to be rather unlikely compared with reaction to give two secondary alkoxy radicals followed by their rapid disproportionation while still in the solvent cage (*9*). Disproportionation in the cage should yield equal amounts of alcohol and ketone. However, in nonviscous solvents a certain fraction of the alkoxy radicals may be expected to escape from the cage (between 20 and 80% perhaps, by analogy with the fraction of alkyl radicals which escape recombination in the cage upon the decomposition of azo compounds) (*19, 31, 32, 39*). If some radicals escape from the cage, more alcohol will be formed than ketone unless the alkoxy radicals undergo a rapid β-scission reaction. Product studies on pulse-radiolyzed cyclohexane saturated with oxygen have shown that about 10% more cyclohexanol is formed than cyclohexanone (*12, 15, 35, 36*). This result lends some support to a reaction *via* alkoxy radicals rather than *via* a highly oriented tetroxide decomposing through a cyclic transition state. On the other hand, the Russell mechanism provides an attractive explanation for the observed low activation energy (and rapid termination) of secondary peroxy radicals compared with tertiary which is not easily accounted for by the Benson mechanism. However, the evidence for either mechanism is not conclusive. If both mechanisms are, in fact, operative, the Russell mechanism would be expected to predominate at lower temperatures and with secondary peroxy radicals containing weakly bound hydrogen atoms on the peroxidic carbon.

A number of critical questions require additional study before the details of the self-reactions of peroxy radicals can be specified with confidence. More precise values of "absolute" rate constants and their temperature coefficients for a variety of radicals under various experimental conditions are required.

A recent report by Bartlett and Guaraldi (5) provides convincing evidence for the existence of the tetroxide as an intermediate in the self-reactions of *tert*-butylperoxy radicals. They estimate ΔH for the formation of tetroxide by dimerization of peroxy radicals to be −6 kcal. per mole and ΔE_{act} for decomposition of the tetroxide to alkoxy radicals and oxygen to be 11 kcal. per mole.

Acknowledgment

We are indebted to F. R. Mayo for the cyclopentenyl hydroperoxide, to H. S. Mosher for the *n*-butyl hydroperoxide, to the Lucidol Corp. for the 1,1,3,3-tetramethyl butyl hydroperoxide, and to the U. S. Peroxygen Corp. for the 2,5-dimethylhexane-2,5-dihydroperoxide.

Literature Cited

(1) Adams, G. E., Boag, J. W., Michael, B. D., *Proc. Roy. Soc. (London)* **289A**, 321 (1966).
(2) Aleksandrov, A. L., Denisov, E. T., *Izv. Akad. Nauk SSSR, Ser. Khim.* **1966**, 1737.
(3) Ayers, C. L., Janzen, E. G., Johnston, F. J., *J. Am. Chem. Soc.* **88**, 2610 (1966).
(4) *Ibid.* **89**, 1176 (1967).
(5) Bartlett, P. D., Guaraldi, G., *Ibid.* **89**, 4801 (1967).
(6) Bartlett, P. D., Günther, P., *Ibid.* **88**, 3288 (1966).
(7) Bartlett, P. D., Traylor, T. G., *Ibid.* **85**, 2407, (1963).
(8) Benson, S. W., *Ibid.* **86**, 3922 (1964).
(9) *Ibid.* **87**, 972 (1965).
(10) Bielski, B. H. J., Saito, E., *J. Phys. Chem.* **66**, 2266 (1962).
(11) Bissing, D. E., Maturak, C. A., McEwan, W. E., *J. Am. Chem. Soc.* **86**, 3824 (1964).
(12) Blackburn, R., Charlesby, A., *Trans. Faraday Soc.* **62**, 1159 (1966).

(13) Blanchard, H. S., *J. Am. Chem. Soc.* **81**, 4548 (1959).
(14) Carlsson, D. J., Howard, J. A., Ingold, K. U., *Ibid.* **88**, 4725 (1966).
(15) Cramer, W. A., *J. Phys. Chem.* **71**, 1171 (1967).
(16) Currie, D. J., Dainton, F. S., *Trans. Faraday Soc.* **61**, 1156 (1965).
(17) Dixon, W. T., Norman, R. O. C., *J. Chem. Soc.* **1963**, 3119.
(18) Factor, A., Russell, C. A., Traylor, T. G., *J. Am. Chem. Soc.* **87**, 3692 (1965).
(19) Hammond, G. S., Sen, J. N., Boozer, C. E., *Ibid.* **77**, 3244 (1955).
(20) Hendry, D. G., Russell, G. A., *Ibid.* **86**, 2368 (1964).
(21) Hendry, D. G., Russell, G. A., *Ibid.*, p. 2371.
(22) Hiatt, R., Clipsham, J., Visser, T., *Can. J. Chem.* **42**, 2754 (1964).
(23) Hiatt, R., Traylor, T. G., *J. Am. Chem. Soc.* **87**, 3766 (1965).
(24) Howard, J. A., Ingold, K. U. *Can. J. Chem.* **42**, 1250 (1964).
(25) *Ibid.* **43**, 2729, 2737 (1965).
(26) *Ibid.* **44**, 1113, 1119 (1966).
(27) *Ibid.* **45**, 785, 793 (1967).
(28) Howard, J. A., Ingold, K. U., unpublished results.
(29) Ingold, K. U., Morton, *J. Am. Chem. Soc.* **86**, 3400 (1964).
(30) Ingold, K. U., Thomas, J. R., unpublished results.
(31) Kodama, S., *Bull. Chem. Soc. Japan* **35**, 652, 658, 824, 827 (1962).
(32) Kodama, S., *et al.*, *Ibid.* **39**, 1009, 1323 (1966).
(33) Lebedev, Ya. S., Tsepalov, V. F., Shlyapintokh, V. Ya., *Dokl. Akad. Nauk SSSR* **139**, 1409 (1961).
(34) Lebedev, Ya. S., Tsepalov, V. F., Shlyapintokh, V. Ya., *Kinetika i Kataliz* **5**, 64 (1964).
(35) Lowever, C. F., *J. Phys. Chem.* **71**, 1112 (1967).
(36) MacLachlan, A., *J. Am. Chem. Soc.* **87**, 960 (1965).
(37) Melville, H. W., Richard, S., *J. Chem. Soc.* **1954**, 944.
(38) Mulcahy, M. F. R., Steven, J. R., Ward, J. C., *Australian J. Chem.* **18**, 1177 (1965).
(39) Nelson, S. F., Bartlett, P. D., *J. Am. Chem. Soc.* **88**, 137, 143 (1966).
(40) North, A. M., "International Encyclopedia of Physical Chemistry and Chemical Physics," C. E. H. Bawn, ed., Topic 17, Vol. 1, p. 84, Pergamon Press, New York, 1966.
(41) Pincock, R. E., Kiovsky, T. E., *J. Am. Chem. Soc.* **87**, 2072, 4100 (1965).
(42) *Ibid.* **88**, 51, 4455 (1966).
(43) Russell, G. A., *Ibid* **79**, 3871 (1957).
(44) Rust, F. F., Youngman, E. A., *J. Org. Chem.* **27**, 3778 (1962).
(45) Thomas, J. R., *J. Am. Chem. Soc.* **85**, 591 (1963).
(46) Thomas, J. R., *Ibid.* **87**, 3935 (1965).
(47) *Ibid.* **88**, 2064 (1966).
(48) Traylor, T. G., Russell, C. A., *Ibid.* **87**, 3698 (1965).
(49) Walling, C., "Free Radicals in Solution," p. 513, Wiley, New York, 1957.
(50) Zwolenik, J. J., *J. Phys. Chem.* **71**, 2464 (1967).

RECEIVED October 9, 1967.

20

Cage Reactions of Acetoxy Radicals

J. C. MARTIN and STEVEN A. DOMBCHIK

University of Illinois, Urbana, Ill. 61803

Double-labeling experiments on acetyl peroxide-^{18}O decomposition rule out the cyclic analog of the Cope rearrangement as a mechanism for the scrambling of label. Hydrolysis of acetyl peroxide, followed by permanganate oxidation, proceeds without label scrambling between carbonyl and peroxidic oxygen. Oxygen thus derived shows the more complete scrambling required by a mechanism involving acetoxy radical intermediates. The rate at which carbonyl label is incorporated into the peroxide oxygen gives a rate constant for label scrambling close to that obtained earlier with carbonyl oxygens. Decompositions in hydrocarbon solvents of increasing viscosity show an increased label scrambling rate. This suggests that the decrease in over-all rate of disappearance of acetyl peroxide with increasing solvent viscosity is correctly ascribed to increased importance of solvent cage recombination of acetoxy radicals.

The decomposition of acetyl peroxide is thought (2, 11, 17, 21, 22) to proceed by a mechanism involving initial O—O bond cleavage to give two acetoxy radicals which can recombine to yield acetyl peroxide (22) or decarboxylate (2, 21) to give methyl radicals with rates competitive with the rate of diffusion from the solvent cage. Evidence based on heavy atom (^{18}O and ^{13}C) kinetic isotope effects has been interpreted (6) in terms of concerted two-bond or three-bond cleavage mechanisms. At least a part of the incompatibility between the observed kinetic isotope effects and a model based on simple O—O bond cleavage is removed by including in the model a large contribution of the cage reaction of acetoxy radicals to regenerate acetyl peroxide (17, 22, 23). The cage recombination postulated to explain the observed (17, 22) scrambling of label is also in keeping with the observation of a near-zero secondary kinetic isotope effect in the decomposition of perdeuterioacetyl peroxide, reported by Koenig and Brewer (11) if one makes the reasonable assumption that the transition state for decarboxylation of the acetoxy radical, a very exothermic reaction (21), is one in which C—C bond cleavage is not appreciably developed.

Few examples have been reported (5, 8, 9, 10, 12, 24) of cage recombination of simple alkoxy or acyloxy radicals to form O—O bonds in isolable molecules. This paper explores further the implications of the observed (17, 22, 23) scrambling of label seen in acetyl peroxide *carbonyl*-^{18}O recovered after partial decomposition.

Experimental

Sodium Acetate-^{18}O. Sodium acetate-^{18}O was prepared according to the procedure described by Drew (4). A mixture of 61 grams (3.38 moles) of H_2O (1.52 atom % excess ^{18}O) and 0.5 ml. of concentrated hydrochloric acid was added slowly with stirring and cooling to 193.9 grams (1.62 moles) of trimethyl orthoacetate. The reaction mixture was heated at 60°C. for 4 hours. A solution of 87.5 grams (1.62 moles) of anhydrous sodium methoxide in 364 ml. of absolute methanol was added dropwise with stirring, and the resulting precipitate (sodium acetate-^{18}O) was filtered and washed. Concentrating the alcohol solution resulted in recovery of additional product to give a total yield of 113.9 grams (1.39 moles, 86%) of sodium acetate-^{18}O.

Acetic Anhydride-^{18}O. To 81.0 grams (0.99 mole) of sodium acetate-^{18}O was slowly added 58.9 grams (0.495 mole) thionyl chloride with stirring and cooling. The reaction mixture was heated at 60°C. for 1.75 hours and then at 70°C. for 5 hours. Distillation gave 46 grams (0.45 mole, 91.4%) of acetic anhydride-^{18}O (1.269 atom % excess ^{18}O per oxygen, b.p., 57°C., 34 mm.). GLC analysis using a 10-foot QF-1 (Dow Corning fluorosilicone) column on 80/90 mesh Analabs ABS indicated about a 5% impurity of acetic acid. The anhydride was used to synthesize acetyl peroxide-*carbonyl*-^{18}O without further purification.

Acetyl Peroxide-*Carbonyl*-^{18}O. Acetyl peroxide-^{18}O was prepared by a modification of the usual procedure (16). An apparatus was devised to allow purification of the peroxide by remote manipulation of stopcocks and valves controlling nitrogen pressure. Recrystallization, filtration, and the addition of fresh solvent were accomplished without exposing the experimenter to the hazard of direct manipulation of crystalline acetyl peroxide.

Acetyl peroxide was recrystallized three times at −78°C. to give a stock solution of peroxide in isooctane, whose concentration was determined either by infrared analysis using the 1790 cm.$^{-1}$ peroxide band or by iodometric titration (13).

Stock solutions in the more viscous solvents were prepared by simply removing ether solvent *in vacuo* and then adding the solvent to the dry crystalline peroxide, omitting the recrystallization steps.

Isotope Scrambling During the Decomposition of Acetyl Peroxide-*Carbonyl*-^{18}O. DETERMINATION OF CARBONYL LABEL. Isotope scrambling in the decomposition of acetyl peroxide-*carbonyl*-^{18}O (synthesized from acetic anhydride-^{18}O containing 1.269 atom % excess ^{18}O) was determined by the process previously described (22, 23), except that the recrystallizations of recovered acetyl peroxide were carried out by remote manipulation. The absence of scrambling or dilution of label during the purification procedure was demonstrated by conversion of acetyl peroxide from stock solution to methyl acetate by treatment with sodium methoxide (22, 23). The resulting methyl acetate showed 1.246 atom % excess ^{18}O per labeled oxygen. The detailed procedure used for ^{18}O analysis has been described (17, 23).

DETERMINATION OF LABEL IN PEROXIDIC OXYGEN. Acetyl peroxide was hydrolyzed to hydrogen peroxide as follows. To a 7.5-ml. aliquot of acetyl peroxide stock solution in isooctane, containing an estimated 0.7 mmole of acetyl peroxide, was added 2 ml. of 2.2N perchloric acid. The mixture was heated with vigorous stirring in a 40°C. oil bath for 5.5 hours. Titration of an aliquot of the aqueous phase with aqueous potassium permanganate indicated a total yield of hydrogen peroxide of 0.73 mmole. A parallel hydrolysis reaction run at room temperature for 4 days gave almost identical results (0.74 mmole).

Thermal Decomposition of Acetyl Peroxide-*Carbonyl*-^{18}O. The isotopic content of the peroxide oxygens of recovered acetyl peroxide was determined by the following procedure. A sample of 12 ml. of isooctane 0.05M in acetyl peroxide (3.57 atom % excess carbonyl ^{18}O) was degassed, sealed in a bomb tube, and heated at 80 ± 0.05°C. for an appropriate time. The bomb tube was opened, and its contents were added to 3 ml. of 2.2N perchloric acid and heated for at least 5.25 hours at 40°C. with rapid magnetic stirring. The aqueous phase was added to one side of a Y-tube. An acidic solution of potassium permanganate (about a three-fold molar excess of that needed to convert all the hydrogen peroxide to oxygen) was added to the second arm of the Y-tube. The solutions were simultaneously degassed and then mixed *in vacuo*. The resulting oxygen was collected after passing it through two traps at −196°C. The oxygen was analyzed for ^{18}O, using both the 34/32 and the 36/34 m/e ratios obtained on a Consolidated-Nier model 21-201 isotope ratio mass spectrometer. Suitable controls showed the reproducibility of the mass spectrometric 34/32 and 36/34 ratios to be within a range of 0.5% over the pressure ranges of samples used. A zero sample analyzed by this procedure showed less than 0.01 atom % excess ^{18}O incorporated in the peroxide oxygens.

The 34/32 and 36/34 ratios of the oxygen samples were calibrated mass spectrometrically by comparison with oxygen samples of known isotopic composition which were blended from samples generated electrolytically (1) from H$_2^{18}$O samples containing 4.63 and 1.39 atom % excess ^{18}O. The H$_2^{18}$O labeling was determined by conversion to CO^{18}O in the usual (16, 23) manner.

Results

The previously reported (22, 23) method for determining the position of ^{18}O label in acetyl peroxide involves multiple recrystallizations of acetyl peroxide, recovered after partial decomposition, followed by treatment with sodium methoxide. Isolating the resulting methyl acetate by GLC provides a compound containing all of the carbonyl oxygen. This is analyzed for total ^{18}O by the usual procedure.

$$CH_3C\overset{O^*}{=}O-O\overset{O^*}{=}CCH_3 \xrightarrow{NaOCH_3} CH_3C\overset{O^*}{\diagdown}OCH_3$$

Samples analyzed by two different workers (this work and Ref. 23) gave satisfactory agreement, but the results showed an objectionable degree of scatter. The data represented by the open circles of Figure 1 define a rate constant for scrambling of label in carbonyl-labeled acetyl peroxide in experiments followed by this analytical method.

A new method has been developed for degrading acetyl peroxide to determine the pattern of labeling based on the isolation of the peroxide

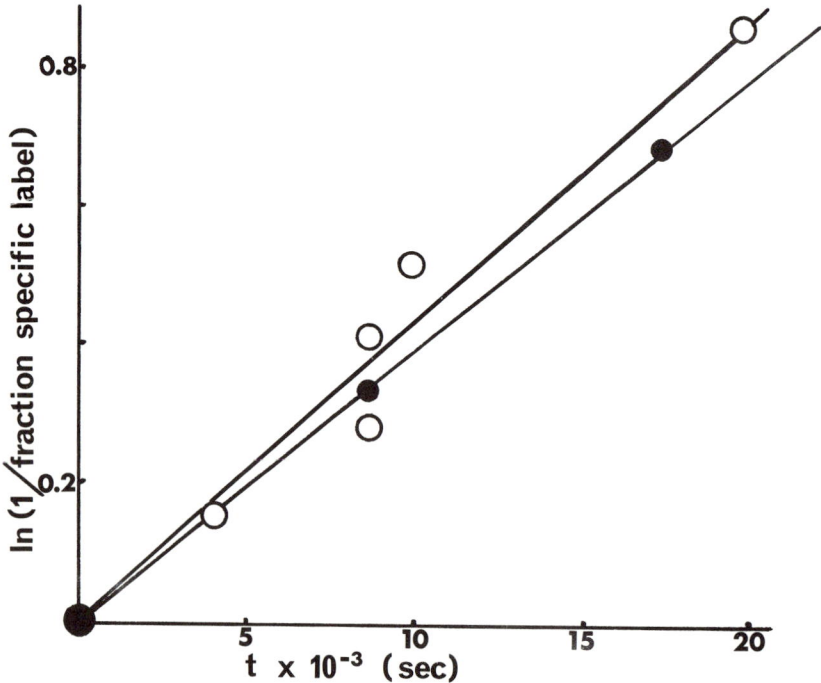

Figure 1. *Scrambling of carbonyl-^{18}O label in recovered acetyl peroxide as measured by disappearance of carbonyl label (open circles) and by appearance of peroxide label (solid circles)*

oxygen rather than the carbonyl oxygen. The method is based on that of Bunton (3), who has shown that no scrambling of label accompanies the over-all process involving the hydrolysis of peracetic acid and conversion of the resulting hydrogen peroxide to oxygen by permanganate oxidation. The hydrolysis of acetyl peroxide is accomplished under comparable conditions, and oxygen derived by this route from specifically carbonyl-labeled acetyl peroxide shows no incorporation of label. The data of Table I (the closed circles of Figure 1) illustrate the much greater precision attained by using this analytical method. These points represent incorporation of label into the peroxide oxygen of acetyl peroxide remaining in solution in isooctane after heating at 80°C. for the indicated time. No purification of the acetyl peroxide is required since none of the decomposition products interfere with the analysis.

Table I. Specificity of Labeling in Acetyl-*carbonyl*-^{18}O Peroxide after Partial Decomposition at 80°C.

Solvent	Time, Sec.	34/32 [a]	Fraction of Carbonyl Specifically Labeled [b]
Isooctane, 0.05M	0	0.00428	1.000
	8700	0.01451	0.712
	17400	0.02228	0.499
Dodecane, 0.08M	0	0.00412	1.000
	8700	0.01578	0.671
	17400	0.02413	0.443
Octadecane, 0.08M	0	0.00412	1.000
	8700	0.01686	0.646
	17400	0.02569	0.401
Mineral Oil, 0.08M	0	0.00413	1.000
	8700	0.01880	0.588
	17400	0.02826	0.331

[a] Mass spectrometric m/e abundance ratios were normalized using oxygen samples from electrolysis of water of known isotopic composition.
[b] The initial level of label in the acetyl-*carbonyl*-^{18}O peroxide was 3.570 atom % excess. The quantities listed represent the excess of label in the carbonyl oxygen over that in the peroxide oxygen. At infinite time equal labeling would be found in the two oxygens, and the fraction of carbonyl specifically labeled would be zero.

The results of this study and similar studies carried out in other solvents are included in Table I and are plotted in Figure 2. Rate constants for scrambling of label, derived from the first-order rate plots of Figure 2, are listed in Table II.

Applying the method to double-labeling studies used acetyl peroxide prepared from acetic anhydride with a high level of labeling (3.57 atom % excess ^{18}O). A sample of this peroxide was recovered after heating at 80°C. for 8700 sec. The oxygen derived from this peroxide showed a

36/34 m/e ratio ($^{18}O^{18}O/^{18}O^{16}O$) of 0.00835. This is to be compared with the values predicted from the two most probable mechanisms for scrambling of label.

Mechanism A:

$$CH_3C\overset{O^*}{\underset{O-O}{\overset{\parallel}{\diagdown}}}\overset{O^*}{\underset{}{\overset{\parallel}{\diagup}}}CCH_3 \rightleftarrows \left[CH_3C\overset{O^*\cdots O^*}{\underset{O\cdots O}{\diagup\diagdown}}CCH_3 \right]^{\ddagger} \rightarrow CH_3C\overset{O^*-O^*}{\underset{O\ \ O}{\diagdown\diagup}}CCH_3$$

Mechanism B:

$$CH_3C\overset{O^*}{\underset{O-O}{\overset{\parallel}{\diagdown}}}\overset{O^*}{\underset{}{\overset{\parallel}{\diagup}}}CCH_3 \rightleftarrows \left[2\ CH_3C\overset{O^*}{\underset{O}{\cdot}} \right] \rightarrow$$

$$\begin{array}{c} CH_3C\overset{O}{\underset{O^*-O}{\diagdown\diagup}}CCH_3 \\ + \\ CH_3C\overset{O}{\underset{O^*-O^*}{\diagdown\diagup}}CCH_3 \end{array}$$

Mechanism A clearly predicts the formation of more double-labeled oxygen, in this case a 36/34 ratio of 0.0149, calculated from statistical considerations. Mechanism B, involving cage recombination of acetoxy radicals, predicts a more random distribution and a 36/34 ratio of 0.00828, closer to the experimentally determined ratio of 0.00835.

Discussion

The double-labeling experiment described above was used to rule out Mechanism A for the scrambling of carbonyl label which is observed in acetyl peroxide recovered after partial decomposition at 80°C. The closely analogous Cope rearrangements (14, 20) are conveniently studied at considerably higher temperatures unless unusual structural features are operative in lowering the energy of activation. The low value of the bond dissociation energy for the O—O bond of diacyl peroxides, relative to that for the central C—C bonds in the 1,5-diene analogs, apparently causes the dissociation to give acetoxy radicals to be preponderantly favored over the cyclic rearrangement.

The close agreement between the 36/34 m/e ratios observed in the oxygen derived from recovered acetyl peroxide in the double-labeling experiment and that calculated for Mechanism B suggests that no more than 1% of the reaction can proceed by Mechanism A.

Previous tracer studies (22, 23) have ruled out the pictured analogous cyclic pathway for formation of methyl acetate.

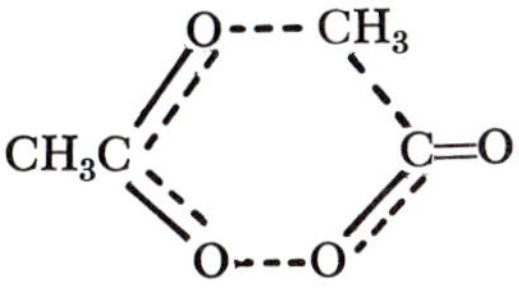

It is clear (23) that three products result from reactions of acetoxy radical pairs within the solvent cage: acetyl peroxide (estimated at 38%), methyl acetate (12.4%), and ethane (2.9%). The implied near equality of rates for decarboxylation of the acetoxy radicals and diffusion from the cage has been given quantitative expression in work of Braun, Rajbenbach, and Eirich (2). These workers studied the variation in the amounts of ethane and methyl acetate formed from acetyl peroxide as a function of solvent viscosity, and they derived a rate constant for the decarboxylation of acetoxy radical at 60°C. of 1.6×10^9 sec.$^{-1}$.

They further noted (2) that the over-all rate constant for disappearance of acetyl peroxide decreases monotonically with increasing solvent viscosity. The attractive hypothesis (19, 22, 23) that the observed rate decrease with increased viscosity in this homologous series of hydrocarbon solvents reflects the increased importance of cage recombination of acetoxy radicals in the more viscous solvents is subject to further test from the data of this paper.

Scheme A

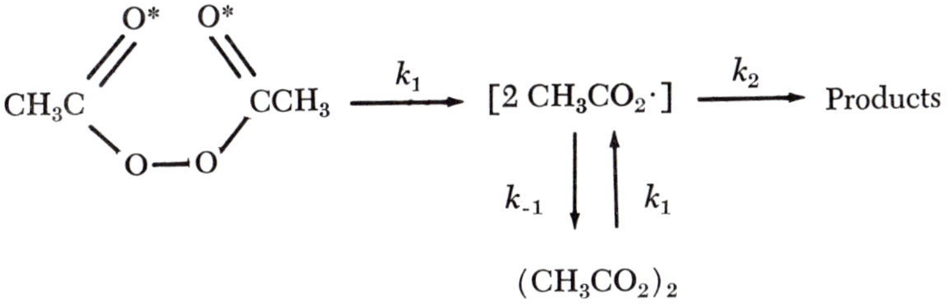

$$CH_3C(O^*)(O-O)CCH_3 \xrightarrow{k_1} [2\ CH_3CO_2\cdot] \xrightarrow{k_2} \text{Products}$$

$$k_{-1} \downarrow \uparrow k_1$$

$$(CH_3CO_2)_2$$

"Scrambled"

In terms of the pictured scheme the rate constants for scrambling of carbonyl label listed in Table II may be written as

$$k_s = \frac{k_1 k_{-1}}{k_{-1} + k_2}$$

The over-all rate constant for disappearance of total acetyl peroxide is

$$k_d = \frac{k_1 k_2}{k_{-1} + k_2}$$

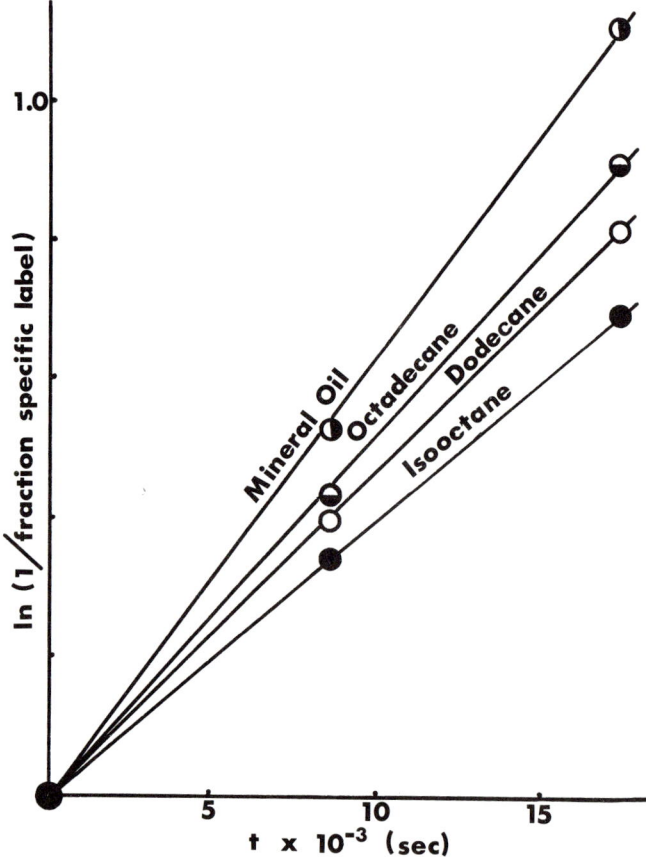

Figure 2. Rates of scrambling of carbonyl-^{18}O label in acetyl peroxide at 80°C. in solvents of varying viscosity

Table II. Rates of Scrambling of Carbonyl Label in Diacetyl Peroxide at 80°C.

Peroxide	Solvent	First-order $k \times 10^5$ (Sec.$^{-1}$)	k_{rel}
Diacetyl	Isooctane	4.00	1.00
	Dodecane	4.68	1.17
	Octadecane	5.25	1.31
	Mineral Oil	6.37	1.59

Let us assume that k_1 is equal to k_g, the rate constant for the gas phase decomposition (15), where no cage effect is expected. This assumption does not always hold (15, 18). For example, it is known (18) that di-*tert*-butyl peroxide (DPB) decomposes about 30% slower in the gas phase than in solution. We can calculate from our value of k_s and the known value of k_g, from the work of Szwarc (7, 21), a value for the fraction of acetoxy radical pairs recombining, f_R, where

$$f_R = \frac{k_{-1}}{k_{-1} + k_2}$$

Since $f_R = 1 - f_D$, where f_D is the fraction of acetoxy radical pairs reacting to give other products, and

$$f_D = \frac{k_2}{k_{-1} + k_2}$$

we can calculate, for a solvent in which k_s has been measured, a value for k_d (calc).

Table III. Solvent Effects on the Decomposition Rate of Acetyl Peroxide

Solvent	Relative k_d (Obs.)	Relative k_d (Calc.)
None	1.55[a]	1.48
Isooctane	1.00[a,b]	1.00
Dodecane	0.79[b]	0.92
Octadecane	0.66[b]	0.85
Mineral Oil[c]	0.44[d]	0.72

[a] From data of Szwarc (7).
[b] From data of Braun, Rajbenbach, and Eirich (2).
[c] American Oil Co., white oil, No. 31, U.S.P.
[d] From our manometric data at 80°C. Isooctane was chosen as reference solvent for all sets of data.

Table III and the plot of Figure 3 compare values of k_d (obs.) and k_d (calc.) in several solvents. The values of k_d (obs.) were taken from the data of Szwarc (7) and Eirich (2) as well as from our laboratory and involve an extrapolation from 60° to 80°C. for the data of Eirich. This was done by making the crude approximation that ΔG^* varies with solvent in the same way at the two temperatures. A close parallel exists between the observed and calculated rates. This suggests that the assumption equating k_1 and k_g is not bad and provides a firmly established case in which the dependence of the rate of decomposition of a radical initiator on solvent viscosity in a homologous series of solvents results from the varying importance of cage recombination in the various solvents. Pryor and Smith (19) using data from *p*-nitrophenylazotriphenylmethane decompositions have suggested that a general criterion for radical cage return to regenerate an initiator molecule might be found in a dependence of decomposition rate on solvent composition parallel to that in the decomposition of acetyl peroxide. They have derived an equation which relates viscosity and the observed first-order rate constant.

$$\log k(\text{obs.}) = \log k_1 + \frac{C_1}{\eta} + C_2$$

where C_1 and C_2 are constants and k_1 is the same as in Scheme A. Kiefer and Traylor (10) have used the method of Pryor and Smith to demonstrate similar behavior for DBP. Using kinetic data they were also able to calculate the fraction of cage collapse for DBP in solvents of varying viscosity. These calculations seem reasonable by comparison with similar results directly observed for di-*tert*-butyl hyponitrite (DBH) and di-*tert*-butyl peroxyoxalate (DBPO) further suggesting the dependence of the rate of radical initiator decomposition on cage recombination. In at least one case, DBH, Kiefer and Traylor have observed that rates of decomposition are 25% faster in Nujol than in isooctane. They explain this in terms of solvent stabilization of DBH by isooctane in excess of that provided by Nujol.

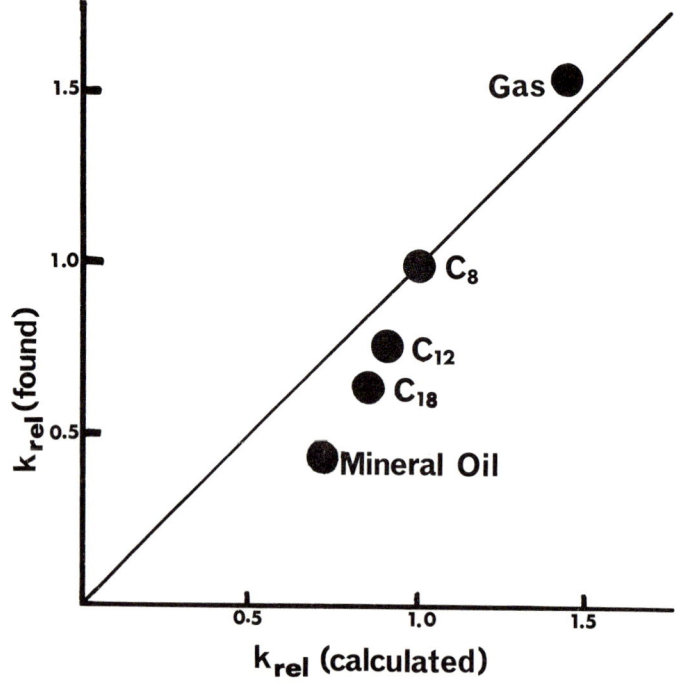

Figure 3. *Rates of over-all disappearance of acetyl peroxide calculated from the rates of scrambling of carbonyl-^{18}O label, assuming the only solvent effect to be the cage effect*

The deviation of points from the theoretical line of Figure 3 could be explained in at least two ways: (a) the crude extrapolation of relative rates from 60° to 80°C. may be in error; new rate data at 80°C. are needed; (b) the assumption that all cage recombinations of acetoxy radicals occur with scrambling of carbonyl label may be in error. If recombination rapid enough to compete with reorientation of the radical fragments originally formed from acetyl peroxide were to give recombination with retention of label specificity in 30% of all recombination events, the agreement between k_d (calc.) and k_d (obs.) would be quite good indeed.

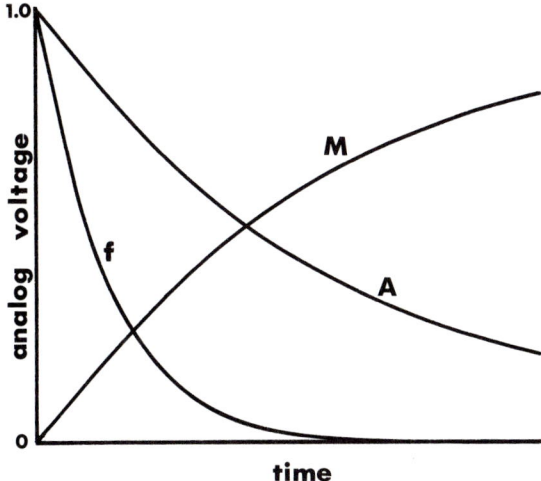

Figure 4. The probability that a radical generated at time zero from acetyl peroxide will be an acetoxy radical (A), a methyl radical (M) or will have diffused from the solvent cage, a quantity proportional to (f). These curves were used to duplicate the observed product yields

With respect to point b above, it is interesting to examine the consequences of an assumption that the three possible cage recombination reactions (those between two acetoxy radicals, between acetoxy and methyl radicals, and between two methyl radicals) proceed with the same (near zero) ΔG^*. With this assumption, a function of time, $f(t)$, is proportional to the probability of cage recombination per unit time between radicals in a pair generated at time zero. We used an analog computer to generate functions representing the composition of radical pairs which were generated at time zero as two acetoxy radicals. Figure 4 shows the first-order decay of the function A, representing the probability that an acetoxy radical generated at time zero will be an acetoxy radical at time

t, the first-order rise of the function M, representing the probability that the radical will have decarboxylated at time t, and one of the possible arbitrary functions $f(t)$, chosen to give the observed product distribution (38% acetyl peroxide, 12.4% methyl acetate, and 2.9% ethane). To these estimates (23) a small correction could be made using our new, more accurate rate constants for scrambling of carbonyl label. The corrections would be smaller, however, than the probable uncertainty existing in rate constants available at 80°C. for the over-all disappearance of the peroxide as followed by infrared spectroscopy. Koenig and Brewer (11) have reported data which suggest that the complete elimination of radical-chain induced decomposition by radical scavenger techniques would give significantly altered decomposition rate constants. The changes would not, however, be sufficiently large to alter the conclusions of this paper. The quantitative extension of these conclusions must await further, more accurate, data on rates and products. Integrals over long time intervals were generated to represent the total observed cage products, where:

$$\text{Acetyl peroxide} = \int fAA \, dt$$
$$\text{Methyl acetate} = 2 \int fAM \, dt$$
$$\text{Ethane} = \int fMM \, dt$$

Attempts were made to fit the product data using various forms of the function $f(t)$. (Early attempts to fit product data using various forms of $f(t)$ were reported in the preprint of this paper, distributed at the Symposium on Oxidation. These were incorrect, owing to a malfunction in our analog computer. The present results require a modification of the conclusion originally drawn.) The plot of $f(t)$ in Figure 4, which accurately reproduces the observed product distribution is of the form

$$f(t) = e^{-k_3 t}$$

The exponential form of $f(t)$ suggests that the assumption equating the rates of the three types of radical coupling for the species involved in this reaction is not bad. The form is approximately that expected for the operation of a diffusion process with k_3 representing the rate of diffusion from the cage. The solution pictured in Figure 4 shows a rate constant for diffusion from the cage 4.3 times larger than that for decarboxylation of the acetoxy radical. Although differences in diffusion rates expected for methyl and acetoxy radicals make it unlikely that a single rate constant should describe the rate of destruction of the caged radical pairs by diffusion, it is interesting that such a simple picture can be used to describe the observed product ratios. The scheme may require elaboration when additional data are available on product ratios in reactions under other conditions and for other peroxides.

Acknowledgment

This research has been supported in part by a grant from the National Science Foundation (NSF-GP6630). Fellowship support for S. A. D. was provided by the National Institute of Health and Standard Oil Co. of California, and for J. C. M. by the John Simon Guggenheim Memorial Foundation.

Literature Cited

(1) Bently, R., *Biochem. J.* **45**, 591 (1949).
(2) Braun, W., Rajbenbach, L., Eirich, F. R., *J. Phys. Chem.* **66**, 1591 (1962).
(3) Bunton, C. A., Lewis, T. A., Llewellyn, D. R., *J. Chem. Soc.* **1956**, 1226.
(4) Drew, E. H., Ph.D. Thesis, University of Illinois, Urbana, Ill., 1961.
(5) Factor, A., Russell, C. A., Traylor, T. G., *J. Am. Chem. Soc.* **87**, 3692 (1965).
(6) Goldstein, M. J., *Tetrahedron Letters* **1964**, 1601.
(7) Herk, L., Feld, M., Szwarc, M., *J. Am. Chem. Soc.* **83**, 2998 (1961).
(8) Hiatt, R., Clipsham, J., Visser, T., *Can. J. Chem.* **42**, 2754 (1964).
(9) Hiatt, R., Traylor, T. G., *J. Am. Chem. Soc.* **87**, 3766 (1965).
(10) Kiefer, H., Traylor, T. G., *J. Am. Chem. Soc.* **89**, 6667 (1967).
(11) Koenig, T. W., Brewer, W. D., *Tetrahedron Letters* **1965**, 2773.
(12) Koenig, T., Deinzer, M., *J. Am. Chem. Soc.* **88**, 4518 (1966).
(13) Kokatnur, V. R., Jelling, M., *J. Am. Chem. Soc.* **63**, 1432 (1941).
(14) Levy, H., Cope, A. C., *J. Am. Chem. Soc.* **66**, 1684 (1944).
(15) Martin, H. Jr., *Angew. Chem. Intern. Ed. Engl.* **5**, 78 (1966).
(16) Martin, J. C., Drew, E. H., *J. Am. Chem. Soc.* **83**, 1234 (1961).
(17) Martin, J. C., Taylor, J. W., Drew, E. H., *J. Am. Chem. Soc.* **89**, 129 (1967).
(18) Molyneux, P., *Tetrahedron* **22**, 2929 (1966).
(19) Pryor, W. A., Smith, K., *J. Am. Chem. Soc.* **89**, 1741 (1967).
(20) Rhoads, S. J., "Molecular Rearrangements," P. deMayo, ed., Vol. I, pp. 684-693, Interscience, New York, 1963.
(21) Szwarc, M., "Peroxide Reaction Mechanisms," J. O. Edwards, ed., p. 173, Interscience, New York, 1962.
(22) Taylor, J. W., Martin, J. C., *J. Am. Chem. Soc.* **88**, 3650 (1966).
(23) *Ibid.*, **89**, 6904 (1967).
(24) Walling, C., Waits, J., *J. Phys. Chem.* **71**, 2361 (1967).

RECEIVED January 29, 1968.

21

Bond Dissociation Energies in the Phenyl Benzoate Molecule and in Related Free Radicals

PETER GRAY

Department of Physical Chemistry, Leeds University, Leeds 2, England

The standard heat of formation of phenyl benzoate as the gaseous species at 25°C. has been determined as: $\Delta H_f°(PhCO_2Ph) = -35 \pm 1$ kcal. per mole. Seven distinct bond dissociation energies are immediately related to this reference basis. Values for these, together with values for the heats of formation of related free radicals, are discussed, and a provisional set is presented. They include the following estimates (kcal. per mole): $D(PhCO_2-Ph) = 94$; $D(Ph-CO_2Ph) = 96$; $D(\cdot CO_2-Ph) = 62$; $D(PhCO-OPh) = 64$. Errors are likely to be around 5 kcal. per mole.

This paper extends previous discussions (6, 7, 8) of the thermochemistry of oxygenated free radicals to the strengths of bonds in phenyl benzoate and in related free radicals and the heats of formation of these free radicals. It is based on our recent redetermination (1) of the heats of combustion and formation of phenyl benzoate—measured principally to establish reliable thermochemistry for the decomposition and explosion (5) of dibenzoyl peroxide, which yields phenyl benzoate as a major product.

Enthalpy of Formation of Phenyl Benzoate

Standard procedures were followed, and experimental details will be published elsewhere. The mean value found for ΔU_c was -31.803 kjoules per gram, with a standard deviation of 0.015%. Washburn corrections and conversion to constant-pressure conditions yielded $\Delta H_c° = -1506.5 \pm 0.5$ kcal. per mole, and $\Delta H_f° = -57.7 \pm 0.5$ kcal. per mole.

Errors quoted are twice the standard deviation, and the new value represents a significant shift from Stohmann's (1892) measurement of 1510.4 kcal. per mole for $-\Delta H_c°$, on which heats of formation in the literature have been previously based.

This value relates to the pure crystalline ester, and to discuss bond dissociation energies it is necessary to have a value for the heat of formation of gaseous phenyl benzoate. The latent heat of sublimation at 25°C. may be derived from separate values for fusion and vaporization. We have measured the latent heats of fusion at 70°C. as $\Delta H_{\text{fus}} = 7.0 \pm 0.3$ kcal. per mole (both electrically and from determining the cryoscopic constant). An average value of the latent heat of vaporization, $\Delta H_{\text{vap}} = 14.2 \pm 0.2$ kcal. per mole, may be evaluated from existing (17) vapor pressure data between 106° and 314°C. To correct these values to 25°C., estimated specific heats must be used. Fortunately, the corrections to ΔH_{fus} and ΔH_{vap} are in opposite directions, and tend to cancel each other; the standard heat of formation of gaseous phenyl benzoate is:

$$\Delta H_f°(\text{PhCO}_2\text{Ph}) = -35 \pm 1 \text{ kcal. per mole.}$$

Bond Dissociation Energies and Radical Thermochemistry

Seven distinct bond dissociation energies are relevant to aromatic esters. They are not independent, and their interrelations are shown below.

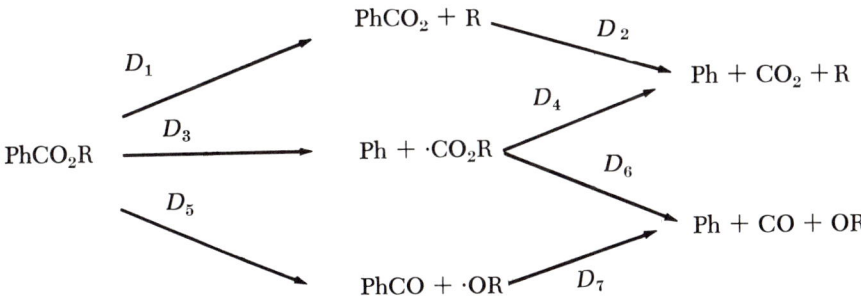

$(D_1 + D_2) = (D_3 + D_4)$ $(D_3 + D_6) = (D_5 + D_7)$

Oxygen-Phenyl Bond in Phenyl Benzoate. By using the new value for the enthalpy of formation of phenyl benzoate together with the most recent values (80 and -21 kcal. per mole, respectively) for the enthalpies of formation of the phenyl (3) and benzoate (11, 12) radicals, we obtain for the strength of the oxygen-phenyl bond in phenyl benzoate a value of

D_1 (PhCO$_2$—Ph) = 94 ± 4 kcal. per mole. The strength of this bond is less than that of the corresponding bond in phenyl acetate D(MeCO$_2$—Ph) = 105 kcal. per mole [a value derived from the enthalpy of hydrolysis (*18*) of phenyl acetate, $\Delta H_{\text{hydr}} = -6.86$ kcal. per mole and the presently accepted (*8*) value for the heat of formation for the acetate radical ΔH(MeCO$_2 \cdot$) = −45]. A difference of 3 to 5 kcal. per mole might be expected to arise from the different inductive effects on the oxygen-phenyl bond in the two parent molecules. It is too great to be readily explicable in terms of superior delocalization in the product radicals, and it contrasts with the similarity found for O—O bond dissociation energies (*11*) in the two parent peroxides—PhCO$_2$CO$_2$Ph and MeCO$_2$CO$_2$Me. In effect, it casts doubts on the literature values (*11*) for the heat of formation of the PhCO$_2 \cdot$ radical.

Phenyl-Carbon Bond in Phenyl-, Methyl-, and Ethyl Benzoate. Full experimental data that would permit the strength of the phenyl-carbon bond in phenyl benzoate to be determined independently are lacking. An estimate can be based on data for methyl benzoate and ethyl benzoate. Values for D_3 in these compounds may be derived from their heats of formation (ΔH_f(PhCO$_2$Me) = −72 kcal. per mole; ΔH_f(PhCO$_2$Et) = −79 kcal. per mole) and from the heats of formation (*7*) of the methoxycarbonyl and ethoxycarbonyl radicals. We take $\Delta H_f(\cdot\text{CO}_2\text{Me}) \simeq -52$ kcal. per mole, and $\Delta H_f(\cdot\text{CO}_2\text{Et}) \simeq -57$ kcal. per mole, revising the original values (*2*) upward in step with the "high" values (*12*) for the corresponding alkyl carbonyl radicals. The bond dissociation energies D_3(Ph—CO$_2$Me) and D_3(Ph—CO$_2$Et) are thus *ca.* 99 kcal. per mole and 101 kcal. per mole, respectively.

In the benzoate esters, inductive effects analogous to those discussed above might be expected to produce a 3 to 5 kcal. weakening from D_3(R—CO$_2$Ph) to D_3(Ph—CO$_2$Ph); on this basis, the strength of the phenyl-carbon bond in phenyl benzoate is estimated to be about 96 kcal. per mole. Although it is not possible to estimate reliably the error in D_3(Ph—CO$_2$Ph), it is not likely to be very large.

Phenyl-Oxygen Bond in the Phenoxycarbonyl Radical. From the relationship D_3(Ph—CO$_2$Ph) + $D_4(\cdot\text{CO}_2$—Ph) = 102 kcal. per mole and the value derived above for D_3 of 96 kcal. per mole, it follows that in the phenoxycarbonyl radical the strength of the bond joining the phenyl group to the OCO group, $D_4(\cdot\text{CO}_2$—Ph) = 102 −96 = 6 kcal. per mole. In turn, the heat of formation of the phenoxycarbonyl radical given by ΔH_f(PhCO$_2$Ph) − ΔH_f(Ph·) + D_3(Ph—CO$_2$Ph) is $\Delta H_f(\cdot\text{CO}_2\text{Ph})$ = −19 kcal. per mole.

Useful comparisons may be made here with the corresponding alkoxycarbonyl radicals, ·CO$_2$Me and ·CO$_2$Et, for which values (*6*) for D_4 of around −10 kcal. per mole have been advanced. These radicals ·CO$_2$R, like the isomeric $RCO_2\cdot$ family, contain an "endothermic" bond. Both $PhCO_2\cdot$ and $\cdot CO_2Ph$ require energy for dissociation.

Some data exist in the literature for the hydroxycarbonyl radical, $\cdot CO_2H$. Back and Sehon (2) concluded that $\Delta H_f(\cdot CO_2H)$ was -62 kcal. per mole, and $D(\cdot CO_2-H)$ was 20 kcal. per mole. This bond strength, which would place ($\cdot CO_2H$) beyond the ($\cdot CO_2Ar$) family thermochemically speaking, seems exceedingly high.

Back and Sehon based their deductions on a kinetic study of the pyrolysis of gaseous phenylacetic acid, which they found to be a homogeneous process obeying first-order kinetics with a velocity constant given by $(k/\text{sec.}^{-1}) = 10^{12.9} \exp(-55000/RT)$. They considered bond fission into benzyl and hydroxycarbonyl radicals to be the rate-determining step and identified their experimental activation energy with the bond dissociation energy $D(PhCH_2-CO_2H)$. Their other thermochemical assumptions were an old experimental value for the heats of formation of the parent acid, $\Delta H_f(PhCH_2CO_2H) = -75$ kcal. per mole and an "intermediate" value for the benzyl radical $\Delta H_f(PhCH_2\cdot) = 42$ kcal. per mole. The former is consistent with empirical estimates based on the heats of formation of toluene and acetic acid, but even if the latter value is revised (3) nearer $\Delta H(PhCH_2\cdot) = 45$, the essential difficulty remains. The mechanism of pyrolysis seems to merit further investigation.

Phenoxyl-Carbon Bond in Phenyl Benzoate. To derive a value for the dissociation energy of the phenoxy-carbon bond in phenyl benzoate, $D_5(PhCO-OPh)$, the heat of formation of the benzoyl and phenoxyl radicals must be known. The currently recommended (12) value for benzoyl is $\Delta H_f(PhCO) = 16$ kcal. per mole. For phenoxyl, an estimate can be based on the heat of formation (9) of gaseous phenol (-23.05 kcal. per mole) and kinetic experiments on the abstraction of hydrogen atoms. Work with methyl radicals (10, 15, 16) suggests that the dissociation energy in phenol is markedly less than in the alcohols. An upper limit of ca. 95 kcal. per mole is placed by Lossing's electron impact studies. New work on substituted phenols (13) suggests that in them D is ca. 88 kcal. per mole. A similar value has been proposed for phenol by Benson (4) on the basis of evidence from inhibition studies involving a chain of resemblance linking $D(PhO-H)$ to $D(HO_2-H)$ and $D(Me_3C-H)$. If the strength of the phenoxyl-hydrogen bond is 88 kcal. per mole, the heat of formation of phenoxyl is 13 kcal. per mole. The strength of the phenoxyl-carbon bond in phenyl benzoate may now be calculated:

$$D_5(PhCO-OPh) = \Delta H_f(PhO\cdot) + \Delta H_f(PhCO) - \Delta H_f^\circ(PhCO_2Ph) = 64 \text{ kcal. per mole}$$

It is the weakest bond in the ester.

In the phenoxycarbonyl radical, the same bond is further weakened by the electronic rearrangements accompanying the formation of the products:

$$D_6(\cdot CO\text{---}OPh) = 7 \text{ kcal. per mole}$$

Reaction Thermochemistry and Mechanisms for Phenyl Esters

Table I summarizes the values recommended here and indicates the magnitude of their uncertainties. These values have many implications outside the chemistry of phenyl benzoate itself. For example, some bond dissociation energies (kcal. per mole) in other species are: $D(Ph\text{---}COPh) = 85$; $D(PhCO\text{---}COPh) = 62$; $D(PhCO_2\text{---}COPh) = 77$; $D(PhO\text{---}CO_2Ph) = 62$; $D(PhCO\text{---}NH_2) = 84$.

Table I. Recommended Values for Bond Dissociation Energies and Radical Thermochemistry in Aromatic Oxidation

Bond Dissociation Energies		Heats of Formation (g) at 25°C.	
Bond	D, (kcal. per mole)	Species	ΔH_f, (kcal. per mole)
$PhCO_2$—Ph	94 ± 4	$PhCO_2Ph$	-35 ± 1
Ph—CO_2Ph	96 ± 5	$PhCO_2$	-21.4 ± 3
PhCO—OPh	64 ± 5	$\cdot CO_2Ph$	-19
PhO—$\dot{C}O$	7 ± 5	$Ph\dot{C}O$	16 ± 5
$O\dot{C}O$—Ph	6 ± 5	PhO·	13 ± 3
Ph—CO_2·	7 ± 3	Ph·	80 ± 2
Ph—$\dot{C}O$	30 ± 5		

Warbentin (19) has indicated how the radical thermochemistry involved can assist in assigning a mechanism for the thermal decomposition of phenyl oxalate. Exclusive initial fission of either the one C—C or the two C—O bonds would lead, via PhOCO or OCCO intermediates, to high CO_2 or high CO yields. In experiments lasting about 75 hours at 500°K. in diphenyl ether, comparable amounts of CO (13%) and CO_2 (9%) are formed. The following steps, possibly concerted:

$$PhOCOCO_2Ph \rightarrow PhO\cdot + \cdot COCO_2Ph$$

$$\cdot COCO_2Ph \rightarrow CO + \cdot CO_2Ph$$

$$\cdot CO_2Ph \rightarrow CO_2 + Ph\cdot$$

could afford a reasonable explanation.

In phenyl benzoate itself, Miller (14) reported results of radiolysis yielding 10 times as much CO as CO_2, which is wholly consistent with predominant split at the weakest bond, PhO—COPh. The situation is similar in phenyl acetate but reversed in benzyl benzoate (and in benzyl acetate), again in agreement with thermochemical requirements.

Conclusions

All the radicals concerned are important in oxidation processes of aromatic molecules, and this paper aims to offer a starting point for the thermochemical dissection of such oxidation processes. It is also hoped that it may stimulate further investigations of radical thermochemistry, especially in the aromatic field. Areas for fruitful work on bond energies include the formate and carbonate esters, including phenyl formate and phenyl carbonate, and the bond strengths in formic acid itself and in benzaldehyde.

Acknowledgments

It is a pleasure to acknowledge the combustion calorimetry carried out by G. P. Adams, D. H. Fine, and P. G. Laye and the extremely useful criticism of S. W. Benson.

Literature Cited

(1) Adams, G. P., Fine, D. H., Gray, P., Laye, P. G., *J. Chem. Soc.* **1967**, B720.
(2) Back, M., Sehon, A., *Can. J. Chem.* **38**, 1261 (1960).
(3) Benson, S. W., Golden, D. M., Rodgers, A. S., personal communication, 1967.
(4) Benson, S. W., *J. Am. Chem. Soc.* **87**, 972 (1965).
(5) Fine, D. H., Gray, P., *Combust. Flame* **11**, 71 (1967).
(6) Gray, P., Thynne, J. C. J., *Nature* **191**, 1357 (1961).
(7) Gray, P., Thynne, J. C. J., Shaw, R., "Progress in Reaction Kinetics," Vol. 4, p. 63, Pergamon Press, New York, 1967.
(8) Gray, P., Williams, A., *Chem. Rev.* **59**, 239 (1959).
(9) Green, J. H. S., *Quart. Rev. Chem. Soc.* **15**, 125 (1961).
(10) Herod, A. A., Ph.D. Thesis, Leeds, 1967.
(11) Jaffe, L., Prosen, E. J., *J. Chem. Phys.* **27**, 416 (1957).
(12) Kerr, J. A., *Chem. Rev.* **66**, 465 (1966).
(13) Mahoney, L. R., Da Rooge, N. A., "Preprint, International Oxidation Symposium," Vol. II, p. 585, 1967.
(14) Miller, A. A., *J. Phys. Chem.* **69**, 1077 (1965).
(15) Mulcahy, M. F. R., Williams, D. J., *Australian J. Chem.* **18**, 20 (1965).
(16) Shaw, R., Thynne, J. C. J., *Trans. Faraday Soc.* **62**, 104 (1966).
(17) Stull, D. R., *Ind. Eng. Chem.* **39**, 533 (1947).
(18) Wadsö, I., *Acta. Chem. Scand.* **14**, 561 (1960).
(19) Warbentin, J., personal communication, 1967.

RECEIVED October 9, 1967.

22

Thermochemistry of Oxidation Reactions

S. W. BENSON and R. SHAW

Department of Thermochemistry and Chemical Kinetics,
Stanford Research Institute, Menlo Park, Calif.

> *The following heats of formation (kcal. per mole) have been calculated, MeO, $+3.5$; EtO, -4.0, tert-BuO, -21.6; and tert-BuO$_2$, 18.6. From the radical heats of formation, the following bond strengths (kcal. per mole) were obtained: RO—H, 104 ± 1; R—OH, 91.5 ± 1; R—OMe, 81 ± 1; R—O, 89 ± 1; R—O$_2$R, 69 ± 1; and R—O$_2$, 26 ± 1, the variance depending on R. Some unusual trends with X in the bond strengths Me—X, Et—X, iso-Pr—X, and tert-Bu—X are discussed. Using group additivity to calculate heats of formation, the bond strengths in some interesting polyoxide molecules and free radicals have been calculated.*

In 1965 Benson (3) discussed the basic thermochemistry and kinetics of combustion reactions. Since then additional thermochemical data have become available and the Benson and Buss Group Additivity Tables (7) have been revised (8).

The object of this paper is to update the earlier calculations (3) of the heats of formation of RO and RO$_2$ and to discuss the resulting implications.

Heats of Formation

Table I gives the recently measured heats of formation of dimethyl, diethyl, and di-*tert*-butyl peroxides. Dimethyl peroxide had not previously been burned, but the values for diethyl and di-*tert*-butyl agreed well with previous work, and the small differences from them were in the expected direction (the heats of formation of unstable compounds, such as peroxides, tend to become more negative with successive measurement). The experimental values were used as standards for the calculation (8) of the group value O—O,C and Table I shows that group additivity is obeyed within the precision of the data.

Table I. Heats of Formation (kcal. per mole)

ROOR	ΔH_f Exptl.	ΔH_f Calcd.	ΔE, Exptl.	ΔH, Exptl.	$\Delta H_f(RO^0)$, Exptl.
MeOOMe	−30.0	−29.2	36.1	36.9	+3.5
EtOOEt	−46.1	−46.2	37.3	38.1	−4
tert-BuOO-tert-Bu	−81.5	−81.6	37.5	38.3	−21.6

The pyrolysis of di-tert-butyl peroxide is one of the most studied (11) and cleanest reactions in gas kinetics. The activation energy is well established as 37.5 kcal. per mole. Assuming that the back-activation energy is zero and adding RT ($T \sim 400°K$.) to convert energy change to enthalpy change, we obtain $\Delta H = 38.3$ kcal. per mole. Di-tert-butyl peroxide has four gauche interactions (8) worth a total of 1.2 kcal. per mole. This strain is relieved when the molecule decomposes. Dimethyl and diethyl peroxides do not have any gauche interactions, so it is reasonable to expect that their activation energies for the decomposition might be 1.2 kcal. per mole higher than di-tert-butyl peroxide. Despite this, the experimental activation energy for dimethyl peroxide is 36.1. Di-tert-butyl peroxide has no easily abstractable hydrogen and so cannot undergo a chain decomposition, but dimethyl peroxide can.

$$CH_3O + CH_3OOCH_3 \rightarrow CH_3OH + CH_2OOCH_3 \quad (1)$$

$$CH_2OOCH_3 \rightarrow CH_2O + CH_3O \quad (2)$$

A small contribution from the chain decomposition would be difficult to detect and would lower both the A factor and the activation energy. A similar argument applies to diethyl peroxide, and the measured (11) low A factors and activation energies are convincing evidence of the chain contribution. Leggett and Thynne (16) recently measured the A factor and activation energy for the decomposition of diethyl peroxide and found them to be normal.

The heats of formation of MeO, EtO, and tert-BuO from the relation $\Delta H_f(RO) = 1/2[\Delta H + \Delta H_f(ROOR)]$ are given in Table I, using experimental and calculated values for ΔH and $\Delta H_f(ROOR)$. The values in Table I are significantly (in the statistical sense) higher than those suggested previously (11, 12). This trend is in keeping with the tendency of "preferred" values of heats of formation of radicals to increase with time. Thus, the "best" heats of formation of CH_3, CF_3, Ph, $PhCH_2$, and allyl have increased by 2 to 10 kcal. per mole in recent years. A consequence of the increased heats of formation of RO is that the RO—H, R—OH, and R—OR bond strengths are correspondingly higher than previously accepted (see e.g., Table II).

Table II. Heats of Formation (kcal. per mole)

ROH	$\Delta H_f(ROH)$, Exptl.	$\Delta H_f(ROH)$, Calcd.	$\Delta H_f(RO^0)$, Exptl.	$D(RO—H)$, Exptl.
MeOH	−48.0	−48.0	+3.5	103.6
EtOH	−56.2	−56.5	−4	104.3
tert-BuOH	−74.7	−74.8	−21.6	105.2

The heat of formation of tert-BuO of −21.6 kcal. per mole gives ΔH_3 = +3.8 for Reaction 3.

$$(CH_3)_3CO \rightarrow CH_3COCH_3 + CH_3 \qquad (3)$$

i.e., ΔE_3 (at 400°K.) = 3.0 kcal. per mole. This is consistent with E_3 = 13 kcal. per mole (11) and E_{-3} = 10 kcal. mole^{-1} (2).

The group value of O-(O)(C) calculated from the dialkyl peroxides gives a value of −57.1 kcal. per mole for the heat of formation of tert-BuOOH, compared with the measured value (11) of −52.3 kcal. per mole. Since the calculated value is more negative, one cannot account for the difference by the otherwise reasonable assumption that the hydroperoxide decomposed a little prior to combustion. An alternative would be that group additivity did not apply to the hydroperoxides, but Benson's

Table III. Values

X	$\Delta H_f(X)^a$	$D(Me—X)$	$D(Et—X)$	$D(\text{iso-Pr}—X)$
H	52.1	104	98.3	94.5
Ph	80	102	98.9	96.7
Me	34	88.2	84.8	83.8
Et	26	84.8	82.2	80.5
MeCO	−4	81.7	79	
HCO	6.3	80	77.9	
SH	33	72.5	70	68.8
I	25.5	56.2	53.6	52.6
Br	26.7	70.2	67.5	67
Cl	28.9	83.6	81.2	79.4
F	18.9	107.9	106.8^b	105.1
NH$_2$	40	79.5	77	78.2
BF$_2$	−130	103	105	99.6
CN	109	123.9	122.6	121
NO$_2$	7.9	59.8	57.7	59.4
OH	9.4	91.4	91.6	92.2
OMe	3.5	81.5	81.2	81.3
ONO	7.9	57.5	58.1	
ONO$_2$	16.6	79.2	79.4	79.9
MeCO$_2$	−40.1	93.1	92.5	93.4

^{a} All quantities in kcal. per mole.
^{b} Calculated from $\Delta H_f(Pr^nF)$. All other $\Delta H_f(RX)$ are experimentally obtained values.

prediction (5) based on group additivity that the measured (11) activation energy for the decomposition of *tert*-butyl hydroperoxide of 37.8 kcal. per mole was too low and ought to be 42.1 kcal. per mole, has recently been substantiated (13) by Hiatt and Irwin's experimental activation energy of 43 kcal. per mole. Assuming ΔH_f (*tert*-BuOOH) = −57.1 and D(*tert*-BuO$_2$—H) = 90 kcal. per mole (3), then ΔH_f (*tert*-BuO$_2$) = −19.2, D(*tert*-Bu—O$_2$) = 25.9, and D(*tert*-BuO$_2$—*tert*-Bu) = 68.4 kcal. per mole.

Bond Strengths

A feature of the thermochemistry of oxygenated compounds is the lack of variation of RO—R' with varying R. Table III shows the values of D(R—X), where R is methyl, ethyl, isopropyl, and *tert*-butyl calculated from experimental values (8) of ΔH_f(RX), and the following values of ΔH_f(R) (kcal. per mole): Me 34, Et 26, *iso*-Pr 17.6, and *tert*-Bu 6.7. The differences in D(R—X) are also tabulated. There are three broad, slightly overlapping, groups. In the C—C group the differences are fairly large with Δ(Me—*tert*-Bu) greater than 7 kcal. per mole. There is an intermediate group where the differences are still significant but less marked. In the last group, where in every case there is a C—O bond

of D(R—X)

D(*tert*-Bu—X)	Δ(Me—Et)	Δ(Me—*iso*-Pr)	Δ(Me—*tert*-Bu)
90.9	5.7	9.5	13.1
92.1	3.1	5.3	9.9
80.4	3.4	4.4	7.8
76.9	2.6	4.3	7.9
	2.7		
	2.1		
65.8	2.5	3.7	6.7
49.4	2.6	3.6	6.8
64.0	2.7	3.2	6.2
78.6	2.4	4.2	5
	1.1	2.8	
75.3	2.5	1.3	4.2
	−2	3.4	
115.7	1.3	2.9	8.2
	2.1	0.4	
90.8	−0.2	−0.8	0.6
80.2	0.3	0.2	1.3
	−0.6		
	−0.2	−0.7	
	0.6	−0.3	

ΔH_f(Ph) (18); ΔH_f(BF$_2$) and ΔH_f(OH) (14); ΔH_f(OMe), this paper. ΔH_f(ONO$_2$) (8); all other ΔH_f(X) (4).

involved, the differences are negligible within experimental error. This is rather unusual and would suggest that there is little variation in R—O· and R—O$_2$· bond strengths for similar R groups. There is no apparent correlation with the absolute bond strengths or with the Hammett σ-meta function (15).

However, there is a trend with the "ionicity" of the C—X bond. The striking decrease in ionization potential Me 228, Et 201, iso-Pr 182, and tert-Bu 159 kcal. per mole led Benson and Bose (6) and later Benson and Haugen (9) to account successfully for the activation energies of four-center reactions by means of a simple, self-consistent, electrostatic model based on semi-ion pairs. The same principles can be applied in the present case. The basic idea is that the weakening of the C—H bond in the series Me—H, Et—H, iso-Pr—H, and tert-Bu—H is caused by increasing stabilization in the alkyl radical. The stabilizing effect of the methyl group arises from polarization of the CH$_3$ groups by the effective plus charge at the C atom nucleus.

This polarization is greater in the radical than in the parent molecule because of a greater "spreading" of the valence electron density in the radical. Thus, in the transition from R$_3$C—H to R$_3$C the R groups see a larger effective positive charge at the C atom in the radical. Although such a polarization is different in principle from hyperconjugation, in terms of stabilization and charge distribution it appears formally equivalent.

In a substituted hydrocarbon the C—X bond will be polarized.

$$\begin{array}{c} R \\ R\!-\!\!\!\!\!\!>\!\!C\overset{+\delta\quad -\delta}{-\!\!-\!\!-} X \\ R \end{array}$$

Table IV

	Δ(Me—Et)	Δ(Me—iso-Pr)	Δ(Me—tert-Bu)	$r_{CX}{}^a$		$\mu_o{}^b$	$(\mu_o/r_{CX})^2$
		Kcal. per Mole		A	D		
H	5.7	9.5	13.1	1.09	0.4		0.14
I	2.6	3.6	6.8	2.16	1.64		0.58
Br	2.7	3.2	6.2	1.91	1.79		0.88
Cl	2.4	4.2	5.0	1.7	1.94		1.3
F	1.1	2.8		1.69	1.8		1.7
SH	2.5	3.7	6.7	1.82	1.26		0.48
NH$_2$	2.5	1.3	4.2	1.48	1.29		0.76
OH	−0.2	−0.8	0.6	1.42	1.71		1.45

[a] Values of r are for MeX (10).
[b] Values of μ_o for MeX (17).

Table V. Standard Bond Dissociation Enthalpies
DH^0 (X—Y) of Some Oxide and Polyoxide Molecules
and Free Radicals[a]

X–	–O·	–O$_2$·	–O$_3$·	–O$_2$H	–O$_3$H	–O$_4$H	–O$_2$CH$_3$	–O$_3$CH$_3$	–O$_4$CH$_3$
H–	102.4	47	62	90	90	90	90	90	90
CH$_3$–	90	27	42	70	70	70	71	71	71
tert-Bu–	88	26	41	69	69	69	70	70	70
HO–	64	–15	0	28	28	28	28	28	28
CH$_3$O–	56	–23	–7	20	21	21	22	22	22
tert-BuO–	57	–21	–6	22	22	22	22	22	22
HO$_2$–	40	–38	–23	5	5	5	5	5	5
CH$_3$O$_2$–	41	–38	–23	5	5	5	5	5	5

[a] Values based on rule of group additivity of thermochemical properties, using an assumed value, $\Delta H_f°$, for O—(O)$_2$ group of 19 ± 2 kcal., and further assuming that in the molecular series H—O$_n$H, $DH°$(H—O$_n$H) = 90 ± 2 kcal. for $n \geq 2$.

The dipole moment, μ_0, is given approximately by $\mu_0 = \delta r_{CX}$. The partial charge on the carbon atom will induce dipoles in the adjacent R groups with resultant stabilization of the molecule.

If R is methyl, we can write:

$$\text{Stabilization energy (per methyl group)} = \frac{\mu^2_{induced}}{2\alpha_{Me}} = E_{CH_3}$$

where α_{Me} is the polarizability of the methyl group; $\mu_{induced} = (\alpha_{Me})$ (field at the methyl group). The field at the methyl group $\sim \delta/(r_{C-Me})^2$ where r_{C-Me} is the bond distance; hence, $E_{CH_3} = \alpha_{Me}\delta^2/2(r_{C-Me})^4 = \text{const.}(\mu_0/r_{CX})^2$.

At the present time no attempt has been made to deal with the more complex substituents, but Table IV shows that in the simple cases considered there, the difference in R—X bonds is clearly reduced with increasing values of $(\mu_0/r_{CX})^2$.

Table V gives the bond strengths in some interesting polyoxide molecules and free radicals. The calculations were based on the reasonable assumption that group additivity is obeyed.

Literature Cited

(1) Baker, G., Littlefair, J. H., Shaw, R., Thynne, J. C. J., *J. Chem. Soc.* **1965**, 1286.
(2) Benson, S. W., "Foundations of Chemical Kinetics," McGraw-Hill, New York, 1960.
(3) Benson, S. W., *J. Am. Chem. Soc.* **87**, 972 (1965).
(4) Benson, S. W., *J. Chem. Educ.* **42**, 502 (1965).
(5) Benson, S. W., *J. Chem. Phys.* **40**, 1007 (1964).
(6) Benson, S. W., Bose, A. N., *Ibid.* **39**, 3463 (1963).
(7) Benson, S. W., Buss, J., *Ibid.* **29**, 546 (1958).

(8) Benson, S. W., Cruickshank, F. R., Golden, D. M., Haugen, G. R., O'Neal, H. E., Rodgers, A. S., Walsh, R., Shaw, R., unpublished manuscript.
(9) Benson, S. W., Haugen, G. R., *J. Am. Chem. Soc.* **87**, 4036 (1965).
(10) *Chem. Soc. (London) Spec. Pub.* **11** (1958).
(11) Gray, P., Shaw, R., Thynne, J. C. J., *Progr. Reaction Kinetics* **4**, 65 (1967).
(12) Gray, P., Williams, A., *Chem. Rev.* **59**, 239 (1959).
(13) Hiatt, R., Irwin, K. C., *J. Org. Chem.* **33**, 1436 (1968).
(14) "JANAF Thermochemical Tables," Dow Chemical Co., Midland, Mich.
(15) Leffler, J. E., Grunwald, E., "Rates and Equilibria of Organic Reactions," p. 173, Wiley, New York, 1963.
(16) Leggett, C., Thynne, J. C. J., *Trans. Faraday Soc.* **63**, 2504 (1967).
(17) McLellan, A. L., "Tables of Experimental Dipole Moments," W. H. Freeman & Co., San Francisco, Calif., 1963.
(18) Rodgers, A. S., Golden, D. M., Benson, S. W., *J. Am. Chem. Soc.* **89**, 4578 (1967).

RECEIVED October 2, 1967. Work supported in part by Grant AP-00353-03 from the Air Pollution Division, U. S. Public Health Service.

Inhibition

K. U. INGOLD
Session Chairman

23

Inhibition of Autoxidation

K. U. INGOLD

Division of Applied Chemistry, National Research Council, Ottawa, Canada

> *Methods for inhibiting the autoxidation of organic substrates are reviewed. Antioxidants can be divided into two broad groups according to whether they reduce the rate of chain initiation (preventive antioxidants) or interfere with the normal propagation process by reacting with free radicals (chain-breaking antioxidants). Preventive antioxidants can be subdivided into peroxide decomposers, metal ion deactivators, and ultraviolet light deactivators. Chain-breaking antioxidants can be subdivided according to whether they add, donate hydrogen, or donate an electron to the chain-propagating radicals. The detailed mechanisms by which these various classes of antioxidants inhibit oxidation are discussed.*

The autoxidation of most organic substances is a free radical chain process. The over-all oxidation rate depends on the rate of chain initiation, chain propagation, and chain termination. The rate can generally be reduced, and/or the length of the induction period can be increased by adding relatively low concentrations of compounds which contain certain specific functional groups. These compounds have been indiscriminately referred to as antioxidants (primary and secondary), inhibitors, retarders, deactivators, stabilizers, etc. The term "antioxidant" is probably most widely used and I employ it as a general term for all compounds which inhibit oxidation—*i.e.*, for all compounds which reduce the rate of attack of oxygen on a substrate.

In theory, at least, antioxidants may be divided into two broad groups according to whether they reduce the rate of chain initiation (the so-called preventive antioxidants) or interfere with the normal propagation process by reacting with free radicals (the chain-breaking antioxidants). Since the two groups of antioxidants interfere at different points in the oxidation process, they have a mutually reinforcing effect on one another. When they are used together, the over-all inhibiting effect is generally greater than the sum of the individual effects from each antioxidant. This phenomenon is referred to as synergism.

In practice, of course, many antioxidants have both preventive and chain-breaking antioxidant properties. Such compounds exhibit autosynergism, and they are generally very effective antioxidants. Autosynergism is usually achieved by incorporating into a single compound two or more functional groups, one of which is known to act predominantly by a preventive mechanism and the other by a chain-breaking mechanism. Considerable interest in antioxidants is currently centered about empirically discovered compounds such as the dialkyl dithiophosphates which were originally believed to function by a single mechanism but which, in fact, function by two mechanisms. Under many circumstances these compounds must owe much of their efficiency to autosynergism.

Synergistic behavior by two antioxidants is not confined to compounds which inhibit by entirely different mechanisms—for example, two chain-breaking phenolic antioxidants may synergize one another. This homosynergism is caused by the suppression of the unfavorable chain propagation reactions of one phenoxy radical by a hydrogen atom transfer from the second phenol.

It is considerably more difficult to inhibit oxidation in the gas phase than in the liquid phase. At the high temperatures of gas-phase oxidations the rates of the chain-propagating and branching reactions are increased to a greater extent than the rates of the chain-terminating reactions. Initiation by surfaces can also constitute a serious problem. The majority of liquid-phase antioxidants which are effective at high temperatures are too involatile to be useful in the gas phase. However, inhibition can be achieved with aliphatic amines, which are generally rather ineffective inhibitors of low temperature liquid-phase oxidations. The mechanisms by which the different types of antioxidants inhibit oxidation are briefly described below.

Preventive Antioxidants

Preventive antioxidants reduce the rate of chain initiation, and they can be subdivided into three main categories.

Peroxide Decomposers. These convert hydroperoxides (ROOH) to nonradical products.

ROOH + D → molecular products (generally ROH + DO or R'CO + H_2O)

This reduces the importance of the various hydroperoxide decomposition reactions which give free radicals. A peroxide decomposer may convert the hydroperoxide to a product which is itself an antioxidant—

e.g., Lewis acids convert cumene hydroperoxide to phenol and acetone.

$$C_6H_5(CH_3)_2COOH \xrightarrow{H^+} C_6H_5OH + (CH_3)_2CO$$

Metal Ion Deactivators. These generally chelate ions of copper and the transition metals which might otherwise catalyze the decomposition of hydroperoxides to free radicals,

$$ROOH + M^{n+} \rightarrow RO\cdot + OH^- + M^{(n+1)+}$$

$$ROOH + M^{(n+1)+} \rightarrow RO_2\cdot + H^+ + M^{n+}$$

Ultraviolet Light Deactivators. These absorb short-wavelength radiation which might otherwise generate free radicals—*e.g.*,

$$R_2CO \xrightarrow{h\nu} R_2\dot{C}\dot{O}$$

$$ROOH \xrightarrow{h\nu} RO\cdot + \cdot OH$$

The mechanisms of inhibition by peroxide decomposers, metal deactivators, and ultraviolet absorbers are fairly well understood in general terms, although many details of the individual reactions remain to be elucidated. Classifying a preventive antioxidant into one of the three categories above will only rarely describe its entire function. The dual behavior of dialkyl dithiophosphates in the liquid phase has been mentioned. Many other phosphorus- and sulfur-containing antioxidants commonly classified as peroxide decomposers can also act as chain breakers. Similarly, the structure of many metal deactivators and ultraviolet absorbers indicates that they must also have some chain-breaking activity.

Chain-Breaking Antioxidants

Chain-breaking antioxidants which interfere with the normal propagation processes may react with peroxy radicals, $RO_2\cdot$, or, more rarely, with the carbon radical, $R\cdot$. The antioxidant may react with the propagating radical by addition, by hydrogen transfer, or by electron transfer. The chain can be terminated directly, but more commonly a new radical is formed, which either continues the chain at a reduced rate or terminates a second chain.

Among chain-breaking antioxidants the stable dialkyl nitroxides such as di-*tert*-butyl nitroxide (I) and 2,2,6,6-tetramethyl-4-pyridone nitroxide (II) inhibit oxidation by the simplest mechanism and therefore exhibit

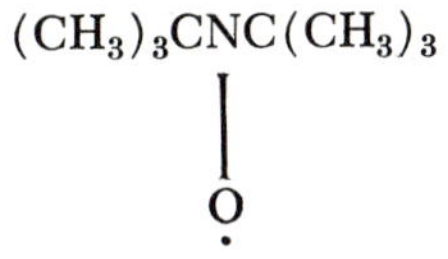

(CH$_3$)$_3$CNC(CH$_3$)$_3$
|
O·

I

[Structure II: 2,2,6,6-tetramethyl-4-oxopiperidine-1-oxyl]

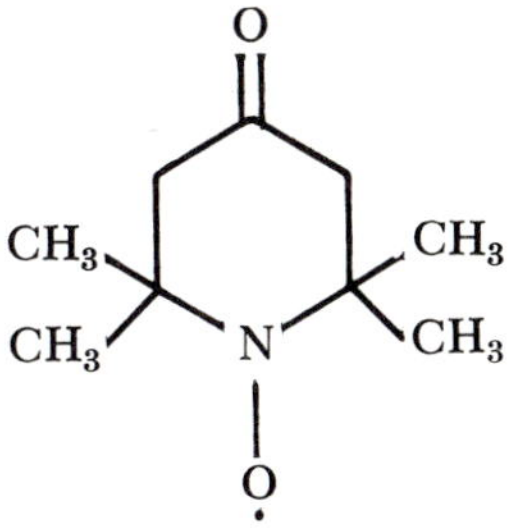

II

the simplest kinetics. These nitroxides react rapidly with R· radicals to give stable molecular products.

$$R_2'NO· + R· \rightarrow R_2'NOR$$

They do not react with RO$_2$· radicals. The inhibited oxidation rate is proportional to the oxygen pressure and inversely proportional to the nitroxide concentration, but it is independent of the substrate concentration—i.e., rate $\propto$ [O$_2$]/[nitroxide]. The nitroxides react slowly with hydrocarbons and hydroperoxides and therefore have little tendency to initiate oxidation. However, they are of disappointingly little practical value compared with more conventional chain-breaking antioxidants because they must compete with molecular oxygen for the R· radicals, and the

$$R· + O_2 \rightarrow RO_2·$$

reaction is extremely rapid.

Stable diaryl nitroxides such as 4,4'-dimethoxydiphenyl nitroxide (III) are somewhat better antioxidants than the dialkyl nitroxides. The III inhibited rate depends on both the oxygen pressure and the substrate

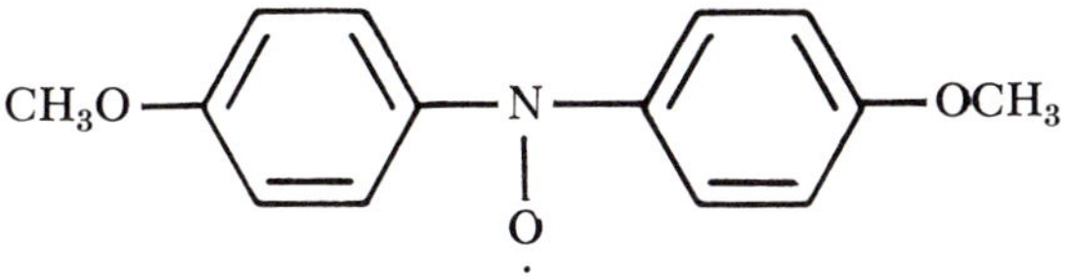

CH$_3$O—⟨C$_6$H$_4$⟩—N(O·)—⟨C$_6$H$_4$⟩—OCH$_3$ (III)

concentration, which implies that III reacts with both R· and RO$_2$· radicals. The reaction with RO$_2$· is much slower than the reaction with R·. Except at rather low oxygen pressures the nitroxide III is a less efficient inhibitor than the corresponding diphenylamine (IV) and hydroxylamine (V).

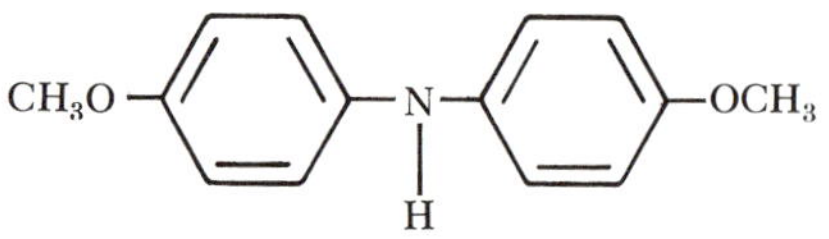

CH₃O—⟨⟩—N(H)—⟨⟩—OCH₃   CH₃O—⟨⟩—N(O—H)—⟨⟩—OCH₃

     IV 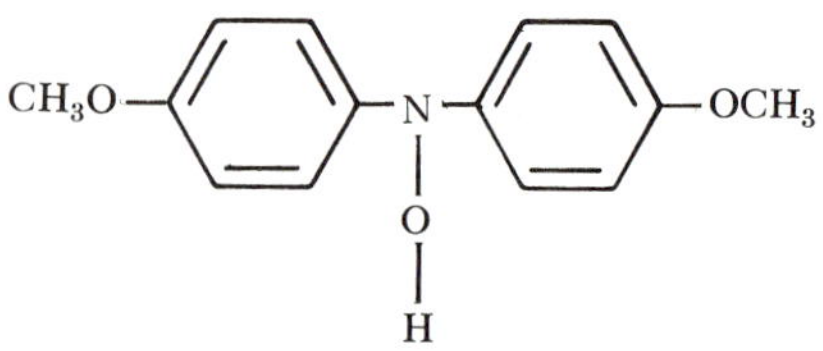   V

Carbon black and many polynuclear hydrocarbons are effective inhibitors of oxidation. Their antioxidant properties are believed to arise from their ability to trap free radicals.

The most useful and most thoroughly studied chain-breaking antioxidants are phenols and aromatic amines. These compounds can generally transfer a hydrogen atom to a peroxy radical.

$$RO_2\cdot + AH \rightarrow ROOH + A\cdot$$

The rate of transfer is accelerated by electron-releasing substituents on the aromatic ring of the antioxidant and retarded by steric protection of the labile hydrogen or its replacement by deuterium. The subsequent fate of the radical A· determines the over-all kinetics of the inhibited reaction and the practical usefulness of the antioxidant. If A· is a fairly stable phenoxy radical, it will probably add a peroxy radical or dimerize.

$$RO_2\cdot + A\cdot \rightarrow ROOA$$

$$A\cdot + A\cdot \rightarrow A_2$$

In this case the kinetics are simple. The inhibited rate is proportional to the substrate concentration and inversely proportional to the antioxidant concentration—*i.e.*, rate $\propto$ [RH]/[AH]. More generally, phenoxy radicals which are not sterically hindered will enter into chain transfer reactions, and the kinetics may become extremely complicated. Chain transfer with the substrate is usually slow, but it can be rapid with hydroperoxides.

$$A\cdot + RH \xrightarrow{slow} AH + R\cdot$$

$$A\cdot + ROOH \xrightarrow{fast} AH + RO_2\cdot$$

Aromatic primary and secondary amines behave generally like the phenols. However, the over-all process is more complex because an additional series of termination and transfer reactions is introduced by the rapid formation of nitroxides in the following reaction.

$$RO_2\cdot + A\cdot \rightarrow RO\cdot + AO\cdot$$

e.g., $RO_2^{\cdot} + \text{Ph–N(H)–Ph} \longrightarrow RO \cdot + \text{Ph–N(O}^{\cdot}\text{)–Ph}$

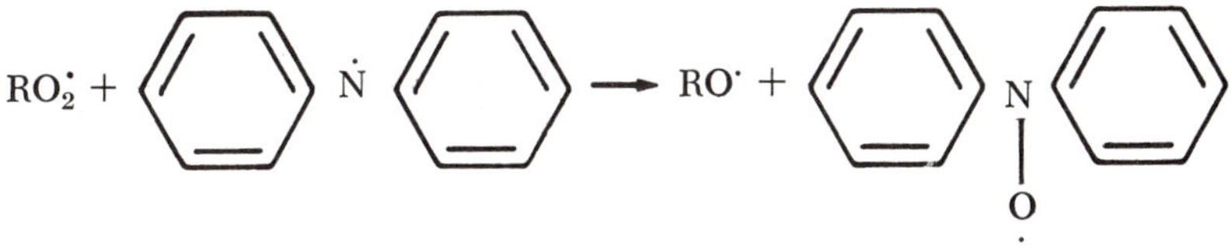

The amine antioxidants therefore give rise to both nitrogen and nitroxide radicals. The reactions of these two species are still not well known.

Several other types of antioxidants can inhibit oxidation by donating a hydrogen atom to a peroxy radical. Howard (15) has described the interesting case of inhibition by a second hydrocarbon or its hydroperoxide. In the particular case he discussed, the inhibition of the oxidation of cumene by Tetralin (TH) or Tetralin hydroperoxide (TOOH) occurs because the tetralylperoxy radical ($TO_2^{\cdot}$) reacts with a cumylperoxy radical ($CO_2^{\cdot}$) much more rapidly than two cumylperoxy radicals react with one another. The inhibition sequence can be represented by the following reactions.

$$CO_2^{\cdot} + TH \xrightarrow{O_2} COOH + TO_2^{\cdot}$$
$$CO_2^{\cdot} + TOOH \rightarrow COOH + TO_2^{\cdot}$$
$$CO_2^{\cdot} + TO_2^{\cdot} \rightarrow COH + \alpha\text{-tetralone} + O_2$$

Triarylmethanes also inhibit oxidation by hydrogen transfer to a peroxy radical. In this case it is the triarylmethyl radical which traps the second peroxy radical.

$$RO_2^{\cdot} + Ph_3CH \rightarrow ROOH + Ph_3C^{\cdot}$$
$$RO_2^{\cdot} + Ph_3C^{\cdot} \rightarrow Ph_3COOR$$

The inhibition efficiency of the triarylmethanes decreases as the oxygen partial pressure is increased because of a decrease in the steady-state concentration of the triarylmethyl radicals.

$$Ph_3C^{\cdot} + O_2 \rightleftarrows Ph_3COO^{\cdot}$$

An inhibition mechanism involving electron transfer between a chain-propagating radical and the antioxidant has frequently been suggested but has rarely been identified with any certainty. This process remains one of the least understood of all inhibition mechanisms. Probably the most clear-cut example of inhibition by one electron transfer (either partial or complete) has come from studies of metal-catalyzed oxidations. Many workers have reported that under certain conditions transition metals may inhibit rather than catalyze oxidations. Cobalt, manganese, and copper are particularly prominent in this respect.

$$RO_2^{\cdot} + Co^{2+} \rightarrow RO_2^{-}Co^{3+} \text{ or } RO_2^{\delta-}Co^{(2+\delta)+}$$

$$RO_2{\cdot} + Mn^{2+} \rightarrow RO_2{^-}Mn^{3+} \text{ or } RO_2{^{\delta-}}Mn^{(2+\delta)+}$$

The inhibition of hydrocarbon oxidation by aromatic tertiary amines which contain no labile hydrogen, such as N,N-dimethylaniline and N,N,N',N'-tetramethyl-p-phenylenediamine, has been assigned to an electron-transfer process. However, this seems rather unlikely as pyridine

Table I. Rate Constants for Some Reactions of

Reaction[a]	Antioxidant (A and AH)
$R{\cdot} + A{\cdot} \rightarrow RA$	Me_4-4-piperidone nitroxide (II) $(MeO)_2$-diphenyl nitroxide (III)
$RO_2{\cdot} + A{\cdot} \rightarrow ROOA$	Phenoxy $(MeO)_2$-diphenyl nitroxide (III)
$A{\cdot} + A{\cdot} \rightarrow A_2$	4-MeO-phenoxy Phenoxy Anilino (?)
$RO_2{\cdot} + A \rightarrow RO_2{^-}A^+$	Manganous stearate Zinc diisopropyl dithiophosphate
$RO_2{\cdot} + AH \rightarrow ROOH + A{\cdot}$	Phenol 4-MeO-phenol 2,6-$tert$-Bu_2-4-methylphenol 2,4,6-$tert$-Bu_3-phenol (VIH) N-Methylaniline Diphenylamine $(MeO)_2$-diphenylamine (IV) $(MeO)_2$-diphenylhydroxylamine (V) Zinc diisopropyl dithiophosphate
$RO_2{\cdot} + R'OOH \rightarrow ROOH + R'O_2{\cdot}$	(Hydroperoxide)
$A{\cdot} + ROOH \rightarrow AH + RO_2{\cdot}$	Phenoxy 4-MeO-phenoxy $tert$-Bu_3-phenoxy (VI)
$A{\cdot} + A'H \rightarrow AH + A'{\cdot}$	$tert$-Bu_3-phenoxy + 4-MeO-phenol $tert$-Bu_3-phenoxy + phenol $tert$-Bu_3-phenoxy + 3-chloroaniline $(MeO)_2$-diphenyl nitroxide + 2,6-$tert$-Bu_2-phenol
$A{\cdot} + RH \rightarrow AH + R{\cdot}$	4-MeO-phenoxy + 9,10-dihydroanthracene Phenoxy + Tetralin $tert$-Bu_3-phenoxy + ethylbenzene $(MeO)_2$-diphenylnitroxide + ethylbenzene

[a] Addition products shown for radical-radical reactions, but disproportionation products may also be formed.

and triphenylamine which also contain no labile hydrogen do not inhibit oxidations under similar conditions.

The radical addition and hydrogen transfer mechanisms of inhibition by chain-breaking antioxidants are now reasonably well understood in both qualitative and quantitative terms. The electron-transfer mechanism of inhibition deserves greater attention.

Chain-Breaking Antioxidants in the Liquid Phase

$\log_{10} k$, (liters/mole/sec.)	Temp., °C.	Ref.
7–8	60, 65	(3, 16)
7–8	65	(3)
~9	57	(23)
4.7	65	(3)
≤ 7.2	60	(19[b])
~9	25	(7, 17)
~9	25	(17)
5.3	60	(8, 9)
1.6	60	(5)
3.7	65	(2, 11)
5.0	65	(2, 11)
4.2	40	(1, 10, 13, 14)
4.2	65	(2, 12)
3.6	65	(3)
4.6	65	(2, 3)
5.5	65	(3)
5.7	65	(3)
2.6	60	(5)
~3	30–56	(15)
~5	57	(23)
≤ 3.3	60	(19,[b] 20[b])
−0.2 ± 0.5	21	(18, 19, 20)
3.8	24	(6, 19, 20)
0.8	24	(6, 19, 20)
−1.2	24	(18)
−3.5	60	(4)
≤ 2.1	60	(19,[b] 20[b])
~0	57	(23)
−5	60	(22)
−6.5	60	(21)

[b] Assuming $k \leq 10^9$ liters per mole per second for reaction of 4-methoxyphenoxy radical with 9,10-dihydroanthracenylperoxy radical (19, 20).

Conclusions

The main chemical types of antioxidants have generally been discovered by a purely empirical approach. Of course, this has meant that our understanding of inhibition chemistry has lagged behind the technology of inhibition. The development of improved antioxidants and synergistic combinations of antioxidants depends increasingly upon a better understanding of the mechanism of inhibition. Fundamental studies of inhibition therefore now tend to concentrate on the determination of the rate constants of the various elementary reactions involved in inhibited systems since this allows the relative importance of the elementary reactions to be estimated quantitatively. Table I lists an arbitrary selection of rate constants for the reactions of a wide variety of chain-breaking antioxidants in the liquid phase. The activation energies for most of the reactions are small (generally <8 kcal. per mole compared with 8 to 15 kcal. per mole for the rate-determining propagation step of oxidation). The rate constants obtained at different temperatures can therefore roughly be compared directly with one another. Only in the reaction

$$A\cdot + RH \rightarrow AH + R\cdot$$

is the structure of R likely to have a very large effect on the rate constants.

The most useful chain-breaking antioxidants react with peroxy radicals. Their rate constants are in the range 10^4 to 10^6 liters per mole per second for this reaction. The radicals formed from the antioxidants (if any) must be reactive towards other free radicals but unreactive in chain transfer.

Literature Cited

(1) Berger, H., et al., ADVAN. CHEM. SER. **75**, 346 (1968).
(2) Brownlie, I. T., Ingold, K. U., *Can. J. Chem.* **44**, 861 (1966).
(3) *Ibid.*, **45**, 2419, 2427 (1967).
(4) Buchachenko, A. L., Sykhanova, O. P., Kaloshnikova, L. A., Neiman, M. B., *Kinetika i Kataliz* **6**, 601 (1965).
(5) Burn, A. J., ADVAN. CHEM. SER. **75**, 323 (1968).
(6) DaRooge, M. A., Mahoney, L. R., *J. Org. Chem.* **37**, 1 (1967).
(7) Dobson, G., Grossweiner, L. I., *Trans. Faraday Soc.* **61**, 708 (1965).
(8) Gol'dberg, V. M., Obukhova, L. K., *Dokl. Akad. Nauk SSSR* **165**, 860 (1965).
(9) Gol'dberg, V. M., Obukhova, L. K., *Izv. Akad. Nauk SSSR, Ser. Khim.* **1966**, 2217.
(10) Howard, J. A., Ingold, K. U., *Can. J. Chem.* **40**, 1851 (1962).
(11) *Ibid.*, **41**, 1744 (1963).
(12) *Ibid.*, **41**, 2800 (1963).
(13) *Ibid.*, **42**, 2324 (1964).
(14) *Ibid.*, **43**, 2724 (1965).

(15) Howard, J. A., Schwalm, W. J., Ingold, K. U., ADVAN. CHEM. SER. **75**, 6 (1968).
(16) Khloplyankina, M. S., Buchachenko, A. L., Neiman, M. B., Vasil'eva, A. G., *Kinetika i Kataliz* **6**, 394 (1965).
(17) Land, E. J., Porter, G., *Trans. Faraday Soc.* **59**, 2016, 2027 (1963).
(18) McGowan, J. C., Powell, T., *J. Chem. Soc.* **1960**, 238.
(19) Mahoney, L. R., *J. Am. Chem. Soc.* **89**, 1895 (1967).
(20) Mahoney, L. R., DaRooge, M. A., *J. Am. Chem. Soc.* **89**, 5619 (1967).
(21) Mamedova, U. G., Buchachenko, A. L., Neiman, M. B. *Izv. Akad. Nauk SSSR, Ser. Khim.* **1965**, 911.
(22) Neiman, M. B., Mamedova, U. G., Blenke, P., Buchachenko, A. L., *Dokl. Akad. Nauk SSSR* **144**, 392 (1962).
(23) Thomas, J. R., *J. Am. Chem. Soc.* **86**, 4807 (1964).

RECEIVED October 9, 1967. Issued as NRC No. 9594.

24

Action of Aliphatic Amines on Slow Oxidation of Acetaldehyde and Ethyl Ether, and on Decomposition of Organic Peroxides in the Gas Phase

P. W. JONES and D. J. WADDINGTON

University of York, York, England

> *Aliphatic amines have much less effect on the later reactions of the gas-phase oxidation of acetaldehyde and ethyl ether than if added at the start of reaction. There is no evidence that they catalyze decomposition of peroxides, but they appear to retard decomposition of peracetic acid. Amines have no marked effect on the rate of decomposition of tert-butyl peroxide and ethyl tert-butyl peroxide. The nature of products formed from the peroxides is not altered by the amine, but product distribution is changed. Rate constants at 153°C. for the reaction between methyl radicals and amines are calculated for a number of primary, secondary, and tertiary amines and are compared with the effectiveness of the amine as an inhibitor of gas-phase oxidation reactions.*

Small amounts of aliphatic amines profoundly affect the reactions between simple organic molecules and oxygen in the gas phase. Inhibition occurs in the slow oxidation region and, at higher temperatures, ignition is either delayed or prevented.

During the slow oxidation of triethylamine, one of the products, diethylamine, appears to inhibit reactions leading to its own formation (9). Moreover, acetaldehyde is formed in the presence of oxygen and is stable at temperatures at which it is normally oxidized readily. This result was ascribed to inhibition of acetaldehyde by both the reactant (triethylamine) and the principal nitrogenous products (diethylamine and ethylamine). These findings preceded several detailed studies of the effect of aliphatic amines on the slow oxidation and the ignition of several organic compounds, principally acetaldehyde, ethyl ether, and paraffin hydrocarbons (Table I).

It is not possible to propose a general mechanism from these studies, for results do not correspond to a definite pattern. Although, in all the systems, secondary amines are the most effective inhibitors, the role played by tertiary amines is confusing. In several systems (Table I, No. 1, 2, and 3) tertiary amines are much more effective than primary amines, but in others they appear to have little or no effect. Again, in acetaldehyde oxidation (Table I, No. 1 and 2) there is generally a linear relationship between the amount of inhibitor added and the induction period before either slow oxidation or ignition of the fuel occurs. In other systems (Table I, No. 3, 4, and 5), however, a much more complex relationship is obtained. Thus, amines may be acting by different mechanisms in different systems.

Table I. Effect of Aliphatic Amines on Gas-Phase Oxidation of Organic Molecules

Study	Fuel	Combustion Region	Temp. Range, °C.	Ref.
1	Acetaldehyde	Slow	124–160	(7, 10)
2	Acetaldehyde	Ignition	218–230	(7, 8)
3	Ethyl ether	Slow	150–160	(28)
4	Ethyl ether	Ignition	175–260	(29)
5	Hexane	Slow and ignition	250–525	(25)

Following studies in which partially oxidized amines were added to the fuel-oxygen mixtures (7, 8), there is little doubt that it is the secondary and primary amines themselves (or a simple condensation product between the amine and the fuel if the latter contains a carbonyl group) that inhibit the oxidation reactions. However, Cullis and Khokhar (8) have suggested that the effectiveness of tertiary amines as inhibitors may, in part, be caused by the corresponding secondary and primary amine formed on oxidation, although one could argue that the reaction between the tertiary amine and oxygen may not occur rapidly enough at the low temperatures at which these compounds can inhibit.

Several mechanisms have been proposed to explain the action of amines as inhibitors. These mechanisms may be divided into three main classes.

Mechanism I. Following work on the slow oxidation of acetaldehyde, during which the surface-volume ratio of the reaction vessel was varied, Cullis and Khokhar (8) interpreted the mechanism in terms of the ability of amines to be absorbed onto the surface of the reaction vessel. Thus, if the chain-initiating processes between the fuel and oxygen occur on the surface, the fuel will not be oxidized until the amine is burned off.

Such a mechanism explains why the inhibiting effect of amines decrease as the extent of the surface increases. Such a mechanism is unlikely in the inhibition of ethyl ether oxidation since there is no correlation between the acidity of the surface of the vessel and the ability of amines to inhibit the reaction (28).

Mechanism II. Amines may be able to stabilize free radicals formed during the reaction between the fuel and oxygen. This could be done either by forming an adduct between a fuel radical, R·, and the amine, followed by reaction to form stable products (28), or by direct hydrogen abstraction (25, 28, 29) to form a radical, I·, from the inhibitor, which does not take any part in the fuel-oxygen chain reaction:

$$R\cdot + IH \rightarrow RH + I\cdot$$

For primary and secondary amines, there is no agreement on the relative importance of the part played by the hydrogen atoms attached to the carbon atoms and those attached to the nitrogen atom if such a mechanism does take place.

Mechanism III. Amines may interact with important molecule intermediates formed during the oxidation of the fuel—*e.g.*, peroxides. If this occurred by a nonchain process, degenerate chain branching would be stopped, and there would be effective inhibition, provided that the initiation reaction between the fuel and oxygen was slow.

This paper is concerned with the effect of amines on intermediates formed during the oxidation of simple organic molecules. Amines were added first to two systems in which peroxides are known to be formed in high yield to see whether they have any effect on the kinetics of the later reactions between the fuel and oxygen; secondly, amines were added to two peroxides under conditions where the amines act as inhibitors of gas-phase oxidation processes.

Experimental

Materials. The gases used (methane, ethane, ethylene, propane, propylene, *n*-butane, 2-methylpropane, and the butenes) were at least 99% pure (Cambrian Chemicals, Ltd.). The purity of each gas was tested by gas chromatography (columns of molecular sieve 5A, silica gel, or Porapak Q).

The purest available commercial samples of acetaldehyde, ethyl ether, and amines were purified by fractional distillation, either by conventional means or by using a spinning band column (Büchi). N-Methyldiethylamine was prepared from diethylamine (*4*) and N,N-dimethylethylamine from ethylamine (*5*).

Ethyl *tert*-butyl peroxide was prepared by reaction of the sodium salt of *tert*-butyl hydroperoxide (purified by recrystallization and frequent washings of acetone to free it from *tert*-butyl peroxide) and diethyl sulfate (*23*). The peroxide was purified by distillation (35–36°C. and 80-mm. pressure). NMR analysis (Perkin Elmer R60) showed that the samples used contained less than 1% *tert*-butyl peroxide and negligible quantities of *tert*-butyl hydroperoxide. A sample of *tert*-butyl peroxide was kindly given by Laporte Industries, Ltd.

In the studies on acetaldehyde and ethyl ether, the samples of amines were premixed with nitrogen (28).

Apparatus. A conventional static vacuum apparatus was used. Reactants were introduced into cylindrical borosilicate glass reaction vessels (250-ml. capacity) suspended in a furnace maintained at a temperature within ±0.1°C. Pressure measurements were made by a transducer (Consolidated Electrodynamics) linked to a recorder (RE 511 Servoscribe).

Analysis. Analysis was done by gas chromatography. All results were determined isothermally using Pye 104 model chromatographs, one equipped with a flame ionization detector, the other with a thermal conductivity cell. Carbon monoxide and other oxygenated products were determined on Porapak Q (17), hydrocarbons on silica gel treated with cupric sulfate and *dl*-alanine (16), and peroxides on both 10% Carbowax 20M on Celite (AW) and 10% PEG 600 on Celite (alk. W). Gases were analyzed using a heated sampling tube attached directly to the reaction vessel (6). Liquid samples were frozen out in traps surrounded by liquid nitrogen, and dissolved in either ethyl ether or dioxane, before being analyzed. Formaldehyde was determined spectrophotometrically using chromotropic acid (1, 2). Concentrations of the reactants and products were determined by a detailed number of internal analyses to ensure that there was no interaction between the products during sampling.

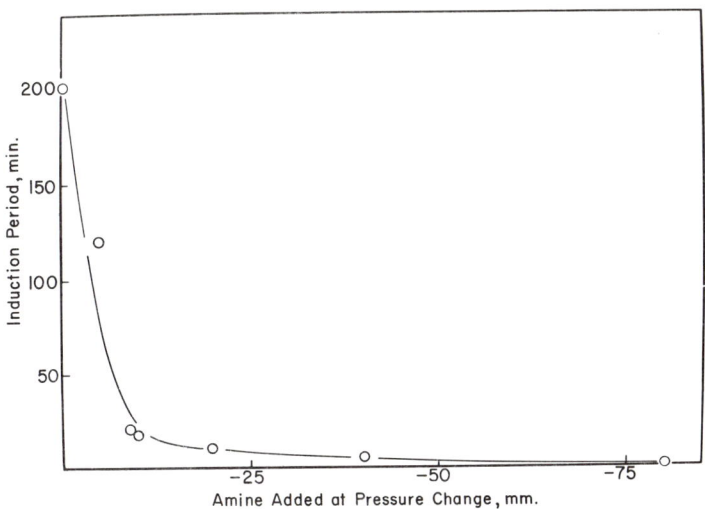

Figure 1. Influence of additions of diethylamine on induction period at 153°C.

Acetaldehyde, 100 mm.
Oxygen, 200 mm.
Diethylamine, 2.5 mm.

Results

When amines are added to the acetaldehyde-oxygen and ethyl ether–oxygen systems, a period of time elapses before oxidation of the fuel begins. The length of this induction period is dependent on the amount of amine added and the structure of the amine. Detailed analysis has shown that the length of the induction period can be used as a parameter for the efficiency of the inhibitor, for it is only when the amine has been consumed that oxidation of the fuel can take place (*28*).

Action of Aliphatic Amines on Oxidation Products of Acetaldehyde. Peracetic acid is formed in large quantities during the early stages of the slow oxidation of acetaldehyde—*i.e.*, during the time of negative pressure change (*12, 20, 21, 22*).

Diethylamine, a powerful inhibitor of acetaldehyde oxidation when added at the start of the reaction (*10*), was added during the oxidation of acetaldehyde. The later the inhibitor is added, the less effect it has on the subsequent reaction, although the length of the induction period still depends on the amount of amine added (Figure 1). Again, the

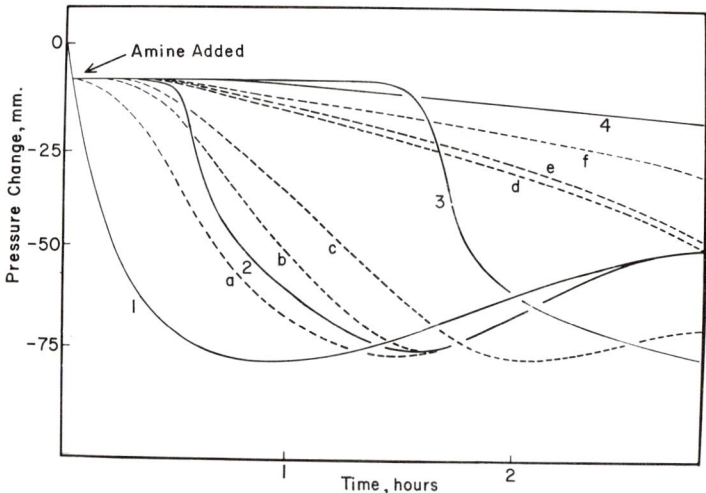

Figure 2.   Influence of ethylamine and diethylamine on reaction at 153°C.

Acetaldehyde,	100 mm.	Ethylamine, mm.	Diethylamine, mm.
Oxygen,	100 mm.	(a) 5.8	(1) —
Amine added at ΔP = −10 mm.		(b) 10.4	(2) 3.6
		(c) 15.0	(3) 4.8
		(d) 20.0	(4) 6.0
		(e) 24.0	
		(f) 31.0	

length of the induction period depends on the structure of the amine added—e.g., diethylamine is considerably more effective than ethylamine (Figure 2).

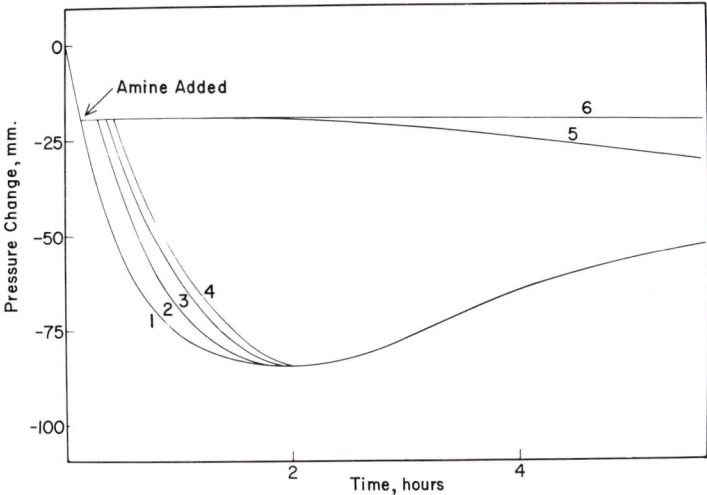

Figure 3. Influence of diethylamine on rate of reaction at 153°C.

Amine added at $\Delta P = -20$ mm.
Acetaldehyde, 100 mm.
Oxygen, 100 mm.

Pressure of Amine, mm.
(1) —
(2) 2.5
(3) 3.9
(4) 6.0
(5) 11.2
(6) 13.5

Similar results are obtained if the inhibitor is added at a later stage (when more peracetic acid is present), but the induction periods are even shorter (Figure 3). Indeed, at the minimum pressure change, when the maximum concentration of peracetic acid is present, it is only when a large amount of diethylamine has been added (over 10% of the total mixture) that decomposition appears to stop (Figure 4).

Action of Aliphatic Amines on Oxidation Products of Ethyl Ether. Amines were added during the oxidation of ethyl ether. Both ethylamine (Figure 5) and triethylamine (Figure 6) are more than three times more effective as negative catalysts when added at the start of reaction than during the reaction, and triethylamine has little effect on the subsequent rate of reaction when added at the minimum pressure change [when the maximum concentration of peracetic acid is present (27)].

Action of Aliphatic Amines on Decomposition of *tert*-Butyl Peroxide. The rate of decomposition of *tert*-butyl peroxide is not greatly affected by adding n-butylamine (Figure 7). However, the pressure change (Figure 7) and the yields of different products are altered (Table II). The most significant changes are the decrease in ethane yield, while the amount of methane formed increases. The yield of *tert*-butyl alcohol also increases. The amount of methyl ethyl ketone formed decreases.

Figure 8 shows how the rate of formation of methane and ethane is altered by adding dimethylamine, while Figure 9 describes how the addition of different amines affects the formation of the two alkanes.

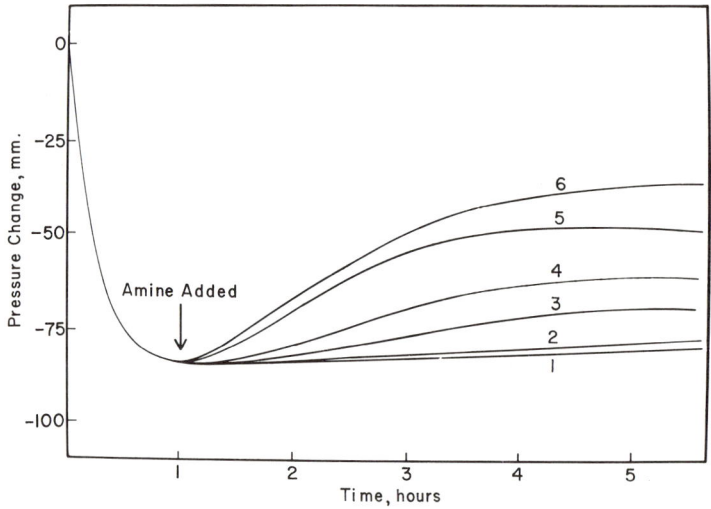

Figure 4. Influence of diethylamine on later stages of reaction at 153°C.

Acetaldehyde, 100 mm. Oxygen, 100 mm.	Pressure of Diethylamine, mm.	Rate of Reaction, mm. hr.$^{-1}$
	(1) —	20
	(2) 5.3	17
	(3) 12.1	10
	(4) 15.2	5
	(5) 17.5	2
	(6) 20.3	1

Action of Diethylamine on Decomposition of Ethyl *tert*-Butyl Peroxide. The rate of decomposition of ethyl *tert*-butyl peroxide is decreased by adding diethylamine (Figure 7), and the yield of products is altered (Table II). Again, the yield of methane is increased at the expense of ethane and *tert*-butyl alcohol is increased at the expense of acetone. Ethanol and acetaldehyde are formed in considerably greater amounts. The yields of carbon monoxide and methyl ethyl ketone are decreased.

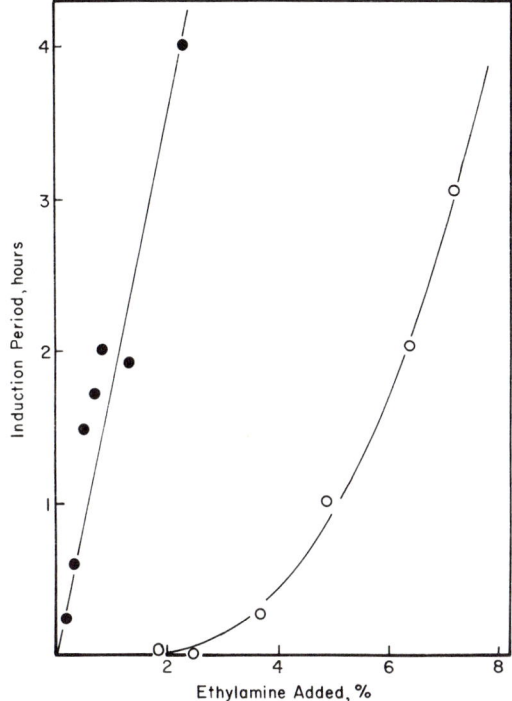

Figure 5. Dependence of induction period on concentration of ethylamine added at 153°C.

Ethyl ether, 100 mm.
Oxygen,   200 mm.
● Amine added at t = 0 min.
○ Amine added at t = 25 min. ΔP = −3 mm.

Discussion

A previous paper described the inhibiting influence of aliphatic amines on the oxidation of ethyl ether in the gas phase (28). It was suggested that one of the principal methods by which amines can act as negative catalysts is by stabilizing free radicals in the gas phase. However, it is necessary to examine the action of amines on the intermediate products formed during oxidation processes.

It is believed that peractic acid is primarily responsible for chain branching during the oxidation of acetaldehyde (12) and ethyl ether (24, 27). Although amines are powerful inhibitors of these oxidation processes, they have much less effect if added during the reaction. The inhibiting power of amines becomes smaller the longer the reaction between the fuel and oxygen has proceeded. Similar behavior is observed on adding aromatic amines to *n*-butane–oxygen and 2-methylpropane–oxygen mixtures (*18*). This result is also supported by circumstantial evidence from the study of the action of triethylamine on acetaldehyde, when Cullis points out that as there is no pressure change during the induction period, no peracetic acid can be formed and thus the action of the amine must be to prevent stages in the reaction cycle earlier than those involving peracetic acid (*8*). Analysis during the induction period in the ethyl ether–oxygen system caused by amines confirmed that no peracetic acid is detectable (*28*). However, it may have been formed in small amounts and destroyed rapidly by the amine. There is little doubt, however, that amines, in large concentrations, appear to retard the decomposition of peracetic acid once the latter has been formed.

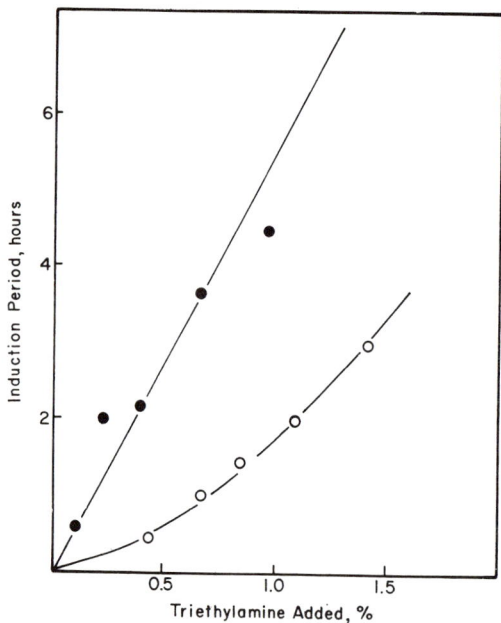

Figure 6. Dependence of induction period on concentration of triethylamine at 153°C.

Ethyl ether, 100 mm.
Oxygen, 200 mm.
● Amine added at t = 0 min.
○ Amine added at t = 25 min. $\Delta p = -3$ mm.

Even large amounts of aliphatic amines do not alter the rate of decomposition of *tert*-butyl peroxide, but they do slow down the rate of decomposition of ethyl *tert*-butyl peroxide.

This retardation of ethyl tert-butyl peroxide decomposition may possibly be caused by competition between the inhibitor and the peroxide for methyl radicals (Reactions 1-4).

$$CH_3 \cdot CH_2 \cdot O \cdot O \cdot C(CH_3)_3 \rightarrow CH_3 \cdot CH_2 \cdot O \cdot + (CH_3)_3 C \cdot O \cdot \quad (1)$$

$$(CH_3)_3 C \cdot O \cdot \rightarrow CH_3 \cdot + CH_3 \cdot CO \cdot CH_3 \quad (2)$$

$$CH_3 \cdot CH_2 \cdot O \cdot O \cdot C(CH_3)_3 + CH_3 \cdot \rightarrow CH_3 \cdot CH \cdot O \cdot O \cdot C(CH_3)_3 + CH_4 \quad (3)$$

$$IH + CH_3 \cdot \rightarrow I \cdot + CH_4 \quad (4)$$

Thus, in the presence of an amine, if $k_4 > k_3$, homolysis will become the principal method of decomposition of the peroxide.

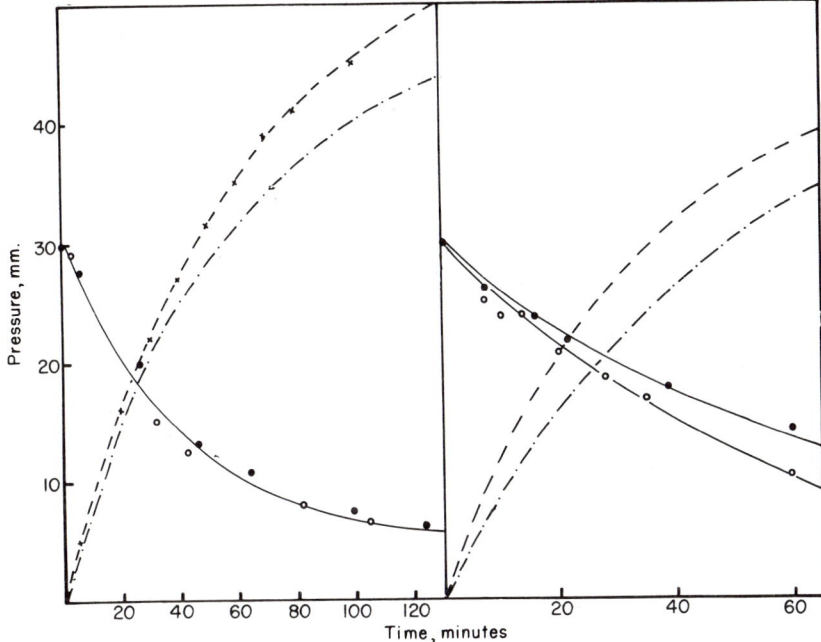

Figure 7. Influence of amines on decomposition of peroxides at 153°C.

(1) tert-Butyl peroxide, 30 mm.
(2) tert-Butyl peroxide, 30 mm. n-Butylamine, 20 mm.
(3) Ethyl tert-butyl peroxide, 30 mm.
(4) Ethyl tert-butyl peroxide, 30 mm. Diethylamine, 20 mm.

Left:
------ Total pressure change (1)
-·-·-· Total pressure change (2)
● Peroxide pressure (1)
○ Peroxide pressure (2)

Right:
------ Total pressure change (3)
-·-·-· Total pressure change (4)
● Peroxide pressure (3)
○ Peroxide pressure (4)

In the decomposition of tert-butyl peroxide, the reaction corresponding to Reaction 3 is less important, and decomposition of the peroxide will be less affected by amine addition.

Table II. Analyses at Final Pressure Change (153°C.)[a]

	tert-Butyl Peroxide, 30 mm.	tert-Butyl Peroxide, 30 mm., n-Butylamine, 20 mm.	Ethyl tert-Butyl Peroxide, 30 mm.	Ethyl tert-Butyl Peroxide, 30 mm., Diethylamine, 20 mm.
Hydrocarbons				
Methane	0.23	0.50	0.39	0.55
Ethane	0.72	0.67	0.49	0.14
Propane	0.03	0.03	—	—
Carbonyl Compounds				
Formaldehyde	—	—	0.04	0.02
Acetaldehyde	—	—	0.07	0.23
Acetone	1.60	1.40	0.78	0.68
Methyl ethyl ketone	0.23	0.10	0.06	0.02
Alcohols				
Ethanol	—	—	0.42	0.71
tert-Butyl alcohol	0.03	0.09	0.18	0.33
Other Compounds				
Carbon monoxide	—	—	0.45	0.25
Parent peroxide	0.07	0.07	0.01	0.02

[a] Moles per initial mole of peroxide.

However, the early stages of the decomposition of *tert*-butyl peroxide have been examined to study the reaction between methyl radicals and both the parent molecule and acetone under these conditions.

$$CH_3\cdot + (CH_3)_3C\cdot O\cdot O\cdot C(CH_3)_3 \rightarrow CH_4 + \cdot CH_2(CH_3)_2C\cdot O\cdot O\cdot C(CH_3)_3 \quad (5)$$

$$CH_3\cdot + CH_3\cdot CO\cdot CH_3 \rightarrow CH_4 + \cdot CH_2\cdot CO\cdot CH_3 \quad (6)$$

A measure of Reaction 2 can be gauged from the formation of methyl ethyl ketone.

$$CH_3\cdot + \cdot CH_2\cdot CO\cdot CH_3 \rightarrow CH_3\cdot CH_2\cdot CO\cdot CH_3 \quad (7)$$

In the presence of an amine, the principal method of forming methane is again by hydrogen abstraction from the amine.

$$IH + CH_3\cdot \rightarrow I\cdot + CH_4 \quad (4)$$

Ethane is formed by combination of methyl radicals formed from *tert*-butoxy radicals.

$$(CH_3)_3C\cdot O\cdot O\cdot C\cdot (CH_3)_3 \rightarrow 2(CH_3)_3CO\cdot \quad (8)$$

$$(CH_3)_3CO\cdot \rightarrow CH_3\cdot + CH_3\cdot CO\cdot CH_3 \quad (2)$$

$$2CH_3\cdot \rightarrow C_2H_6 \quad (9)$$

It has been suggested that ethane (3) may also be formed from an amine.

$$CH_3 \cdot + CH_3 \cdot CH_2 \cdot NH \cdot CH_2 \cdot CH_3 \rightarrow C_2H_6 + \cdot CH_2 \cdot NH \cdot CH_2 \cdot CH_3 \quad (10)$$

However, such a reaction does not appear to be particularly important under the conditions studied.

Thus, assuming Reactions 5, 6, and 4 are responsible for producing methane, and Reaction 9 responsible for ethane:

$$\frac{R_{CH_4}}{R^{1/2}_{C_2H_6}} = \frac{k_5[P] + k_6[A] + k_4[IH]}{k_9^{1/2}}$$

where R_{CH_4} and $R_{C_2H_6}$ are the rates of formation of methane and ethane, and [P], [A], and [RH] are the initial pressures of peroxide, acetone, and amine.

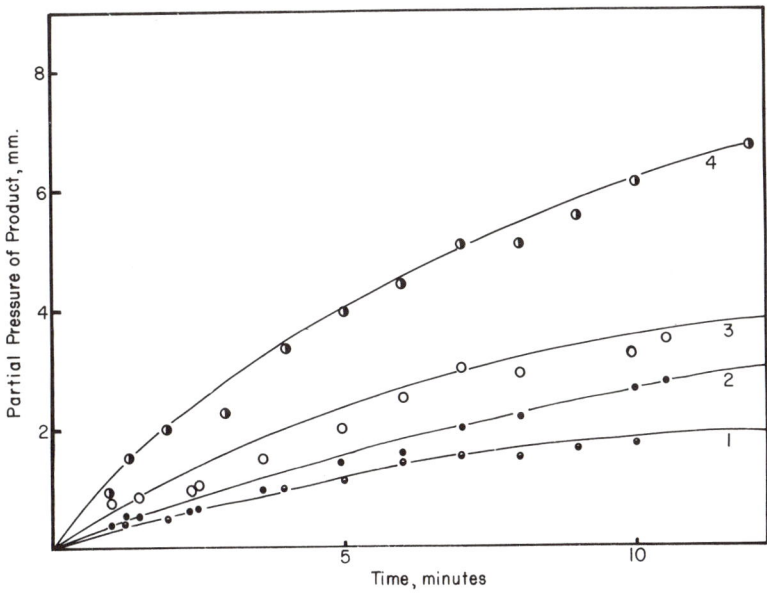

Figure 8. Influence of dimethylamine on formation of methane and ethane from tert-butyl peroxide at 153°C.

tert-Butyl peroxide, 30 mm. Dimethylamine, 20 mm.
(1) Methane. D.t.B.P. 30 mm.
(2) Methane. D.t.B.P. 30 mm. Amine 20 mm.
(3) Ethane. D.t.B.P. 30 mm. Amine 20 mm.
(4) Ethane. D.t.B.P. 30 min.

The rate constant, k_4, for different amines (calculated from Figure 9) is given in Table III. The value for k_9 was taken to be $10^{12.34}$ cc. moles^{-1} sec.$^{-1}$ (26).

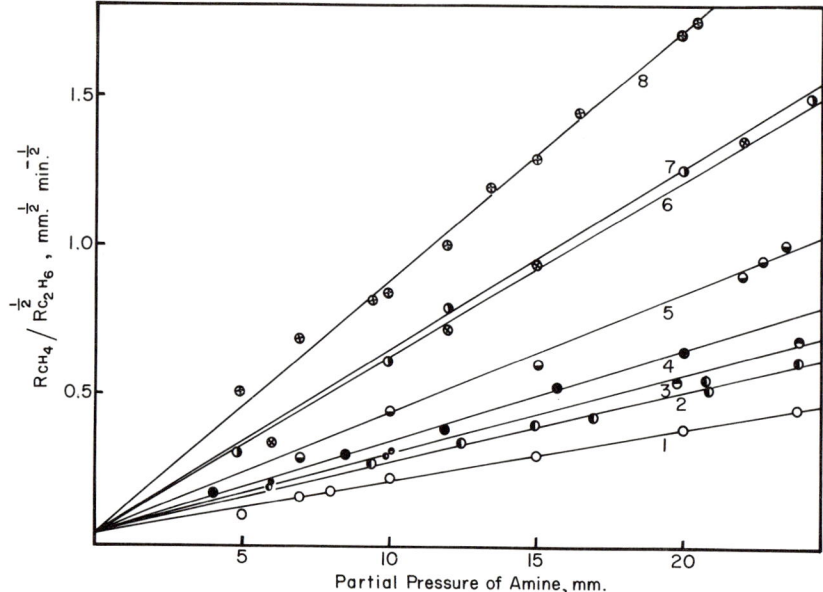

Figure 9. Influence of amines on formation of methane and ethane from tert-butyl peroxide (30 mm.) at 153°C.

(1) Trimethylamine
(2) Ethylamine
(3) n-Butylamine
(4) N,N-Dimethylethylamine
(5) N-Methyldiethylamine
(6) Triethylamine
(7) Dimethylamine
(8) Diethylamine

The values of k_4 for ethylamine and dimethylamine agree with those determined by Gray and co-workers, and we favor lower values for k_4 for other amines. One of the principal causes for the discrepancies is in the value taken for k_9. Gray (15) has pointed out that rate constants for many abstraction reactions calculated in this way may not have a precise physical meaning since there is no direct evidence concerning which hydrogen atom is being abstracted from an amine molecule. For example, at 150°C., with methylamine, the ratio of hydrogen atoms abstracted from the methyl and amino groups is 1 to 2, in favor of the hydrogen atoms attached to nitrogen (15). However, the ratio is reversed for ethylamine (13), since 67% of the hydrogen atoms are abstracted from the methylene group and 29% from the imino group, the ratio for each hydrogen atom being 2 to 1 in favor of the hydrogen atoms attached to the α-carbon atom.

There is much less difference in the abstraction rate constants, k_4, between ethylamine and n-butylamine (Table III) than that found between methylamine and ethylamine. This is expected if the hydrogen atoms in the methylene group adjacent to the nitrogen atom are most prone to attack by methyl radicals. The relationship between the rate of abstraction from methylamine, ethylamine, and n-butylamine closely parallels their efficiency as inhibitors of the slow oxidation of acetaldehyde (10) and ethyl ether (28).

Table III. Rate Constants of Reactions between Methyl Radicals and Amines at 153°C.

Reactions	$\log k_4$ k, cc. moles^{-1} sec.$^{-1}$	Ref.
$CH_3 \cdot + CH_3NH_2$	6.50	(3)
$CH_3 \cdot + CH_3NH_2$	6.87	(15)
$CH_3 \cdot + CH_3NH_2$	6.48	(15)
$CH_3 \cdot + \overline{CH_3NH_2}$	6.60	(15)
$CH_3 \cdot + CH_3CH_2NH_2$	7.5	(3)
$CH_3 \cdot + CH_3CH_2NH_2$	7.11	(13)
$CH_3 \cdot + CH_3CH_2NH_2$	6.91	Figure 9
$CH_3 \cdot + CH_3CH_2NH_2$	6.85	(13)
$CH_3 \cdot + \overline{CH_3CH_2NH_2}$	6.61	(13)
$CH_3 \cdot + n\text{-}C_4H_9\overline{NH_2}$	6.96	Figure 9
$CH_3 \cdot + (CH_3)_2NH$	8.0	(3)
$CH_3 \cdot + (CH_3)_2NH$	7.57	(15)
$CH_3 \cdot + (CH_3)_2NH$	7.32	Figure 9
$CH_3 \cdot + (CH_3)_2NH$	7.5	(15)
$CH_3 \cdot + (C_2H_5)_2NH$	8.1	(3)
$CH_3 \cdot + (C_2H_5)_2NH$	7.47	Figure 9
$CH_3 \cdot + (CH_3)_3N$	7.2	(19)
$CH_3 \cdot + (CH_3)_3N$	7.3	(11)
$CH_3 \cdot + (CH_3)_3N$	6.76	Figure 9
$CH_3 \cdot + C_2H_5N(CH_3)_2$	7.02	Figure 9
$CH_3 \cdot + (C_2H_5)_2NCH_3$	7.14	Figure 9
$CH_3 \cdot + (C_2H_5)_3N$	7.7	(19)
$CH_3 \cdot + (C_2H_5)_3N$	7.31	Figure 9

The difference in reactivities of dimethylamine and diethylamine toward radical attack is not as marked as between methylamine and ethylamine (Table III), and this suggests that attack on the imino group is most important; hence, adding methylene groups on passing from dimethylamine to diethylamine will not have such a pronounced difference on the rate of abstraction as between methylamine and ethylamine. The importance of the imino hydrogen atom is seen when comparing the rates of abstraction from dimethylamine and trimethylamine and comparing diethylamine with N-methyldiethylamine (Table III). It has been calculated that the ratio of hydrogen abstraction from dimethylamine is 1 to 20 in favor of the hydrogen atom in the imino group compared with a hydrogen atom in the methyl group (14).

A comparison of the four tertiary amines studied gives unequivocal evidence of the importance of the methylene group in the amine (Table III), the difference in k_4 between trimethylamine and N,N-dimethylethylamine being larger than the difference between the latter amine and N-methyldiethylamine or between N-methyldiethylamine and triethylamine. This effect again resembles the differences found between methylamine, ethylamine, and n-butylamine. It is of interest to compare the C_4 isomers, n-butylamine, diethylamine, and N-methyldiethylamine. The order of values for k_4 are: secondary > tertiary > primary, which corresponds to the order of inhibitor efficiency in the slow oxidation region for acetaldehyde-oxygen (10) and ethyl ether–oxygen (28) mixtures.

The linear relationships in Figure 9 indicate that Reaction 10 is relatively unimportant under these conditions. If it were important, one would expect (3) a decrease in the ratio of $R_{CH_4}/R^{1/2}{}_{C_2H_6}$ as the concentration of amine is increased for:

$$\frac{R_{CH_4}}{R^{1/2}{}_{C_2H_6}} = \frac{k_4[IH]}{k_9^{1/2}} \left[1 + \frac{k_{10}[IH]}{k_9[CH_3]} \right]^{-1/2}$$

The results given in this paper show that aliphatic amines do not catalyze the decomposition of peroxides, and compared with their effect at the start of reaction, they have much less effect on the later stages of oxidation, although they appear to retard the decomposition of peracetic acid. The reactions of radicals with aliphatic amines indicate that an important mode of inhibition is most probably by stabilization of free radicals by amine molecules early in the chain mechanism, possibly radicals formed from the initiation reaction between the fuel and oxygen. For inhibition to be effective, the amine radical must not take any further part in the chain reaction set up in the fuel-oxygen system. The fate of the inhibitor molecules is being elucidated at present.

Acknowledgment

The authors thank Shell Research, Ltd., for the provision of a studentship (P.W.J.). They also thank M. A. Warriss for helpful technical assistance.

Literature Cited

(1) Altshuller, A. P., Miller, D. L., Sleva, S. F., *Anal. Chem.* **33**, 621 (1961).
(2) Bricker, C. E., Johnson, H. R., *Ibid.* **17**, 400 (1945).
(3) Brinton, R. K., *Can. J. Chem.* **38**, 1339 (1960).
(4) Clarke, H. T., Gillespie, H. B., Weisshaus, S. Z., *J. Am. Chem. Soc.* **55**, 4571 (1933).

(5) Copp, J. L., *Trans. Faraday Soc.* **51,** 1056 (1955).
(6) Crawforth, C. G., Waddington, D. J., *J. Gas Chromatog.* **6,** 103 (1968).
(7) Cullis, C. F., Khokhar, B. A., Seventh Symposium (International) on Combustion, p. 171, Butterworths, London, 1959.
(8) Cullis, C. F., Khokhar, B. A., *Trans. Faraday Soc.* **56,** 1235 (1960).
(9) Cullis, C. F., Waddington, D. J., *Proc. Roy. Soc.* **244A,** 100 (1958).
(10) Cullis, C. F., Waddington, D. J., *Trans. Faraday Soc.* **53,** 1317 (1957).
(11) Edwards, D. A., Kerr, J. A., Lloyd, A. C., Trotman-Dickenson, A. F., *J. Chem. Soc.* (A) **1966,** 621.
(12) Farmer, J. B., McDowell, C. A., *Ibid.* **48,** 624 (1952).
(13) Gray, P., Jones, A., *Ibid.* **62,** 112 (1966).
(14) Gray, P., Jones, A., Thynne, J. C. J., *Ibid.* **61,** 474 (1965).
(15) Gray, P., Thynne, J. C. J., *Ibid.* **59,** 2275 (1963).
(16) Hart-Davis, A. J., private communication.
(17) Hollis, O. L., *Anal. Chem.* **38,** 309 (1966).
(18) Ingold, K. U., Puddington, I. E., *Can. J. Chem.* **37,** 1376 (1959).
(19) Kozak, P. J., Gesser, H., *Ibid.* **1960,** 448.
(20) McDowell, C. A., Thomas, J. H., *J. Chem. Phys.* **17,** 588 (1949).
(21) McDowell, C. A., Thomas, J. H., *J. Chem. Soc.* **1950,** 1462.
(22) McDowell, C. A., Thomas, J. H., *Trans. Faraday Soc.* **46,** 1030 (1950).
(23) Rust, F. F., Seubold, F. H., Vaughan, W. E., *J. Am. Chem. Soc.* **72,** 338 (1950).
(24) Salooja, K. C., *Combustion and Flame* **9,** 33 (1965).
(25) Salooja, K. C., *J. Inst. Petrol.* **49,** 58 (1963).
(26) Shepp, A., *J. Chem. Phys.* **24,** 939 (1956).
(27) Waddington, D. J., *Proc. Roy. Soc.* **252A,** 260 (1959).
(28) Waddington, D. J., *Ibid.* **265A,** 436 (1962).
(29) Waddington, D. J., Seventh Symposium (International) on Combustion, p. 165, Butterworths, London, 1959.

RECEIVED October 9, 1967.

Discussion

H. S. Olcott (University of California, Berkeley, Calif.): We have studied the effects of aliphatic amines on the autoxidation of a fish oil and squalene in air at moderate temperatures. There was little protection unless phenolic-type inhibitors were also added, in which case secondary amines were more effective than primary or secondary amines. However, at 70°C. trioctylamine alone protected the fish oil, whereas at lower temperatures it did not (*2*). Further study revealed that peroxides react with trioctylamine to yield some dioctylhydroxylamine which has antioxidant properties (*1*). These and other observations (*3*) indicate that peroxides formed during the initial stages of autoxidation may react with other constituents to form new compounds with enhanced antioxygenic activity. Additional work on the effects of substituted hydroxylamines and nitroxides on the oxidation of unsaturated lipids is in progress.

Literature Cited

(1) Harris, L., Olcott, H. S., *J. Am. Oil Chemists' Soc.* **43**, 11 (1966).
(2) Olcott, H. S., "Lipids and Their Oxidation," H. W. Schultz, ed., p. 173, Avi Publishing Co., New York, 1962.
(3) Olcott, H. S., Van der Veen, J., Brown, W. D., *Nature* **191**, 1201 (1961).

25

Mechanism of Oxidation Inhibition by Zinc Dialkyl Dithiophosphates

A. J. BURN

British Petroleum Co., Ltd., BP Research Centre, Chertsey Road, Sunbury-on-Thames, Middlesex, England

> *The kinetics of the zinc diisopropyl dithiophosphate–inhibited oxidation of cumene at 60°C. and Tetralin at 70°C. have been investigated. The results cannot be accounted for solely in terms of chain-breaking inhibition by a simple electron-transfer mechanism. No complete explanation of the Tetralin kinetics has been found, but the cumene kinetics can be explained in terms of additional reactions involving radical-initiated oxidation of the zinc salt and a chain-transfer step. Proposed mechanisms by which zinc dialkyl dithiophosphates act as peroxide-decomposing antioxidants are discussed.*

The mechanism of hydrocarbon autoxidation and its inhibition have been widely studied over the last two decades (1, 19), and as a result two principal types of antioxidant have been recognized: chain-breaking inhibitors which interrupt the autoxidation propagation cycle by removing the chain-propagating peroxy radicals in a reaction involving hydrogen or electron transfer, and peroxide decomposers which prevent the accumulation of chain-initiating hydroperoxides by causing their decomposition to nonradical products.

Although zinc dialkyl dithiophosphates, $[(RO)_2PS_2]_2Zn$, have been used as antioxidants for many years, the detailed mechanism of their action is still not known. However, it is certain that they are efficient peroxide decomposers. The effect of a number of organic sulfur compounds, including a zinc dithiophosphate, on the rate of decomposition of cumene hydroperoxide in white mineral oil at 150°C. was investigated by Kennerly and Patterson (13). Each compound accelerated the hydroperoxide decomposition, the zinc salt being far superior in its activity to the others. Further, in each case the principal decomposition product was phenol, indicating an ionic mechanism (*10*). Zinc diisopropyl dithiophosphate has been shown (*7*) to accelerate the decomposition of *tert*-butyl hydroperoxide in carbon tetrachloride at 20°C. The decomposition of cumene hydroperoxide in benzene at room temperature is also accelerated by the related nickel and ferric diisopropyl dithiophosphates (*11*) at a concentration of only 1% of that of the hydroperoxide, which was completely decomposed. The principal products were acetone and phenol, again those typical of ionic decomposition (*10*).

The inhibition of hydrocarbon autoxidation by zinc dialkyl dithiophosphates was first studied by Kennerly and Patterson (*13*) and later by Larson (*14*). In both cases the induction period preceding oxidation of a mineral oil at 155°C. increased appreciably by adding a zinc dialkyl dithiophosphate. In particular, Larson (*14*) observed that zinc salts containing secondary alkyl groups were more efficient antioxidants than those containing primary groups. In these papers the inhibition mechanism was discussed only in terms of peroxide decomposition.

More recently it has been shown (*6, 7*) that zinc dialkyl dithiophosphates also act as chain-breaking inhibitors. Colclough and Cunneen (*7*) reported that zinc isopropyl xanthate, zinc dibutyl dithiocarbamate, and zinc diisopropyl dithiophosphate all substantially lowered the rate of azobisisobutyronitrile-initiated oxidation of squalene at 60°C. Under these conditions, hydroperoxide chain initiation is negligible, and it was therefore concluded that inhibition resulted from removal of chain-propagating peroxy radicals. Also, consideration of the structure of these zinc dithioates led to the conclusion that no suitably activated hydrogen atom was available, and it was suggested that inhibition could be accounted for by an electron-transfer process as follows:

$$[R_2NCS_2]_2Zn + RO_2 \cdot \rightarrow (R\ddot{O}_2)^- (R_2NCS.\dot{S})^{+}Zn\ \ddot{S}CSNR_2$$

This scheme appears to involve a free thiyl type radical ($R_2NCS.\dot{S}$).

The conclusion that chain-breaking inhibition by zinc dialkyl dithiophosphates involves electron transfer was reached independently by Burn (*6*) following a more detailed qualitative study of the inhibition of the azonitrile-initiated oxidation of squalane and cumene and the noninitiated oxidation of indene by metal dialkyl dithiophosphates and related compounds (I to IV):

$$[(RO)_2PS_2]_xM \qquad\qquad [Pr\text{-}i\text{-}OCS_2]_2Zn$$
$$\text{I} \qquad\qquad\qquad\qquad \text{II}$$

$$[(C_{12}H_{25}O)_2PO_2]_2Zn \qquad\qquad [(Pr\text{-}i\text{-}O)_2PS_2]_2$$
$$\text{III} \qquad\qquad\qquad\qquad \text{IV}$$

Only compounds of type I and II, which contain both metal and sulfur, were found to act as inhibitors. Compounds III and IV, for example, were ineffective. The inhibition of indene oxidation by zinc dithiophosphates was considered to be a key result in this work since it rules out the intermediate formation of free thiyl radicals. There is adequate evidence (2, 9, 16, 17, 22) that thiyl radicals, including $(RO)_2PS_2\cdot$, add rapidly to olefins and that in the presence of oxygen the following sequence of reactions would occur.

$$RS\cdot + \text{[olefin]} \longrightarrow \underset{\cdot}{\overset{RS}{\bigvee}} \xrightarrow{O_2} \underset{\cdot O_2}{\overset{RS}{\bigvee}}$$

The radical $(RO)_2PS_2\cdot$ in indene would thus effectively act as a chain-carrying radical and is therefore not considered to be an intermediate in chain-breaking inhibition by dithiophosphates. A two-stage mechanism was therefore proposed, involving a stabilized zinc salt–peroxy complex, either a radical or an ion pair as illustrated in structures V and VI.

$$\underset{V}{\overset{R-O-O}{\underset{|}{\underset{(RO)_2P\diagup{\overset{S\cdot}{\|}}\diagdown_{S-Zn-S}\diagup P(OR)_2}{}}}} \qquad \underset{VI}{\overset{RO_2^-}{\underset{(RO)_2P\diagup{\overset{S^+\cdot}{\|}}\diagdown_{S-Zn-S}\diagup P(OR)_2}{}}}$$

On attack of a second peroxy radical at the remaining thionosulfur atom electron transfer is completed to yield disulfide as illustrated in VII and VIII.

The importance of ionization of the metal-sulfur bond in the electron-transfer process accounts for inhibitory properties being observed only for compounds of types I and II. Support for this mechanism was also obtained by the isolation of the disulfide (IV) from a reaction between peroxy radicals and zinc diisopropyl dithiophosphate.

The present paper reports the results of a kinetic study of the inhibition of the azobisisobutyronitrile-initiated autoxidation of cumene at 60°C. and of Tetralin at 70°C. by zinc diisopropyl dithiophosphate, undertaken to test the validity of the chain-breaking inhibition mechanism proposed above. In addition, the effectiveness of several metal dialkyl dithiophosphates as antioxidants in the autoxidation of squalane at 140°C. has been studied. In this system we might now expect both chain breaking and peroxide decomposition to be important. Previous proposals on the mechanism of peroxide decomposition are discussed in detail, incorporating the limited results of the present work which are relevant to this type of antioxidant action.

$$\begin{array}{ccc} \text{VII} & & \text{VIII} \end{array}$$

(Scheme showing structures VII, VIII and a disulfide product with (RO)$_2$P groups, S—Zn—S bridges, and R—O—O interactions)

Experimental

Materials. Cumene and Tetralin were purified by extraction with concentrated sulfuric acid, until the extracts were colorless, then with 2N caustic soda and distilled water, and finally dried and distilled. Both were stored in darkness under N_2 and percolated through silica gel immediately before use. *tert*-Butylbenzene (99.9% by GLC) was used as an inert diluent for cumene and Tetralin where indicated. Squalane (M and B Embaphase) was used as received. AIBN was recrystallized from ether and had a melting point of 102°–103°C. Metal dialkyl dithiophosphates were prepared as described previously (6); zinc diisopropyl dithiophosphate was finally recrystallized twice from *n*-heptane and had a melting point of 146°C.

Rate Measurements. Oxidation rates, which were normally carried out under pure oxygen at 760-mm. total pressure, and azonitrile decomposition rates were measured using apparatus previously described.

Induction Periods. Induction periods for squalane oxidation were measured using an automatic recording oxygen absorption apparatus (23).

Results and Discussion

Chain-Breaking Inhibition Mechanism. According to the mechanism proposed earlier (6), the inhibition of the autoxidation of a hydrocarbon

(RH) by a zinc dialkyl dithiophosphate (ZnP) should ideally be representable by the following simplified scheme (Reactions 1 to 6).

$$\text{AIBN} \rightarrow 2e\text{I}^{\cdot} \qquad (1)$$

$$\text{R}^{\cdot}\ (\text{I}^{\cdot}) + \text{O}_2 \rightarrow \text{RO}_2^{\cdot}\ (\text{IO}_2^{\cdot}) \qquad (2)$$

$$\text{RO}_2^{\cdot} + \text{RH} \rightarrow \text{RO}_2\text{H} + \text{R}^{\cdot} \qquad (3)$$

$$\text{RO}_2^{\cdot} + \text{RO}_2^{\cdot} \rightarrow \text{inactive products} \qquad (4)$$

$$\text{RO}_2^{\cdot} + \text{ZnP} \rightarrow \text{complex (V or VI)} \qquad (5)$$

$$\text{Complex (V or VI)} + \text{RO}_2^{\cdot} \rightarrow \text{inactive products} \qquad (6)$$

Scheme 1

AIBN represents azobisisobutyronitrile, the initiator used in this work, and e is the efficiency with which it initiates oxidation chains. The experimental conditions were such that bimolecular self-termination (Reaction 4) is negligible compared with chain termination by reaction of peroxy radicals with zinc dialkyl dithiophosphate.

Assuming that Reaction 5 is not reversible and that it is the rate-determining step in the proposed termination mechanism, the rate of oxidation predicted on application of the steady-state theory is given by Equation A.

$$R = -\frac{dO_2}{dt} - (2e - 1)k_1[\text{AIBN}] = \frac{ek_1k_3[\text{AIBN}][\text{RH}]}{k_5[\text{ZnP}]} \qquad (A)$$

A correction to the observed rate $(-dO_2/dt)$ is made by subtracting the term $(2e - 1)k_1[\text{AIBN}]$, which allows for oxygen absorption and nitrogen evolution by AIBN.

If Reaction 5 is reversible and a fast equilibrium with complex is set up, the mechanism is similar kinetically to that proposed by Hammond and his co-workers (3, 5) for inhibition by phenol, and Equation B is derived.

$$R = k_3[\text{RH}] \left[\frac{ek_1k_{-5}[\text{AIBN}]}{k_5k_6[\text{ZnP}]}\right]^{1/2} \qquad (B)$$

The results of a study of the zinc diisopropyl dithiophosphate–inhibited oxidation of cumene at 60°C. are shown in Figures 1 to 3. The initial oxidation rate is directly proportional to the AIBN concentration, but the dependence of initial rate on the cumene concentration or the reciprocal of the zinc salt concentration, although reasonably linear, is not in direct proportion.

The results of a similar study of the inhibited oxidation of Tetralin at 70°C. are shown in Figures 4 to 6. Again, the initial oxidation rate is directly proportional to the AIBN concentration. In this case unusual dependence of oxidation rate on both the Tetralin concentration and reciprocal of zinc salt concentration was observed. As Figure 6 shows, the oxidation rate, at two different AIBN concentrations, is independent of the Tetralin concentration until this falls below about 1.5 moles per liter. Also, although increasing the ZnP concentration (Figure 5) initially decreases the oxidation rate, the rate is ultimately increased. These results cannot arise from the occurrence of termination reactions other than 5 and 6 (such as those involving R·) since oxidation rates in this system were independent of the oxygen pressure in the range 350 to 760 mm.

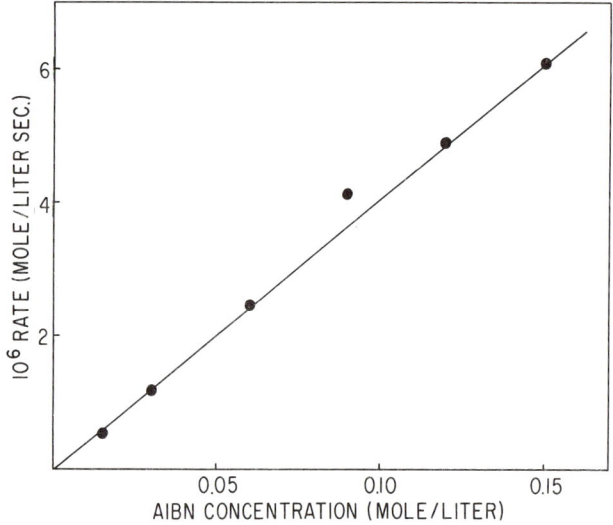

Figure 1. Initial oxidation rate of 7.2M cumene at 60°C. containing 0.02M ZnP as a function of AIBN concentration

From these results, it is clear that neither Equation A nor B represents the kinetics of the zinc diisopropyl dithiophosphate–inhibited autoxidation of cumene or Tetralin. This does not immediately indicate that the mechanism in Scheme 1 is wrong since it is highly idealized and takes no account of possible side reactions. A similar situation occurs in the inhibition of hydrocarbon autoxidation by phenols (AH), for which a basic mechanism similar to that in Scheme 1 is accepted. Termination occurs *via* Reactions 7 and 8 instead of Reactions 5 and 6.

$$RO_2\cdot + AH \rightarrow RO_2H + A\cdot \tag{7}$$

$$A\cdot + RO_2\cdot \rightarrow \text{inactive products} \tag{8}$$

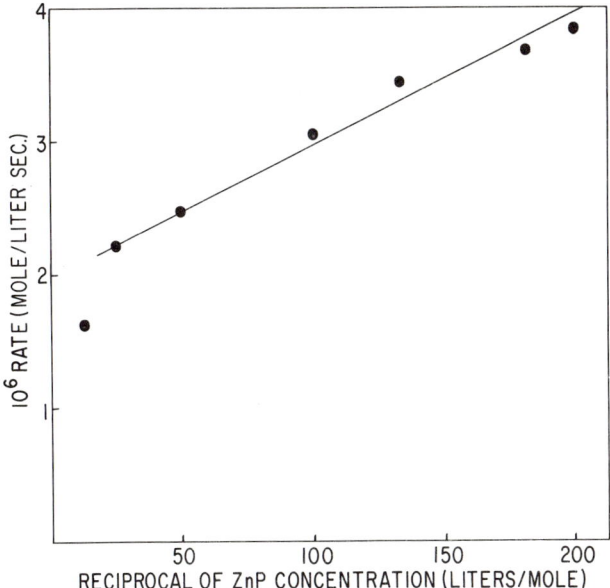

Figure 2. Initial oxidation rate of 7.2M cumene at 60°C. containing 0.06M AIBN as a function of ZnP concentration

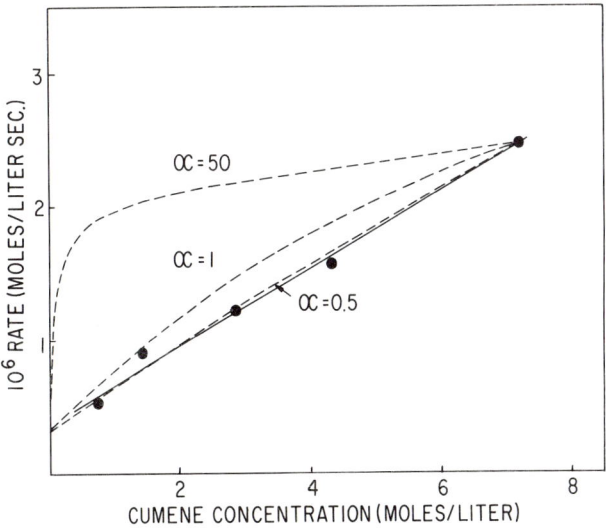

Figure 3. Initial oxidation rate as a function of concentration of cumene in tert-butylbenzene at 60°C. containing 0.06M AIBN and 0.02M ZnP

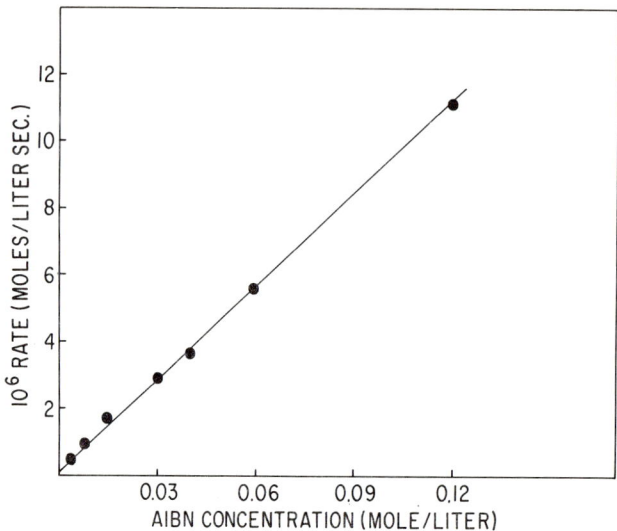

Figure 4. Initial oxidation rate of 7.25M Tetralin at 70°C. containing 0.02M ZnP as a function of AIBN concentration

Equation C (analogous to Equation A) can thus be derived for the theoretical rate of phenol-inhibited oxidation.

$$R = \frac{ek_1k_3[\text{RH}][\text{AIBN}]}{k_7[\text{AH}]} \quad (C)$$

Although Equation C often adequately describes the kinetics of hydrocarbon oxidation inhibition by phenols, more complicated kinetics are common (12), particularly for hydrocarbons which yield hydroperoxide as the primary oxidation product. The complications arise mainly from reversibility of Reaction 7 but can also result from a chain-transfer reaction:

$$\text{A}\cdot + \text{RH} \rightarrow \text{AH} + \text{R}\cdot \quad (9)$$

or from bimolecular phenoxy radical termination:

$$\text{A}\cdot + \text{A}\cdot \rightarrow \text{inactive products} \quad (10)$$

One possible problem peculiar to a quantitative study of the inhibition of oxidation of aromatic hydrocarbons by zinc dialkyl dithiophosphates is that peroxide decomposition could yield a phenol during the initial-rate measurement. Rate curves for the zinc diisopropyl dithiophosphate–inhibited oxidation of cumene are shown in Figure 7. In the initial presence of hydroperoxide the uninhibited rate is never reached, and the reaction soon exhibits autoinhibition, presumably caused by the formation of phenol. In the initial absence of hydroperoxide, on the other hand, a more clearly defined induction period is obtained, and auto-inhibition occurs only after a period during which the uninhibited rate was reached and also after a higher oxygen absorption than was necessary in the initial presence of hydroperoxide. It is unlikely therefore that phenol was formed before the end of the induction period and hence can be assumed not to affect the initial rate. Furthermore, hydroperoxide slightly increases the initial rate while phenol production would be expected to reduce it. An increase in initial rate owing to hydroperoxide was unexpected, and if the zinc salt inhibits oxidation by electron rather than hydrogen transfer, an explanation different from that put forward to account for a similar effect in phenolic inhibition—i.e., reversibility of Reaction 7—must be proposed. Possibly a coordination complex of zinc salt with hydroperoxide is formed, which cannot react with peroxy radicals.

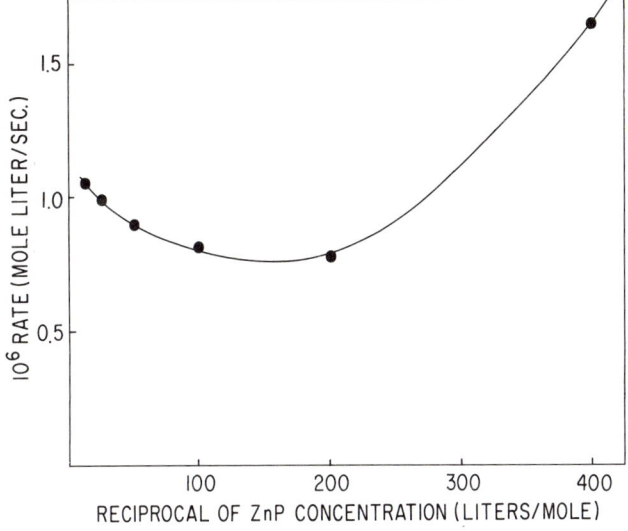

Figure 5. Initial oxidation rate of 7.25M Tetralin at 70°C. containing 0.008M AIBN as a function of ZnP concentration

The initial rates in Figures 1 to 6 were normally linear over the period required to measure them, and the effect of hydroperoxide on rate is therefore probably small in the concentration ranges used. However, this problem is being investigated in more detail. From the induction period in Figure 7, determined by the method described by Hammond and his coworkers (4), a stoichiometric factor, n, the number of peroxy radicals reacting with each zinc salt molecule, can be calculated. Values of n, shown in parentheses, for several metal diisopropyl dithiophosphates were similarly obtained: Zn (1.81), Cd (2.38), Pb (1.90), Ni (1.94), Fe (1.25), assuming in each case a rate of radical production of $2ek_1$-[AIBN] where $e = 0.60$ and $k_1 = 1.0 \times 10^{-5}$ sec.$^{-1}$ at 60°C. The value of n predicted by the postuated electron-transfer mechanism is 2, and the generally low experimental results might be interpreted as indicating a side reaction in which zinc salt is consumed in a nontermination reaction.

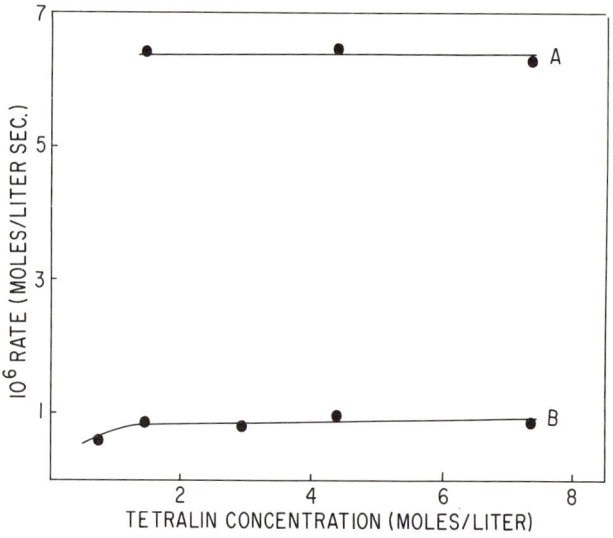

Figure 6. Initial oxidation rate as a function of concentration of Tetralin in tert-butylbenzene at 70°C. containing 0.02M ZnP

A: 0.06M AIBN
B: 0.008M AIBN

Further indication of side reactions of the zinc salt is obtained by extrapolating the graphs in Figures 3 and 6 to zero hydrocarbon concentration. An intercept on the rate axis suggests that oxygen absorption would be expected even in the absence of an oxidizable hydrocarbon. Experimental confirmation of this point has been obtained by studying the rate of oxygen absorption by a series of dithiophosphates at 70°C. in the presence of AIBN, using as solvent *tert*-butylbenzene which is inert to autoxidation under these conditions (Table I). In the absence of AIBN no oxidation occurs, and it might be argued that these results could be accounted for by an increase in the decomposition rate of AIBN or in the efficiency of radical production. The rate of decomposition of

AIBN is, however, virtually unaffected by the presence of dithiophosphates (Table II). Further, with specific reference to the oxidation of the disulfide in Table I, which has no effect on the rate of AIBN-initiated autoxidation of cumene (6), it is unlikely that the efficiency of radical production from AIBN increases since this would produce a prooxidant effect in cumene. Thus, the zinc salt inhibitor is being oxidized in competition with the main chain reaction.

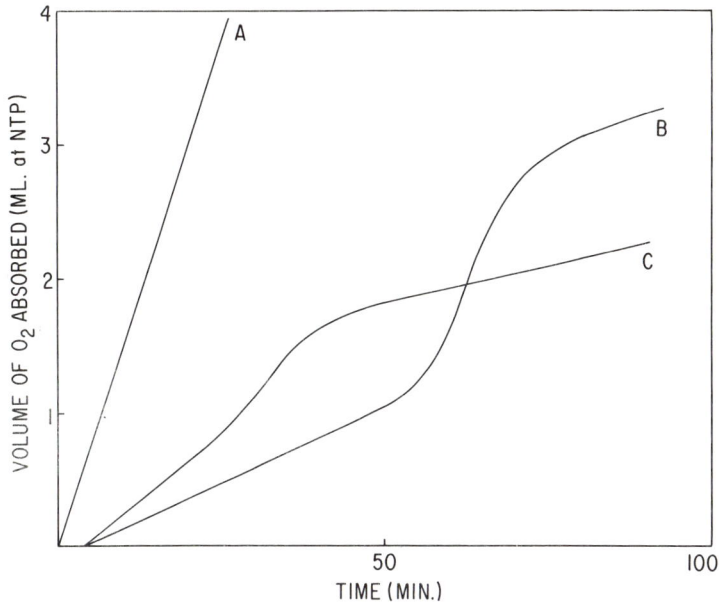

Figure 7. Oxidation of 5 ml. of cumene at 60°C. containing 0.06M AIBN

A: No inhibitor
B: Zinc diisopropyl dithiophosphate, 0.0013M
C: Zinc diisopropyl dithiophosphate, 0.0013M, plus cumene hydroperoxide, 0.017M

Because of the lack of information in the literature on the radical reactions of compounds of quinquevalent phosphorus, it is impossible to postulate a readily acceptable mechanism for the oxidation of zinc dialkyl dithiophosphates. Colclough and Cunneen (7) rejected immediately the possibility of hydrogen abstraction, but in view of the present results serious consideration has been given to this reaction. During this work it was shown (15) that abstraction of hydrogen from trialkyl phosphates, trialkyl phosphonates, and sodium dialkyl phosphates can occur at room temperature in an aqueous medium in the presence of hydroxy radicals.

Electron spin resonance spectra of the resulting phosphoroalkyl radicals were obtained. Absolute rate constants for this process are not known, and similar abstraction by a peroxy radical in nonaqueous media is not necessarily possible. However, in a zinc salt–inhibited hydrocarbon oxidation the additional reactions in Scheme 2 might be envisaged; part molecules are shown for convenience.

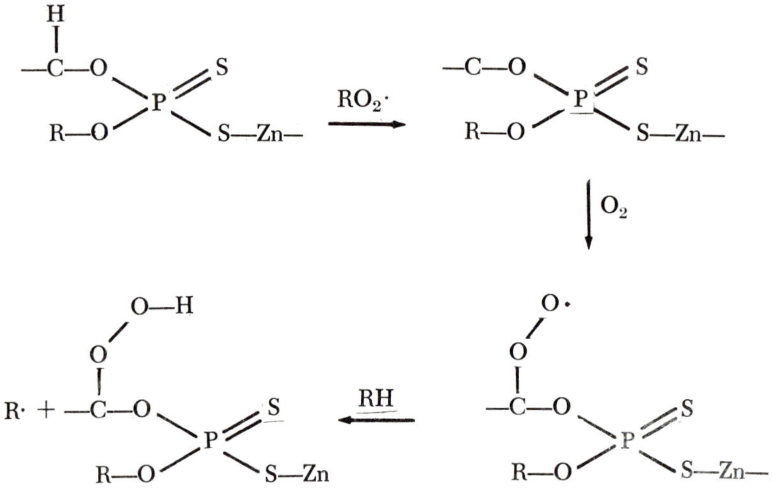

Scheme 2

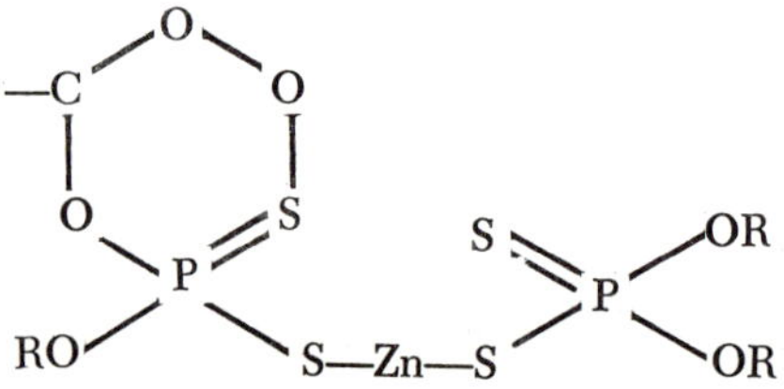

If hydrogen abstraction from the zinc salt is fast enough to compete with that from the hydrocarbon substrate, then Scheme 2 amounts to a chain-transfer reaction similar to that illustrated in Reaction 9 for phenolic inhibition. Also, an alternative termination *via* the peroxyalkyl-ester radical (X) is conceivable since this radical might be expected to cyclize to the form shown in XI, which is analogous to the radical V mentioned above. Further reaction with a peroxy radical at the remaining thionic sulfur atom would again lead to nonradical products. These reactions are summarized in Scheme 3.

$$RO_2\cdot + ZnP \rightarrow ZnP\cdot + RO_2H \quad (11)$$

$$ZnP\cdot + O_2 \rightarrow ZnPO_2\cdot \quad (12)$$

$$ZnPO_2\cdot + RH \rightarrow ZnPO_2H + R\cdot \quad (13)$$

$$ZnPO_2\cdot \rightarrow \text{complex (XI)} \quad (14)$$

$$\text{Complex (XI)} + RO_2\cdot \rightarrow \text{inactive products} \quad (15)$$

Scheme 3

Assuming that Reactions 5 and 14 are irreversible, combining Schemes 1 and 3 and applying the steady-state theory leads to a theoretical rate of oxygen absorption given by Equation D:

$$R = \frac{ek_1[\text{AIBN}]}{[\text{ZnP}]} \left[\frac{k_3[\text{RH}] + k_{11}[\text{ZnP}](1 + 2\alpha[\text{RH}])/(1 + \alpha[\text{RH}])}{k_5 + k_{11}/(1 + \alpha[\text{RH}])} \right] \quad (D)$$

where $\alpha = k_{13}/k_{14}$. If Reactions 5 and 14 are reversible and a fast equilibrium is set up in each case, then the rate of oxygen absorption is given by:

$$R = (k_3[\text{RH}] + 2k_{11}[\text{ZnP}]) \left[\frac{2ek_1[\text{AIBN}]}{[\text{ZnP}](k_5k_6/k_{-5} + k_{11}k_{14}k_{15}/k_{13}k_{-14}[\text{RH}])} \right]^{1/2}$$

Since the results in Figures 1 and 4 show that the zinc salt–inhibited oxidation is first order in AIBN, it is reasonable to rule out reversibility in Reactions 5 and 14, which demands half order in AIBN (*see also* Equation B).

In the absence of chain transfer (Reaction 13) $\alpha = 0$, and Equation D reduces to Equation E.

$$R = \frac{ek_1[\text{AIBN}]}{[\text{ZnP}]} \left[\frac{k_3[\text{RH}] + k_{11}[\text{ZnP}]}{k_5 + k_{11}} \right] \quad (E)$$

Equation E is consistent with linear dependence of oxidation rate on [AIBN], 1/[ZnP], and [RH] as shown in Figures 1, 2, and 3, respectively, but it does not adequately describe the observed kinetics since it requires that the intercepts on the rate axis in Figures 2 and 3 should be equal. Equation D cannot be similarly ruled out since it is also clearly consistent with the linear dependence of rate on [AIBN] and 1/[ZnP], while the dependence on [RH] is less obvious. Because of the illogical nature of the oxidation rate of Tetralin as a function of the zinc salt concentration (Figure 5), it is impossible to fit the Tetralin results to any of the rate equations derived so far. No complete explanation has yet been found for these results, and we will consider only the cumene results in further detail.

Table I. AIBN-Initiated Oxidation of Organic Phosphorus Compounds in *tert*-Butylbenzene at 70°C.

Compound Formula	Concn., Mole/Liter	AIBN Concn., Mole/Liter	Oxidation Rate[a] Mole/Liter Sec.
$[(Pr^iO)_2PS_2]_2Zn$	0.02	0	0
	0.02	0.03	7.4×10^{-7}
	0.02	0.06	1.5×10^{-6}
	0.09	0.06	1.3×10^{-6}
$[(EtO)_2PS_2]_2Zn$	0.02	0.06	1.5×10^{-6}
$[Et_2PS_2]_2Zn$	0.02	0.06	1.5×10^{-6}
$[(C_6H_{11}O)_2PS_2]_2Zn$	0.02	0.06	1.5×10^{-6}
$[(Pr^iO)_2PS_2]_2$	0.02	0.06	9.8×10^{-7}
$(Bu^nO)_3PO$	0.02	0.06	7.4×10^{-7}
2,6-Di-*tert*-butyl-4-methylphenol	0.02	0.06	0

[a] Not corrected for N_2 evolution from AIBN.

Table II. Rate of Thermal Decomposition of AIBN (0.3 Mole per Liter) in the Presence of Organic Phosphorus Compounds (0.1 Mole per Liter) at 70°C.

Compound Added	Solvent	Rate Constant, Sec.$^{-1}$
—	PhCl	4.6×10^{-5}
$[(Pr^iO)_2PS_2]_2Zn$		4.4×10^{-5}
$[(EtO)_2PS_2]_2Zn$		5.7×10^{-5}
$[(Pr^iO)_2PS_2]_2$		4.4×10^{-5}
$(Bu^nO)_3P=O$		4.9×10^{-5}
—	Ph-*tert*-Bu	4.1×10^{-5}
$[(Pr^iO)_2PS_2]_2Zn$		4.0×10^{-5}
$[(EtO)_2PS_2]_2Zn$		3.8×10^{-5}

From Equation D, the slope (β_1) and the intercept (β_2) of a plot of rate *vs.* 1/[ZnP] are given by:

$$\beta_1 = \frac{ek_1k_3[\text{AIBN}][\text{RH}]}{k_5 + k_{11}/(1 + \alpha[\text{RH}])} \tag{F}$$

and

$$\beta_2 = \frac{ek_1k_{11}[\text{AIBN}](1 + 2\alpha[\text{RH}])/(1 + \alpha[\text{RH}])}{k_5 + k_{11}/(1 + \alpha[\text{RH}])} \tag{G}$$

Values of $\beta_1 = 1.0 \times 10^{-8}$ and $\beta_2 = 2.0 \times 10^{-6}$ can be derived from Figure 2, the data in which were obtained using fixed concentrations of AIBN = 6.0×10^{-2} mole per liter and cumene = 7.2 moles per liter. Thus, using values of $e = 0.60$, $k_1 = 1.0 \times 10^{-5}$ sec.$^{-1}$, $k_3 = 0.5$ mole per liter sec., and any arbitrary value of α, constants k_5 and k_{11} can be evaluated from the simultaneous Equations F and G. By knowing e, k_1, k_3, k_5, k_{11} and α, the variation of oxidation rate with [RH] at fixed concentrations of ZnP = 2.0×10^{-2} mole per liter and AIBN = 6.0×10^{-2} mole per liter can be derived from Equation D. As expected, a nonlinear dependence is obtained, but it can readily be shown that at a value of $\alpha = 0.5$, a very close fit to the observed data is obtained (*see* broken lines in Figure 3). When $\alpha = 0.5$, values of $k_5 = 42$ l. mole^{-1} sec.$^{-1}$, and $k_{11} = 404$ l. mole^{-1} sec.$^{-1}$ are obtained, and from these the slope of the plot of oxidation rate *vs.* [AIBN], at fixed concentrations of ZnP = 2.0×10^{-2} mole per liter and cumene = 7.2 moles per liter can be calculated as 41.7×10^{-6}. The experimental value (*see* Figure 1) is 40.6×10^{-6}.

Equation D is thus shown, by choice of a single independent variable α, to represent closely the observed kinetics of zinc diisopropyl dithiophosphate–inhibited oxidation of cumene. The value $\alpha = 0.5$ is reasonable since it requires only that the rate of chain transfer (Reaction 13) be of the same order as termination by Reactions 14 and 15. We have no independent evidence that hydrogen abstraction from zinc dialkyl dithiophosphates by peroxy radicals is possible, but otherwise Scheme 3 seems acceptable. It is unfortunate that the thermal instability of the di-*tert*-butyl derivative, the hydrolytic instability of the diphenyl derivative (which gives phenol), and the total insolubility of $[Ph_2PS_2]_2Zn$ precluded the use of these compounds in this kinetic study. Perhaps oxidation of the zinc salt follows a mechanism different from that in Scheme 3 but that the over-all reaction is kinetically similar. The results using Tetralin also strongly suggest that the kinetic complexities may depend on a reaction not yet considered in detail, such as coordination with hydroperoxide, now being investigated.

Peroxide Decomposition Mechanism. Since virtually no work has been reported which concerns only the mechanism by which zinc dialkyl dithiophosphates act as peroxide decomposers, it is pertinent to discuss metal dialkyl dithiophosphates as a whole. The mechanism has been studied both by investigating the products and the decomposition rates of hydroperoxides in the presence of metal dithiophosphates and by measuring the efficiency of these compounds as antioxidants in hydrocarbon autoxidation systems in which hydroperoxide initiation is significant.

In our own work no direct attempt has yet been made to adopt the former approach; however, some relevant qualitative information has emerged from the kinetic study. The autoinhibitive type of behavior exhibited in the zinc dialkyl dithiophosphate–inhibited oxidation of cumene, referred to above (Figure 7) and presumed to be caused by phenol formation, has been investigated further. Using several metal salts the phenomenon of autoinhibition has been almost universally observed for this system (Figure 8). Differences in degree depending on the metal are noticeable, and of particular significance is the complete absence of autoinhibition for the potassium salt. This compound therefore probably does not decompose cumene hydroperoxide to give phenol and may not even be a peroxide decomposer. The effect of the initial presence of hydroperoxide can be seen (Figure 9) for the inhibited oxidation of cumene which was used without prior percolation through silica gel to remove hydroperoxide. Inhibition by salts of the transition metals iron and nickel is particularly affected. For ferric iron no inhibition is observed until the autoinhibition phase begins. Assuming that autoinhibition in this system is caused by phenol production by hydroperoxide decomposition, the nature of the metal appears to be important in this process.

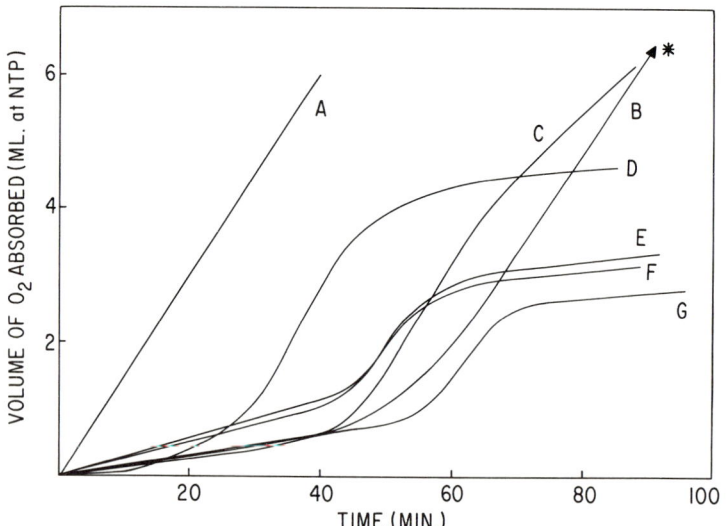

Figure 8. Oxidation of 5 ml. of cumene at 60°C. containing 0.06M AIBN

* Still linear up to 12 ml. of O_2 absorbed
A: No inhibitor
B: Potassium di(4-methyl-2-pentyl)dithiophosphate, 0.005M
C, D, E, F, G: 0.001M lead, ferric, nickel, zinc, and cadmium diisopropyl dithiophosphates, respectively

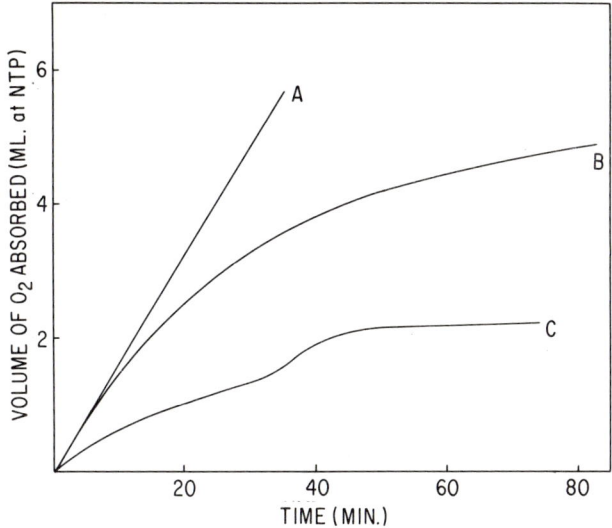

Figure 9. Oxidation of peroxidized cumene 5 ml. at 60°C. containing 0.06M AIBN

A: No inhibitor
B: Ferric diisopropyl dithiophosphate, 0.001M
C: Nickel diisopropyl dithiophosphate, 0.001M

We have carried out a limited study of the effect of metal dialkyl dithiophosphates on a hydroperoxide-autocatalyzed oxidation system. Table III summarizes induction periods for the oxidation of squalane at 140°C. These results do not unambiguously reflect the peroxide-decomposing property of each dithiophosphate; radical capture also occurs.

The basic importance of the metal in the compounds tested is once more emphasized by the lack of oxidation inhibition in the presence of the disulfides $[(RO)_2PS_2]_2$. Also, the pro-oxidant effect of nickel and iron dithiophosphates shows that under appropriate conditions, radical rather than the reported (11) ionic decomposition can occur. Regarding the effect of alkyl group structure in the metal salts on the induction period, general agreement has been found with the results of Larson (14), who investigated the inhibition of oxidation of a refined white mineral oil at 150°C. by zinc dialkyl dithiophosphates—i.e., for a given metal, secondary alkyl dithiophosphates are more effective antioxidants than those containing primary alkyl groups. In agreement with others (11, 13), Larson considered that a reaction product of the zinc salt, rather than the original compound, is the active antioxidant. Thus, noting also that the secondary alkyl zinc salts are thermally less stable than the primary alkyl compounds, he suggested that the active peroxide decomposer is a thermal decomposition product. This product is presumably acidic, formed in a β-elimination reaction from the zinc salt, and the effect of alkyl group structure on the rate of this reaction has been reported more recently (18). Although an acidic thermal decomposition product would account (10) for the products of the zinc salt–catalyzed decomposition, it is unlikely to be the effective peroxide-decomposing species for three reasons. Peroxide decomposition by zinc salts has been observed at room temperature (7, 11) when their thermal decomposition could not be responsible. The disulfide, $[(Pr\text{-}i\text{-}O)_2PS_2]_2$, has practically the same thermal stability (18) as zinc diisopropyl dithiophosphate, but compounds of the disulfide type do not inhibit squalane oxidation. Finally, the dithiophosphinate, $[Et_2PS_2]_2Zn$, which cannot undergo β-elimination (because it is not an ester) still strongly inhibits the oxidation of squalane (Table III).

Table III. Effect of Metal Dialkyl Dithiophosphates, $[(RO)_2PS_2]_xM$, (at 4×10^{-5} gram atoms of Phosphorus per liter) on the Oxidation of Squalane at 140°C.

Metal (M)	Alkyl Group (R)	X	Induction Period, Min.
Zinc	4-Methyl-2-pentyl		119 ± 9
	n-Hexyl		67 ± 5
	Isopropyl	2	80 ± 12
	n-Propyl		65 ± 10
	Ethyl[a]		70
Antimony	4-Methyl-2-pentyl	3	95 ± 11
Cadmium	4-Methyl-2-pentyl	2	79 ± 5
	n-Hexyl		30 ± 2
Lead	4-Methyl-2-pentyl	2	67 ± 6
	n-Hexyl		25 ± 0.5
Bismuth	4-Methyl-2-pentyl	3	45 ± 0.5
	n-Hexyl		49 ± 2
Iron	4-Methyl-2-pentyl	3	2.5
Nickel	4-Methyl-2-pentyl	2	2.5
	n-Hexyl		2.5
	Isopropyl		2.5
None (disulfide)	4-Methyl-2-pentyl		5 ± 0.5
	n-Hexyl	2	4 ± 0.5
	Isopropyl		4 ± 0.5
None	None	–	4 ± 0.5

[a] For $(Et_2PS_2)_2Zn$ (5×10^{-4} mole/liter) at 160°C.

Kennerly and Patterson (*13*) studied the effect of several organic sulfur compounds, including thiols, sulfides, a disulfide, sulfonic acids, and a zinc dialkyl dithiophosphate, on the decomposition rate of cumene hydroperoxide in white mineral oil at 150°C. In each case they found phenol as the major product. They suggested that the most attractive mechanism by which to explain these results involves ionic rearrangement catalyzed by acids or other electrophilic reagents (*10*) as

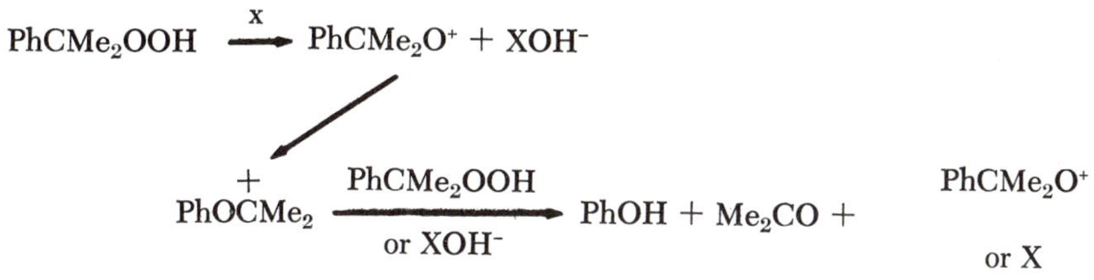

$$PhCMe_2OOH \xrightarrow{X} PhCMe_2O^+ + XOH^-$$

$$PhOCMe_2 + \xrightarrow[\text{or XOH}^-]{PhCMe_2OOH} PhOH + Me_2CO + PhCMe_2O^+ \text{ or } X$$

Scheme 4

in Scheme 4. Since no obvious process existed for converting the sulfur compounds tested to strong acids which was compatible with the observed structure-activity relationship, they further concluded that an electrophilic species other than H^+ must be involved. However, no such species derived from the zinc dialkyl dithiophosphate was defined. In a kinetic study (*13*), first-order dependence on both hydroperoxide and sulfur compound was observed, and approximate rate constants were derived from the following expression:

$$\frac{-dP}{dt} = k[D][P] \qquad (H)$$

where D is the peroxide decomposer, and P is the hydroperoxide. Incorporating the following simplified peroxide decomposition reactions,

$$P + D \rightarrow Y + D'$$
$$P + D' \rightarrow Z + D'$$
$$P + D' \rightarrow \text{inactive products}$$

where D' is the electrophilic active decomposer derived from D, into the usual hydrocarbon autoxidation mechanism and using Equation H, they derived an expression for the induction period (t_{ind}) of the form shown in Equation I

$$t_{ind} = KD_o^2 \qquad (I)$$

where D_o is the initial concentration of the peroxide decomposer. In this derivation it was assumed, to a first approximation, that the peroxide decomposer D did not interfere with the normal oxidation chains—an assumption now known to be unacceptable for a zinc dialkyl dithiophosphate. However, Equation I was verified experimentally, although only two different induction periods were cited for the oxidation of mineral oil at 155°C. containing zinc di(4-methyl-2-pentyl)dithiophosphate over a narrow concentration range. We have been unable to confirm this equation and find instead that for the oxidation of squalane at 160°C. in the presence of zinc diisopropyl dithiophosphate, $t_{ind} = KD_o^{0.8}$ as shown in Figure 10.

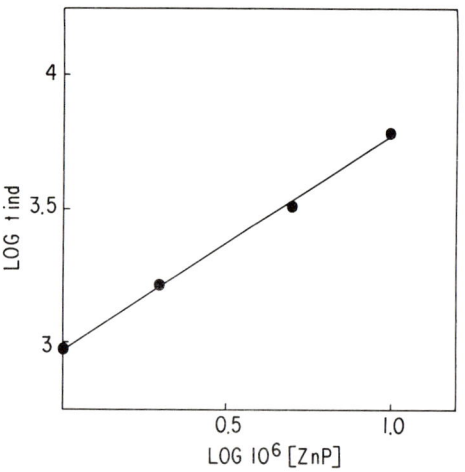

Figure 10. Induction period for squalane oxidation at 160°C. as a function of ZnP concentration

Shopov and his co-workers have recently published two papers on hydroperoxide decomposition by barium dialkyl dithiophosphates. The decomposition rate of cumene hydroperoxide at 140°C. in the presence of barium dibenzyl dithiophosphate was found (20) not to be described by Equation H. A mechanism, similar to that of Kennerly and Patterson (13) but slightly more detailed was proposed as follows:

$$ROOH + D \rightarrow RO\cdot + OH^- + D^{\cdot+}$$
$$ROOH + D^{\cdot+} \rightarrow RO_2\cdot + H^+ + D$$
$$RO_2\cdot + R'H \rightarrow ROOH + R'\cdot$$
$$RO_2\cdot + D^{\cdot+} \rightarrow \text{inactive products}$$

Scheme 5

where D is the original peroxide decomposer, and R'H is the hydrocarbon solvent. This mechanism suffers from the obvious disadvantage of producing chain-carrying radicals at a rate presumably greater than that from thermal decomposition; thus:

$$2ROOH \rightarrow RO\cdot + RO_2\cdot + H_2O$$

and although the peroxide-decomposing intermediate is also postulated to capture radicals, it is difficult to believe that Scheme 5 would account for oxidation inhibition. Nevertheless, from this scheme an expression for the rate of decomposition of hydroperoxide of the form shown in Equation J was derived:

$$\ln \frac{[ROOH]_o}{[ROOH]_t} = k_1[D]t + k_2 t^2 \qquad (J)$$

which was shown to fit their experimental results. Because of the thermal instability of barium dibenzyl dithiophosphate at 140°C. mentioned in this paper, it is doubtful, however, that these results can represent unambiguously peroxide decomposition by the barium dibenzyl salt itself or that other metal dialkyl dithiophosphates would behave similarly. Shopov (21) has also investigated the effect of barium salts on the decomposition of cumene hydroperoxide at 90°C. in liquid paraffin. The rates (W) were considered as the sum of the thermal decomposition, the dithiophosphate-induced decomposition, and the radical-induced decomposition as in Equation K.

$$W = W_T + W_D + W_{RO_2}\cdot \qquad (K)$$

By studying the reaction in the presence and absence of a radical acceptor, 2,6-di-*tert*-butyl-4-methylphenol, radical-induced decomposition was selectively eliminated and the separate terms in Equation K were evaluated. Using barium dibenzyl dithiophosphate the following ratio was obtained:

$$W_T : W_D : W_{RO_2}\cdot = 1 : 21.2 : 1940$$

Hence, radical-induced decomposition predominates, rather surprisingly since this will not account for high yields of phenol obtained using zinc salts, assuming that these compounds behave similarly. Some confusion also arises in deriving these figures by using maximum rates and initial rates. In the absence of radical acceptor and dithiophosphate a measure of $W_T + W_{RO_2}\cdot$ is obtained. This is given as a maximum rate after one hour and is approximately 40 times the initial rate. Comparing this maximum rate with initial rates measured over a few minutes in other cases in the presence of inhibitors gives undue bias to the importance of radical-induced decomposition.

The catalytic nature of the action of metal dialkyl dithiophosphates in the decomposition of cumene hydroperoxide at room temperature has been clearly shown by Holdsworth, Scott, and Williams (11). They investigated the effect of a metal xanthate, metal dithiocarbamates, and some nonmetallic sulfur compounds as well as metal dithiophosphates. These compounds, at a concentration of 1 mole % of that of the cumene hydroperoxide in benzene, caused its complete decomposition, giving high yields of phenol and acetone. Zinc dinonyl and diethyl dithiocarbamates also readily gave sulfur dioxide as a product. Since sulfur dioxide was also shown to catalyze the decomposition of cumene hydroperoxide to phenol and acetone, it was suggested that the sulfur compounds tested represent a potential reservoir of sulfur dioxide and that this is the active peroxide decomposer. However, no evidence was advanced to suggest either that dithiocarbamates yield sulfur dioxide in the presence of the other hydroperoxides mentioned in the same paper, or in particular that any metal dialkyl dithiophosphate correspondingly yielded sulfur dioxide. It is almost certainly a mistake to assume that the active intermediates involved in hydroperoxide decomposition by sulfur compounds are identical even for such formally similar compounds as dithiophosphates and dithiocarbamates.

No readily acceptable mechanism has been advanced in reasonable detail to account for the decomposition of hydroperoxides by metal dialkyl dithiophosphates. Our limited results on the antioxidant efficiency of these compounds indicate that the metal plays an important role in the mechanism. So far it seems, at least for the catalytic decomposition of cumene hydroperoxide on which practically all the work has been done, that the mechanism involves electrophilic attack and rearrangement as shown in Scheme 4. This requires, as commonly proposed, that the dithiophosphate is first converted to an active form. It does seem possible, on the other hand, that the original dithiophosphate could catalyze peroxide decomposition since nucleophilic attack could, in principle, lead to the same chain-carrying intermediate as in Scheme 4; thus,

$$PhCMe_2OOH + X \rightarrow PhCMe_2O\overset{+}{X} + OH^-$$
$$\downarrow$$
$$PhO\overset{+}{C}Me_2 \leftarrow PhCMe_2\overset{+}{O} + X$$

Nucleophilic attack has been shown (8), for phosphines, to occur at the hydroxy, not alkoxy, oxygen. However, a possibility with a metal dithiophosphate is that formation of a coordination complex occurs and that subsequent nucleophilic attack on the alkoxy oxygen involves a six-membered transition state.

Competing reactions can also be expected to result ultimately in oxidation of the zinc salt to other products such as sulfate, with corresponding loss of catalytic activity.

Acknowledgment

The author thanks J. F. Ford for many useful discussions and R. C. Palmer for his part in the derivation and application of oxidation rate equations. Permission to publish this paper has been given by The British Petroleum Co., Ltd.

Literature Cited

(1) "Autoxidation and Antioxidants," W. O. Lundberg, ed., Vol. I, Interscience, London, 1961.
(2) Bacon, W. E., LeSuer, W. M., *J. Am. Chem. Soc.* **76**, 670 (1954).
(3) Boozer, C. E., Hammond, G. S., *Ibid.* **76**, 3861 (1954).
(4) Boozer, C. E., Hammond, G. S., Hamilton, C. E., Sen, J. N., *Ibid.* **77**, 3233 (1955).
(5) *Ibid.*, p. 3238.
(6) Burn, A. J., *Tetrahedron* **22**, 2153 (1966).
(7) Colclough, T., Cunneen, J. I., *J. Chem. Soc.* **1964**, 4790.
(8) Denney, D. B., Goodyear, W. F., Goldstein, B., *J. Am. Chem. Soc.* **82**, 1393 (1960).
(9) Ford, J. F., Pitkethly, R. C., Young, V. O., *Tetrahedron* **4**, 325 (1958).
(10) Hawkins, E. G. E., "Organic Peroxides," p. 90, Spon, London, 1961.
(11) Holdsworth, J. D., Scott, G., Williams, D., *J. Chem. Soc.* **1964**, 4692.
(12) Howard, J. A., Ingold, K. U., *Can. J. Chem.* **42**, 2324 (1964).
(13) Kennerly, G. W., Patterson, W. L., *Ind. Eng. Chem.* **48**, 1917 (1956).
(14) Larson, R., *Sci. Lubrication* **10**, 12 (August 1958).
(15) Lucken, E. A. C., *J. Chem. Soc.* **1966**, 1354.
(16) Oswald, A. A., Griesbaum, K., Hudson, B. E., Division of Petroleum Chemistry Preprints, Vol. 8 (1), 5, 144th ACS Meeting, Los Angeles, March, April 1963.
(17) Oswald, A. A., Griesbaum, K., Naegele, W., *J. Am. Chem. Soc.* **86**, 3791 (1964).
(18) Rowe, C. N., Dickert, J. J., American Chemical Society Symposium, Detroit **10** (2), D-71 (1965).
(19) Scott, G., "Atmospheric Oxidation and Antioxidants," Elsevier, London, 1965.
(20) Shopov, D., Ivanov, S. K., *Neftekhimiya* **5**, (3), 410 (1965).
(21) Shopov, D., Ivanov, S. K., Kateva, J., *Erdol Kohle* **19**, 732 (1966).
(22) Thaler, W. A., Oswald, A. A., Hudson, B. E., *J. Am. Chem. Soc.* **87**, 311 (1965).
(23) Young, V. O., *Chem. Ind., London* **1967**, 658.

RECEIVED November 6, 1967.

26

A New Method for Determining the Absolute Rate Constants of Autoxidation of Some Hydrocarbons

H. BERGER, A. M. W. BLAAUW, M. M. AL, and P. SMAEL

Koninklijke/Shell-Laboratorium, Amsterdam, The Netherlands

> *When a slow steady-state autoxidation of a suitable hydrocarbon is disturbed by adding either a small amount of inhibitor or initiator, a new stationary state is established in a short time. The change in velocity during the non-steady state can be followed with sensitive manometric apparatus. With the aid of integrated equations describing the non-steady state the individual rate constants of the autoxidation reaction can be derived from the results. Scope and limitations of this method are discussed. Results obtained for cumene, cyclohexene, and Tetralin agree with literature data.*

The autoxidation at moderate temperatures of hydrocarbons, including inhibition by a hindered phenol, for example, is generally described by the following mechanism:

			Rate
Initiation	$\rightarrow$ R·		R_i
Propagation	R· + O_2 $\rightarrow$ ROO·		Fast
	ROO· + RH $\rightarrow$ R· + ROOH		$k_p[\text{RH}][\text{RO}_2\cdot] = \dfrac{dO_2}{dt}$
Termination	$2\text{RO}_2\cdot \rightarrow$ Non-radical products		$2k_t[\text{RO}_2\cdot]^2$
Inhibition	$\text{RO}_2\cdot + \text{AH} \rightarrow \text{ROOH} + \text{A}\cdot$ A· + $\text{RO}_2\cdot \rightarrow$ AOOR	$\Big\}$	$2k_A[\text{AH}][\text{RO}_2\cdot]$ Fast

(We adhere to the convention, also adopted by Walling (5) that rate constants of reactions which consume two radicals are written as $2k$. To avoid confusion, the factor 2 is retained in Table I and in the equations.)

Only ratios of rate constants of autoxidation can be determined by ordinary steady-state velocity measurements. Non-steady state methods, notably the sector method, have been used to evaluate the individual or absolute rate constants. We have devised a method for measuring the change in rate of autoxidation under non-stationary conditions directly, by means of sensitive manometric apparatus, and for deriving individual rate constants from these measurements.

In our method reproducible non-stationary states are effected as follows: the low stationary-state rate of an autoxidizing hydrocarbon is decreased by a factor of 2 to 5 by adding an appropriately small amount of inhibitor. Under the conditions outlined below, the time required to establish the new stationary state at the inhibited rate is not immeasurably small, as it would be in conventional measurements, but of the order of 100–300 sec. With sufficiently sensitive apparatus a number of determinations of the decreasing velocity can be made, which delineate the course of the non-steady state. Similarly a non-steady state with an increasing velocity can be realized by introducing a small amount of initiator.

Experimental

Apparatus and Procedure. The reaction vessel is a cylindrical borosilicate glass flask with ground glass joint, height 30 mm., diameter 75 mm. (Figure 1). It is equipped with four side baffles projecting 15 mm. into the flask with a 45° slant upward, relative to the direction of stirring. Connections to the pressure transducer are made first by means of a solid ground glass stopper with a wide (5 mm.) channel, so as to avoid blocking by liquid drops, and further with capillary tubing, submerged as far as possible in the thermostat bath. Also provided are capillary side connections to a conventional gas buret (closed during non-steady state measurements), to a well-lagged Metrohm piston buret, equipped with a step-geared synchronous motor, and to a closely fitting rubber serum cap which allows insertion of the 250-mm. needle of a 50 μliter Hamilton syringe down to a few millimeters over the magnetic stirring bar. The magnetic stirring motor is fed from a constant-voltage transformer since varying the stirring speed varies pressure, presumably owing to changes in centrifugal pressure on the gas bubbles. The thermostat is a lagged glass basin, the water temperature being regulated by means of a Thermotrol controller and an efficient stirrer. Temperature constancy in our runs was 0.002°C. (range 20–50°C.).

The differential pressure transducer is a model PID-0.1PSID with model CD11 carrier demodulator (Pace Engineering Co., North Hollywood, Calif.). It is operated at 0.4 maximum sensitivity and is connected to a standard potentiometric recorder using the 200-mv. range. The reference side of the transducer is connected to an empty flask submerged in the thermostat (Figure 2).

The reaction vessel is filled with hydrocarbon, leaving only enough gas volume to allow formation of a vortex down to the stirrer, so that gas bubbles are swept through the liquid in both horizontal and—because of the 45° baffles—vertical swirls, enabling a rapid exchange of gas between vortex and bubbles. The volume of the vortex and the required length of the flask's neck determine the volume of the gas space (*ca.* 20 ml.), including capillary lines and Metrohm buret. A rate of oxygen uptake of 0.002 ml./min. produces a 20-mv. change of output in 20 sec. The "noise" from the stirred liquid determines the limit of sensitivity; the latter was influenced favorably also by the baffles which contribute to a smooth pattern of flow.

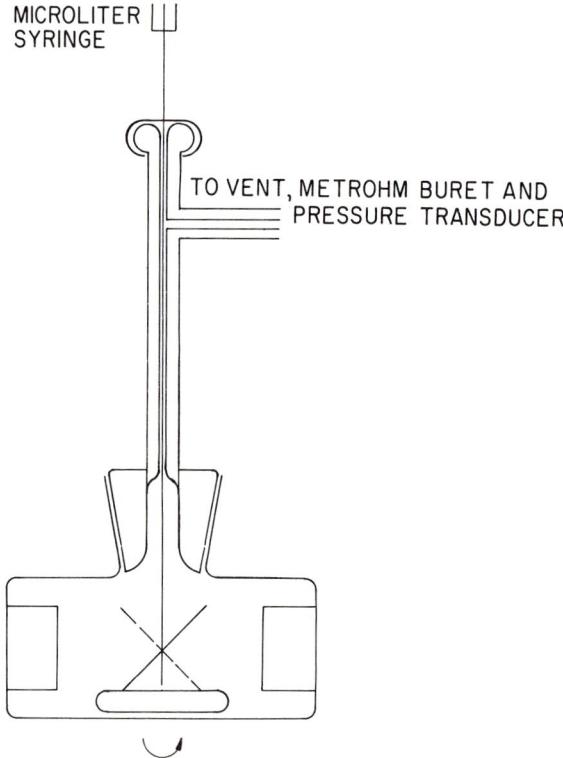

Figure 1. Reaction vessel with 45° baffles

During measurement of the reaction rates the pressure drop is recorded in 20-mv. steps, the original pressure being rapidly re-established by the Metrohm buret. In this way one measures under essentially constant pressure; this is necessary since otherwise the change in equilibrium oxygen content of the solution would interfere with the measurement. The setup is calibrated by allowing the Metrohm buret (with geared-down motor) to introduce oxygen at a known rate of the same order of magnitude as that which is being measured. From the apparent rate so measured one obtains a factor for the conversion from mv./sec. to ml.

O_2/sec. If rates do not come near the limit of sensitivity a narrow gas buret may also be used, but this is more time-consuming.

To measure a reaction in its non-steady state, the injection needle is mounted through the serum cap, the syringe being filled with the required amount of inhibitor or 2,2'-azoisobutyronitrile, dissolved in the hydrocarbon. The initial steady state velocity is measured, reference and reaction vessel are vented, the reference closed, the syringe emptied, and the reaction vessel closed. Within 5 sec. measurement can be resumed; three to ten 20-mv. steps of the non-steady state velocity can be obtained, depending on the duration of the non-steady state and the velocities involved. The final velocity is also measured. From the results, graphs of velocity vs. time are drawn as well as graphs of the appropriate logarithmic expression, whose slope is converted to the absolute rate constant required.

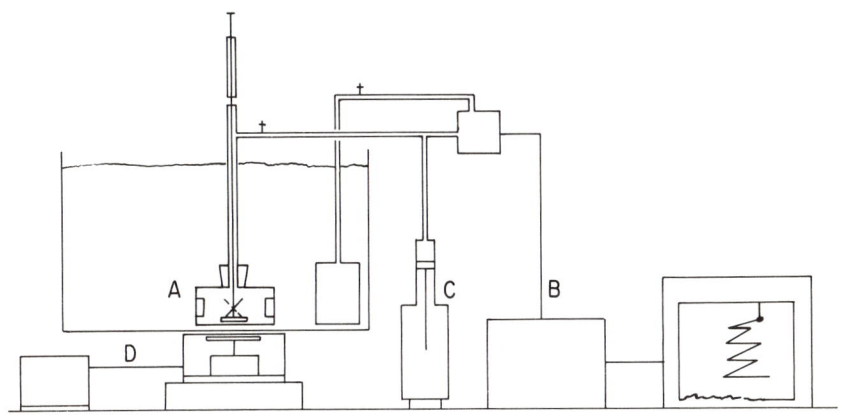

Figure 2. Apparatus for non-steady state autoxidation measurements

A: *Reaction vessel with syringe*
B: *Differential pressure transducer to recorder*
C: *Piston buret (synchronous motor)*
D: *Magnetic stirrer (constant-voltage transformer)*

The rate constant for physical gas absorption γ_1 (*see* Appendix) was determined by measuring the logarithmic decrease in rate of oxygen uptake by a hydrocarbon slightly unsaturated with respect to oxygen (not taking up oxygen chemically). It was found that γ_1 values as high as 0.08 sec.$^{-1}$ could be obtained.

Starting Materials. Cumene and Tetralin were washed with sulfuric acid, aqueous sodium hydroxide, and water, dried over magnesium sulfate and over sodium, and fractionated from sodium. Cyclohexene (Eastman Kodak grade) was distilled only. All hydrocarbons were stored under nitrogen at −15°C. and percolated over alumina before use. Rate constant ratios obtained on these hydrocarbons from conventional measurements are listed in Table I.

Table I. Absolute Rate Constants

Hydrocarbon ([RH])	T, °C.	$\Delta R_i \times 10^{-11}$ Mole/Liter/ Sec.	$[AH] \times 10^{-7}$ Mole/Liter	v_0 10^{-8} Mole/Liter/ Sec.	v_∞
Cumene (7.2)	40	4.7	—	1.70	5.25
		4.7	5	6.56	0.25
			—	0.75	1.55
			4	3.90	0.37
			4	9.18	1.29
			4	8.74	1.22
Cyclohexene (10)	30		2	2.70	1.05
			4	3.34	1.15
			2	2.34	0.97
Tetralin (7.34)	40		5	11.7	4.97
			5	11.5	0.02
			5	5.90	1.10

[a] Determined according to non-steady state Equations 1 or 2.
[b] From k_p and $k_p/(2k_t)^{1/2}$.
[c] From k_p and $2k_A/k_p$.
[d] Determined in conventional rate measurements.

2,2′-Azoisobutyronitrile (AIBN) (purchased from Fluka) was recrystallized from ether, m.p., 103°C. 2,6-Di-*tert*-butyl-*p*-cresol (Ionol) was recrystallized from ethanol, m.p., 71°C.

Principles of Measuring

The steady state condition for the concentration of $RO_2\cdot$ radical in an autoxidation is

$$\frac{d[RO_2\cdot]}{dt} = 0 = R_i - 2k_t[RO_2]^2 \qquad (1)$$

When a small amount of an inhibitor, AH, is added, $[RO_2\cdot]$ decreases according to

$$0 \neq \frac{d[RO_2\cdot]}{dt} = R_i - 2k_t[RO_2\cdot]^2 - 2k_A[AH][RO_2\cdot] \qquad (2)$$

of Hydrocarbon Autoxidation

Abs. Rate Constant Liter/Mole/Sec.			Rate Constant Ratios[a]	
k_p[a]	$2k_t \times 10^{-6}$[b]	$2k_A \times 10^{-4}$[c]	$\dfrac{k_p}{(2k_t)^{1/2}} \times 10^{-3}$	$\dfrac{2k_A}{k_p} \times 10^{-3}$
0.42	0.02[5]	2.4		
0.44	0.03	2.5		
0.50	0.03[6]	2.9		
0.40	0.02	2.3	2.65	57.5
0.37	0.02	2.15		
0.33	0.01[6]	1.90		
av. 0.41	0.024	2.4		
(0.5 at 50°C.)[e,g]	(0.03 at 50°C.)[e,g]			
6.65	2.6	6.05		
5.95	2.1	5.4	1.30	9.10
7.6	3.4	6.9		
av. 6.7	2.65	6.1		
(6.1)[f]	(5.6)[f]	(2.7)[f]		
5.7	4.0	3.4	3.0	6.0
11.3	14	6.6		
10.7	13	6.6		
av. 9.2	9.5	5.5		
(8.5)[f]	(10.0)[f]	(5)[f,g]		

[e] From Melville and Richards (3).
[f] From Howard and Ingold (2).
[g] Not directly comparable.

Since at $t = 0$, $R_i - 2k_t[RO_2·]^2 = 0$, the initial relative decrease of $[RO_2·]$ is

$$\frac{1}{[RO_2·]_o} \frac{d[RO_2·]_o}{dt} = -2k_A[AH]. \qquad (3)$$

For experimental reasons (discussed below) the relative rate of decrease should be smaller than 0.01 sec.$^{-1}$. Since k_A is of the order of 10^4 liter per mole per sec., [AH] must be of the order of 10^{-6} mole per liter. This low concentration of inhibitor will only have an appreciable effect on the rate of autoxidation if the rate of initiation—and consequently the rate of oxidation—is low. To measure the decrease in rate during the short-lived non-steady state, one must be able to determine these low velocities within short periods of time. From the usual inhibition formulas one can compute, for instance, that in order to obtain a ratio of original to inhibited rate of about 5 with [AH] = 10^{-6} mole per liter, one requires an initiation rate of $\sim 10^{-11}$ mole per liter per sec., resulting in a rate of oxygen consumption of ~ 0.01 ml. per min. for an 80-ml. reaction volume.

By restricting the gas space over the reaction mixture and by using a sensitive pressure transducer, it was possible to measure rates of $\sim 2.10^{-8}$ mole per liter per sec. of O_2 (*i.e.*, 0.002 ml. O_2 per min. in an 80-ml. reaction volume) within 20-sec. intervals, and this is sufficient for measuring a number of velocities in non-steady states of 100–300-sec. duration.

A prerequisite for the measurement is that the rate of stirring should be sufficiently high to ensure that the rate constant of physical gas absorption does not become limiting; this problem is dealt with in the appendix. This condition is not to be confused with that applying at high rates—namely, that the physical rate of absorption should not become limiting; here we have the more stringent requirement that a rate of change in oxygen consumption of 0.01 sec.$^{-1}$ can be followed without physical limitations. The required stirring efficiency was obtained with a baffled reaction vessel described above.

With the low rates of initiation and oxidation used, the difficulties usually encountered in attempts to obtain reproducible data from radical reactions are aggravated. The main problem is that with hydrocarbons showing the expected linear dependence of the oxidation rate on $R_i^{1/2}$ in the usual range of R_i (10^{-8}), deviations occurred at R_i 10^{-10} owing to either spontaneous initiation or spurious inhibition or both. The difficulties arising from these effects were circumvented by incorporating the latter into the integrated formulas describing the non-steady state. It is necessary to assume then that the inhibition can be described as a first-order termination: $\dfrac{d[RO_2\cdot]}{dt} = -X[RO_2\cdot]$ with X a constant during the 500–1000-sec. span of the measurement.

Adding inhibitor or initiator to create the non-steady state also introduces a physical disturbance of the system, owing to differences in temperature and/or in content of dissolved gas. These effects were minimized by adding microliter quantities of solutions only.

Results

Originally we attempted to determine initial slopes of $\dfrac{dO_2}{dt}$ vs. time graphically, but this procedure (though feasible) was abandoned in favor of using integrated formulas which allow determination of the rate constants from the slope of a straight line. Since the low steady state velocities measured often deviated from the theoretically expected values, we finally used formulas which accounted for both spontaneous initiation and

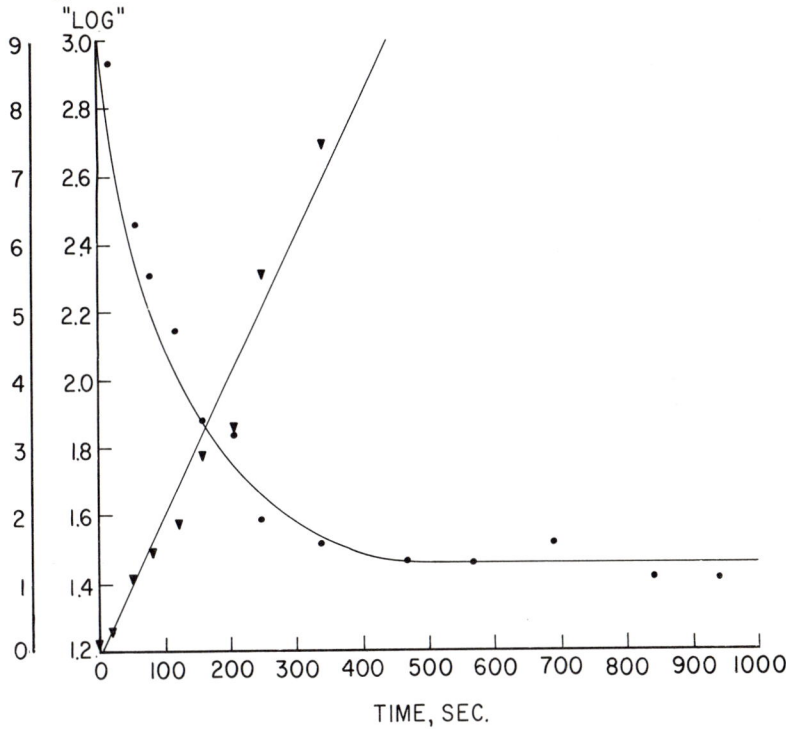

Figure 3. Autoxidation of cumene at 40°C.

Non-steady state after adding 4 × 10⁻⁷M 2,6-di-tert-butyl-p-cresol. Values of "Log" on ordinate defined according to Equation 6 in text

first-order termination (discussed above and in Appendix). The formulas are:

$$\Delta v = v_t - v_o, \quad c_1 = \frac{2k_A[\text{AH}]}{k_p[\text{RH}]} \quad c_2 = \frac{2k_t}{(k_p[\text{RH}])^2} \quad (4)$$

For adding a small amount of initiator, effect ΔR_i:

$$\log \frac{\Delta v + \dfrac{\Delta R_i}{c_2(v_\infty - v_o)}}{v_\infty - v_t} = 0.4343 \left\{ \frac{\Delta R_i}{v_\infty - v_o} + c_2(v_\infty - v_o) \right\} k_p[\text{RH}] t + C \quad (5)$$

For adding a small amount of inhibitor AH ($\Delta v < 0$):

$$\log \frac{\dfrac{c_1}{c_2} \dfrac{v_o}{v_o - v_\infty} + \Delta v}{v_t - v_\infty} = 0.4343 \left\{ \frac{c_1 v_o}{v_o - v_\infty} - c_2(v_o - v_\infty) \right\} k_p[\text{RH}] t + C \quad (6)$$

The values of absolute rate constants of propagation given in Table I have been obtained by using these formulas.

Discussion

Except for cumene (discussed below), no great accuracy is obtained in the rapid individual measurements of the low velocities employed, but the deviations are smoothed out considerably in the logarithmic straight lines, so that the resultant slopes are reasonably accurate (Table I and Figure 3). A fundamental drawback of the method is the limited range of conditions under which measurements can be made (except for cumene); the deviations observed are therefore regarded as a measure of the reproducibility of the method rather than of the accuracy of values obtained.

Since the duration of the non-steady state can be shown (4) to be inversely related to $k_t^{1/2}$ and $R_i^{1/2}$, a high value of k_t is unfavorable because it reduces both the time available for measurement and the rate to be measured. For this reason cumene, having a low k_t, is the obvious choice for demonstrating the method. Cyclohexene and Tetralin, on the other hand, probably represent the limit of what can be measured; their relatively high k_t is partially offset by high values of k_p [RH], which increase the velocities, hence also the accuracy obtainable at a given R_i. The method proved ineffective with fluorene, which must be measured in solution (*e.g.*, 1.8M in chlorobenzene which reduces its k_p [RH] by a factor of 4 relative to Tetralin), *sec*-butylbenzene and cyclohexylbenzene, which probably combine a relatively low k_p with considerably higher k_t than for cumene (*1*).

Concerning the results, the values obtained for cumene, Tetralin, and cyclohexene agree well with those reported in the literature.

Appendix

Non-Steady State Equations with Correction for Spontaneous Initiation and First-Order Termination. Thoroughly purified hydrocarbons should exhibit a square-root dependence of oxidation rate on initiation rate, R_i; we found, however, that even if this behavior is obtained with R_i of the order of 10^{-8} mole per liter per sec., deviations may occur with the low rates of initiation used in the non-steady state measurements ($R_i \sim 10^{-11}$). Also, spontaneous initiation of the order of $R_i \sim 10^{-12}$ may occur. If we assume that the deviations can be described as a constant first-order termination, we can derive corrected formulas for the non-steady state behavior upon adding a small amount of inhibitor AH or initiator ΔR_i, as follows.

Initial condition: spontaneous initiation $R_i°$, first-order termination constant X:

$$R_i° - 2k_t[RO_2·]_o^2 - X[RO_2·]_o = 0 \tag{A1}$$

Non-steady state after adding ΔR_i or AH:

$$\frac{d[RO_2·]}{dt} = R_i° + \Delta R_i - 2k_t[RO_2·]^2 - X[RO_2·] \neq 0$$

or (A2)

$$\frac{d[RO_2·]}{dt} = R_i° - 2k_t[RO_2·]^2 - X[RO_2·] - 2k_A[AH][RO_2·] \neq 0$$

Final condition:

$$R_i° + \Delta R_i - 2k_t[RO_2·]_\infty^2 - X[RO_2·]_\infty = 0$$

or (A3)

$$R_i° - 2k_t[RO_2·]_\infty^2 - X[RO_2·]_\infty - 2k_A[AH][RO_2·]_\infty = 0$$

By subtracting Equation 1 from Equations 2 and 3, one can eliminate $R_i°$. Subtracting Equation 1 from Equation 3 ($R_i°$ eliminated) gives an expression for X to be substituted in Equation 2 ($R_i°$ eliminated) which can then be rearranged to give a quadratic differential equation in $\Delta RO_2· = [RO_2·] - [RO_2·]_o$:

$$-\frac{d[RO_2·]}{dt} = -\frac{d\Delta RO_2·}{dt} = 2k_t(\Delta RO_2·)^2 +$$

$$-\Delta RO_2· \left\{ 2k_t([RO_2·]_\infty - [RO_2·]_o) - + \frac{\Delta R_i}{[RO_2·]_\infty - [RO_2·]_o} \right\} - \Delta R_i \tag{A4}$$

or

$$-\frac{d\Delta RO_2·}{dt} = 2k_t(\Delta RO_2·)^2 - \Delta RO_2· \left\{ 2k_t([RO_2·]_\infty - [RO_2·]_o) \right.$$

$$\left. + \frac{2k_A[AH][RO_2·]_o}{[RO_2·]_\infty - [RO_2·]_o} \right\} + 2k_A[AH][RO_2·]_o \tag{A5}$$

The square roots in the expression

$$\int \frac{dx}{ax^2 + bx + c} = \frac{1}{\sqrt{b^2 - 4ac}} \ln \frac{ax + b - \sqrt{b^2 - 4ac}}{ax + b + \sqrt{b^2 - 4ac}} + C$$

can be extracted. After converting to rates of oxygen consumption according to $v = k_p[RH][RO_2·]$ and introducing the constants $c_1 = 2k_A[AH]/k_p[RH]$ and $c_2 = 2k_t/[k_pRH]^2$, whose values are available from steady state measurements, we obtain the following expressions for the non-steady state when a constant spontaneous initiation and/or first-order termination are involved ($\Delta v = v_t - v_o$):

$$\log \frac{\Delta v + \dfrac{\Delta R_i}{c_2(v_\infty - v_o)}}{v_\infty - v_t} = 0.4343 \left\{ \frac{\Delta R_i}{v_\infty - v_o} + c_2(v_\infty - v_o) \right\} k_p[\text{RH}]t + C \tag{A6}$$

AH ($\Delta v < 0$)

$$\log \frac{c_1/c_2 \dfrac{v_o}{v_o - v_\infty} + \Delta v}{v_t - v_\infty} = 0.4343 \left\{ \frac{c_1 v_o}{v_o - v_\infty} - c_2(v_o - v_\infty) \right\} k_p[\text{RH}]t + C \tag{A7}$$

Oxygen Uptake by a Solution in which the Oxygen Consumption is Rapidly Changing According to a First-Order Law. DEFINITIONS. The symbols C and $\dfrac{dC}{dt}$ refer to oxygen concentration in the solution and C_∞ is the saturation value; v refers to the chemical reaction which consumes oxygen; $\dfrac{dO_2}{dt}$ is the rate of absorption of oxygen gas by the solution.

The rate of oxygen absorption by an unsaturated solution is given by:

$$\frac{dO_2}{dt} = \gamma_1 (C_\infty - C) \tag{A8}$$

If the chemical reaction consumes oxygen at a rate $v = v_o e^{-\gamma_2 t}$ we have

$$\frac{dC}{dt} = -v_o e^{-\gamma_2 t} + \frac{dO_2}{dt} \tag{A9}$$

$$\frac{dC}{dt} + \gamma_1 C = -v_o e^{-\gamma_2 t} + \gamma_1 C_\infty \tag{A10}$$

This is a linear differential equation whose solution is:

$$C = \frac{1}{\gamma_1 - \gamma_2} v_o e^{-\gamma_2 t} + C_\infty + a e^{-\gamma_1 t} \tag{A11}$$

The value of a follows from $C_o = -\dfrac{v_o}{\gamma_1 - \gamma_2} + C_\infty + a$, so that we obtain an expression giving $\dfrac{dO_2}{dt}$ directly—viz.,

$$-\gamma_1(C - C_\infty) = \frac{dO_2}{dt} = \frac{\gamma_1}{\gamma_1 - \gamma_2} v - v_o e^{-\gamma_1 t}\left(\frac{\gamma_1}{\gamma_1 - \gamma_2} - 1\right) \tag{A12}$$

It is seen that, when $\gamma_1 \gg \gamma_2$, the oxygen consumption follows v exactly; any discrepancies are governed by $\dfrac{\gamma_1}{\gamma_1 - \gamma_2}$. By computing curves for different values of γ_1 and γ_2, we find that there are large differences between observed and actual rate if $\gamma_2 > 0.3\, \gamma_1$. However, if $\gamma_1 = 0.05$ and $\gamma_2 = 0.01$, for example, we find that $\dfrac{dO_2}{dt}$ and v take the form of two apparently congruent curves which are separated by 15 seconds.

Literature Cited

(1) Alagy, J., Clement, G., Balaceanu, C., *Bull. Soc. Chim. France* **1961**, 1792.
(2) Howard, A., Ingold, K. U., *Can. J. Chem.* **44**, 1119 (1966).
(3) Melville, H. W., Richards, S., *J. Chem. Soc.* **1954**, 944.
(4) North, A. M., "Kinetics of Free Radical Polymerization," p. 25, Pergamon Press, Oxford, 1966.
(5) Walling, C., "Free Radicals in Solution," p. 67, Wiley, New York, 1957.

RECEIVED October 9, 1967.

INDEX

A

Absolute rate constants for
  hydrocarbon autoxidation ... 6, 346
Abstraction by HO_2 109
Acenaphthene 194
Acetaldehyde 97
  aliphatic amines vs. oxidation of . 310
  diethylamine vs. oxidation of ... 310
  ethylamine vs. oxidation of 310
Acetoxy radicals, cage reactions of . 269
Acetone 102
Acetyl peroxide 269
Acridan 184
Activated complexes 154
Activation energies 158
Acyl peroxide radicals 5
AIBN 327, 350
Alcohol, solvation by 237
Alcohols, C—H bond energies in .. 107
Alcohols, ionic catalysis in chain
  oxidation of 112
Aliphatic amines
  vs. decomposition of *tert*-butyl
    peroxide 311
  vs. decomposition of ethyl *tert*-
    butyl peroxide 314
  as inhibitors 307
  vs. oxidation of acetaldehyde ... 310
  vs. oxidation of ethyl ether 311
  reaction of methyl radicals with . 316
n-Alkanethiols 221
1-Alkenes, liquid-phase oxidation of
  high molecular weight 78
Alkenyl hydroperoxides 78
Alkoxyhydroperoxides 247
Alkylene oxides 5
Alkylperoxy radicals 38
Allylic hydroperoxides 93
  isomerization of 99
Alpha-olefins 79
Amines as antioxidants 168
Anthracene 203
  dianion 241
Anthraquinone 203
Anthrone 184, 207
Antioxidants
  aromatic amine 300
  amines and phenols as 168
  chain-breaking 298
  electron transfer 301
  nitroxide radicals as 298
  phenolic 300
  rate constants for reactions of ... 304
Aralkyl hydrocarbons 26

Aromatic amine antioxidants 300
Arrhenius plots 236
Autoxidation
  absolute rate constants for
    hydrocarbon 6
  of carbanions 167
  of chloroprene 138-9
  of electron deficient molecules .. 171
  of Grignard reagents 172
  of 9, 10-dihydroanthracene,
    base-catalyzed 203
  hydrocarbon 323
  of hydrocarbons, absolute rate
    constants of 346
  inhibition of 296
  of mercaptans 169
  of phosphorus compounds 170
  reactivities of hydrocarbons in.. 39, 55
Azelaic acid 245
2, 2'-Azobisisobutyronitrile 140, 327
2, 2'-Azoisobutyronitrile 350

B

Base-catalyzed autoxidation of
  9,10-dihydroanthracene 203
Benzhydrol 184
Benzhydrylamine 184
Benzhydryl chloride 184
Benzophenone ketyl 186
Benzopinacol 187
Benzyltrimethylammonium
  hydroxide 204
Bis(dimethylamino)methane 226
Bond dissociation energies 282
Bond energies in alcohols, C—H.. 107
Bond strengths 291
Br^- 115
Brassylic acid 245
Butadiene with hydrocarbons,
  co-oxidation of 24
tert-Butanethiol 220
Butene hydroperoxide 93
2-Butene-1-hydroperoxide 94
1-Butene-3-hydroperoxide 94
3-Butene-2-hydroperoxide, decomposition of 97
Butene, photosensitized oxidation of 94
n-Butylamine vs. decomposition of
  tert-butyl peroxide 312
n-Butylamine, reaction of methyl
  radicals with 318
sec-Butylbenzene 354
tert-Butyl peroxide, aliphatic
  amines vs. decomposition of.. 311

tert-Butyl peroxide, n-butylamine
   vs. decomposition of 312
tert-Butylperoxy radicals 259, 265

C

Cage reactions of acetoxy radicals.. 269
Cage recombination 275
Carbanions 208
   autoxidation of 167
   oxidation of 174
Catalysis
   by HCO_3^- ions 117
   by transition metals 112
   cobalt 220
   copper 220
   in chain oxidation of alcohols,
      ionic 112
   metal ion 219
   nickel 220
Catalyst for decomposition 243
Catalyzed autoxidation 248
Cerium(IV) oxidation 246
Chain-breaking antioxidants 298
Chain-breaking inhibitors 324
Chain degradation 246, 250
Chain oxidation 150
   of alcohols, ionic catalysis in.... 112
   of cyclohexanol 117
Chain propagation 159
Chain reactions, oxidative 2
Chain termination 159
   kinetic parameters for 264
   in the oxidation of cumene 21
C—H bond energies in alcohols ... 107
Chemiluminescence of tetrakis
   (dimethylamino)ethylene ... 225
Chloroprene
   autoxidation of 138
   dimerization of 140
   peroxide 141
Chromatographic purification of
   ozonization products 247
Cobalt 97
   catalysis 220
Collisions, kinetic theory of....... 2
Combinations of reactions 4
Co-oxidation of
   butadiene with hydrocarbons ... 24
   cumene and Tetralin 7, 9, 46
   hydrocarbons 38
   styrene and cumene 44
   styrene and Tetralin 43
   substituted styrenes 41
   Tetralin and cis-Decalin 7, 9, 49
Copper catalysis 220
Critical levels, oxidation at 4
Cumene 6, 349
   chain termination in the
      oxidation of 21
   co-oxidation of styrene and 44
   co-oxidation of Tetralin and...7, 9, 46
   hydroperoxide 261, 323
Cumylperoxy radicals 259, 265
Cyclic olefins 32

Cyclohexanol 112
   chain oxidation of 117
Cyclohexene 349
Cyclohexylbenzene 354
Cyclohexenyl hydroperoxide 261

D

cis-Decalin, co-oxidation of
   Tetralin and 7, 9, 49
Decarboxylation 280
Decomposition
   of 3-butene-2-hydroperoxide ... 97
   catalyst for 243
   of methyl ethyl ketone peroxy
      radical 163
   of peroxide radicals 4
Degenerate chain branching 3
α-Deuteriobenzhydrol 185
9-Deuteriofluorenol 191
Dialkyl peroxides, polymeric 78
Diarylcarbinols 174
Diarylmethanes 174
2,6-Di-tert-butyl-p-cresol 350
α,ω-Dicarboxylic acids 245
α,α-Dideuteriodiphenylmethane ... 180
9,9-Dideuteriofluorene 181
Dielectric constant 155, 235
Diethylamine
   vs. decomposition of ethyl tert-
      butyl peroxide 312
   vs. oxidation of acetaldehyde ... 310
   reaction of methyl radicals with . 319
Diethyl malonate 196
9,10-Dihydroanthracene 184
   base-catalyzed autoxidation of .. 203
   oxidation of 205
9,10-Dihydrophenanthrene 193
Diisopropyl dithiophosphate,
   kinetics of oxidation
      inhibition by zinc 327
Dimerization of chloroprene 141
Dimethylamine 226
N,N-Dimethylethylamine, reaction
   of methyl radicals with 318-9
2,5-Dimethylhexane-2,5-
   dihydroperoxide 263
Diphenylacetic acid 184
1,1-Diphenylacetone 184
Diphenylacetonitrile 184
1,2-Diphenylethane-1,2-diol 187
1,1-Diphenylethylene 241
Diphenylmethane 176
Diphenylmethide ion 180
Diphenylmethyl hydroperoxide ... 196
1,1-Diphenyl-2-(methylsulfinyl)
   ethanol 176
Disulfides, liquid-phase oxidation
   of thiols to 216
Dithiophosphates
   hydrogen abstraction from 333
   oxidation of 332
   peroxide decomposition of
      dialkylmetal 337
Donor-acceptor complexes 244

INDEX

E

Electron deficient molecules, autoxidation of	171
Electron transfer	182, 217
antioxidants	301
one-	210
oxidation inhibition mechanism	324
Enthalpy of formation of phenyl benzoate	282
Epoxidizing agents	5
ESR spectroscopy	182
self-reactions of peroxy radicals by	258
Ethanethiol	220
Ethylamine	
vs. oxidation of acetaldehyde	310
vs. oxidation of ethyl ether	311
Ethyl tert-butyl peroxide, aliphatic amines vs. decomposition of	314
Ethyl tert-butyl peroxide, diethylamine vs. decomposition of	312
reaction of methyl radicals with	318
Ethyl ether	
aliphatic amines vs. oxidation of	311
ethylamine vs. oxidation of	311
triethylamine vs. oxidation of	311

F

Fatty acids	2
Fatty alcohols	2
Flow systems	5
9-Fluorenol	189
Free radical mechanism	3
Free thiyl type radical	324
First-order termination	354
Fluorene	354

G

Grignard reagents, autoxidation of	172
Gamma-radiation-induced oxidation of 2-propanol	102
$G_W(OH)$	110
Group additivity	288

H

HCO_3^- ions, catalysis by	117
Heats of formation	288
Heteroatoms in oxidation	166
1-Hexadecene	79
oxidation, products of	89
Hexanitratoammonium cerate	246
High molecular weight 1-alkenes, liquid-phase oxidation of	78
Hindered phenol, inhibition by a	346
HO_2, abstraction by	109
HO_2 radical	103
Homolytic decomposition of hydrogen peroxide	113
Hydrated electron	103
Hydrocarbon autoxidation, absolute rate constants for	6
Hydrocarbon autoxidation and inhibition	323
Hydrocarbons	
absolute rate constants of autoxidation of	346
aralkyl	26
in autoxidation, reactivities of	39, 55
co-oxidations of	38
butadiene with	24
hydrogen atom abstraction from	20
to peroxy radicals, hydrogen atom transfer from	19
reactivities of peroxy radicals toward	6
reactivity of	38
relative reactivities of	34
Hydrogen	
abstraction from dithiophosphates	333
atom	103
abstraction from hydrocarbons	20
transfer from hydrocarbons to peroxy radicals	19
transfer from hydroperoxides to peroxy radicals	13
bonds	155
peroxide, homolytic decomposition of	113
transfer, intramolecular	214
Hydroxycarbonyl radical	284
4-Hydroxydiphenylamine	194
Hydroxyl radical	102
Hydroxymethyl radicals	105
Hydroperoxide	
butene	93
2-butene-1-	94
1-butene-3-	94
chain mechanism	166
conversion products	5
cumene	261
cyclohexenyl	261
decomposition of 3-butene-2-	97
effect on inhibited oxidation rate	331
2-phenylbutyl-2-	261
Tetralin	261
Hydroperoxides	3, 115, 260, 290
alkenyl	78
allylic	93
isomerization of allylic	99
reactivities of peroxy radicals toward	6
to peroxy radicals, hydrogen atom transfer from	13
9-Hydroperoxyfluorene	196

I

Inhibited oxidation	
rate, hydroperoxide effect on	331
of squalane	342
of Tetralin	327
Inhibiters, chain-breaking	324
Inhibition	
of autoxidation	296
by a hindered phenol	346
hydrocarbon	323

Inhibition *(Continued)*
 mechanism, electron transfer
  oxidation 324
 mechanism of oxidation 323
 of oxidation of chloroprene 140
 by zinc diisopropyl dithio-
  phosphate, kinetics of
  oxidation 327
Inhibitors, aliphatic amines as.... 307
Inhibitors, oxidation 5
Initiated oxidation of chloroprene.. 140
Intermediate ion triplet 243
Intramolecular hydrogen transfer .. 214
Ionic catalysis in chain
 oxidation of alcohols 112
Ionic intermediate 227
Ionicity 292
Ionol 350
Ion triplet, intermediate 243
Isomerization
 of allylic hydroperoxides 99
 of methyl ethyl ketone peroxy
  radical 163
 of peroxide radicals 4
Isopropylbenzene 2
Isotope effect 179

K

Kinetic parameters for chain
 termination 264
Kinetic theory of collisions 2
Kinetics of oxidation inhibition
 by zinc diisopropyl
 dithiophosphate 327

L

Liquid-phase
 co-oxidations 40
 olefin oxidation 78
 oxidation 78
 of high molecular weight
  1-alkenes of organic
  compounds, trends of
  research on the thiols
  to disulfides 216

M

Mechanism of oxidation inhibition. 323
Mechanisms for phenyl esters 286
Mercaptans 216
 autoxidation of 169
Metal dialkyl dithiophosphates,
 peroxide decomposition by ... 337
Metal ion catalysis 219
Metal ion deactivators 298
p-Methoxystyrene 41
Methylamine, reaction of methyl
 radicals with 318
N-Methyldiethylamine, reaction
 of methyl radicals with...... 320
Methyl ethyl ketone 158
 peroxy radical, decomposition of. 163
 peroxy radical, isomerization of.. 163

4,5-Methylenephenanthrene 184
Methyl radicals 269
 correlation of reactivity of 34
 with aliphatic amines, reaction of 316
 with n-butylamine, reaction of.. 318
 with diethylamine, reaction of .. 319
 with dimethylamine, reaction of. 318
 with ethylamine, reaction of 318
 with methylamine, reaction of .. 318
 with N-methyldiethylamine,
  reaction of 320
 with N,N-dimethylethylamine,
  reaction of 319
 with triethylamine, reaction of.. 320
 with trimethylamine, reaction of. 319
9-(Methylsulfinylmethyl)-9-
  hydroxyfluorene 189
Methyl vinyl carbinol 97
Methyl vinyl ketone 97
Monomers, structure *vs.* rate
 of oxidation of 146

N

Nickel catalysis 220
Nitrobenzene 181
2-Nitropropane 196
m-Nitrostyrene 41
p-Nitrostyrene 41
Nitroxide radicals as antioxidants .. 298
Non-steady state methods 347

O

O_2^- 110
1-Octene ozonide 247
Olefin oxidation, liquid-phase 78
One-electron transfer 210
Organic compounds, trends of
 research on liquid-phase
 oxidation of 1
Oxidation
 of carbanions 174
 at critical levels 4
 of 9,10-dihydroanthracene 205
 of dithiophosphates 332
 inhibition mechanism, electron
  transfer 324
 inhibitors 5
 of ozonization products 245
 reactions, thermochemistry of .. 288
 of tetrakis(dimethylamino)
  ethylene 225
Oxidative chain reactions 2
Oxides, alkylene 5
Oxygen-phenyl bond 283
Oxygen pressure 233
Ozonization products, chromatographic purification of 247
Ozonization products, oxidation of. 245

P

Performic acid oxidation 245
Peroxide
 decomposers 297, 323

INDEX

Peroxide
  decomposition by metal dialkyl
    dithiophosphates 337
  polymeric 138
  radicals, isomerization and
    decomposition of 4
  thermal stabilities 91
Peroxides 288
  polymeric dialkyl 78
Peroxy radicals 150
  correlation of reactivity of 34
  decomposition of a methyl ethyl
    ketone 163
  by ESR spectroscopy, self-
    reactions of 258
  toward hydrocarbons and
    hydroperoxides,
    reactivities of 6
  hydrogen atom transfer from
    hydroperoxides to 13, 19
  isomerization of methyl ethyl
    ketone 163
Phenalene 192
Phenalenyl radical 193
Phenol, hindered 346
Phenolic antioxidants 300
Phenols as antioxidants 168
Phenoxycarbonyl radical 284
Phenoxyl-carbon bond 285
Phenoxyl-hydrogen bond 285
Phenyl benzoate 282
2-Phenylbutyl-2-hydroperoxide ... 261
Phenyl-carbon bond 283
Phenyl esters, thermochemistry and
  mechanisms for 286
Phenyl oxalate, thermal
  decomposition of 286
Phenyl-oxygen bond 284
Phenyl-2-pyridylmethane 192
Phenyl-4-pyridylmethane 192
Phorphorus compounds,
  autoxidation of 170
Photosensitized oxidation of butenes 94
Platinum black 253
Polymeric dialkyl peroxides 78
Polymeric peroxide 138
Polyoxide molecules 293
Potassium peroxide 199
Potassium superoxide 186
Pre-exponential factors 158
Products of 1-hexadecene oxidation 89
Propagation, chain 159
2-Propanol 112
  gamma-radiation-induced
    oxidation of 102
Propionaldehyde 97

R

Radical thermochemistry 283
Radicals, acyl peroxide 5
Radiolysis 286

Rate constants
  of autoxidation of hydrocarbons,
    absolute 6, 346
  for reactions of antioxidants 304
Rate of oxidation of monomers,
  structure vs. 146
Rates and routes of oxidation reactions, solvents vs. 150
Reaction vessel surface 4
Reactivities
  of hydrocarbons in autoxidation. 39
  of hydrocarbons, relative 34
  of peroxy radicals toward
    hydrocarbons and
    hydroperoxides 6
Reactivity of hydrocarbons 38
  in autoxidation 55
Relative reactivities of hydrocarbons 34
Research on liquid-phase oxidation
  of organic compounds,
  trends of 1
ROOH 115
Rotating sector 266
Routes of oxidation reactions,
  solvents vs. rates and 150

S

Self-reactions of peroxy radicals
  by ESR spectroscopy 258
Solvation by alcohol 237
Solvent cage 269
Solvent effects 206, 237
Solvent viscosity 275
Solvents vs. rates and routes of
  oxidation reactions 150
Spontaneous initiation 354
Squalane, inhibited oxidation of... 342
Stabilization energy 293
Structure vs. rate of oxidation
  of monomers 146
Styrene 27, 41
  and cumene, co-oxidation of 44
  and Teteralin, co-oxidation of .. 43
Styrenes, co-oxidations of
  substituted 41, 42
Substituted styrenes, co-oxidations
  of 41, 42
Surface, reaction vessel 4
Synergism 297

T

Termination, chain 159
Termination in the oxidation of
  cumene, chain 21
Tetraaminoethylene 225
Tetrakis(dimethylamino)ethylene,
  oxidation and
  chemiluminescence of 225
Tetralin 6, 349
  co-oxidation of 7, 9
    styrene and 43
    cumene and 46
    cis-Decalin and 49

Tetralin *(Continued)*
   hydroperoxide 261
   inhibited oxidation of 327
Tetramethylbutylperoxy radicals .. 265
Tetramethylhydrazine 226
Tetraphenylcyclopentadiene 192
sym-Tetraphenylethane 184
Tetroxide formation 265
Thermochemistry of oxidation
   reactions 288
Thermochemistry of phenyl esters. 286
Thermal decomposition of phenyl
   oxalate 286
Thiols to disulfides, liquid-phase
   oxidation of 216
Thiophenol 220
Thioxanthene 184
Thiyl type radical, free 324
TMAE 225
   fluorescence 233
Transition metals, catalysis by 112
Trends of research on liquid-phase
   oxidation of organic
   compounds 1
Trialkyl phosphites 170
m-Trifluoromethylnitrobenzene ... 181
Triethylamine, reaction of methyl
   radicals with 320

Triethylamine *vs.* oxidation of
   ethyl ether 311
Trimethylamine, reaction of methyl
   radicals with 319
1,1,2-Triphenylethane 184
   1,2-diol 187
Triphenylmethane ...,......... 174
Triphenylmethide ion 174
Triphenylmethyl 174
Triplet, intermediate ion 243

U

Ultraviolet light deactivators 298

V

Valence-shell expansion 170
Vanadium 97

W

Water radiolysis 110

X

Xanthene 184
9-Xanthenol 189

Z

Zinc dialkyl dithiophosphates 323